Fatty Acids: A Comprehensive Study

Fatty Acids: A Comprehensive Study

Editor: Jack Fielding

STATES
ACADEMIC PRESS
www.statesacademicpress.com

States Academic Press,
109 South 5th Street,
Brooklyn, NY 11249, USA

Visit us on the World Wide Web at:
www.statesacademicpress.com

ISBN: 978-1-63989-197-9 (Hardback)

Cataloging-in-Publication Data

Fatty acids : a comprehensive study / edited by Jack Fielding.
p. cm.
Includes bibliographical references and index.
ISBN 978-1-63989-197-9
1. Fatty acids. 2. Carboxylic acids. 3. Biochemistry. I. Fielding, Jack.
QP752.F35 F38 2022
612.397--dc23

Table of Contents

Preface

Fatty acids are carboxylic acids, having a long aliphatic chain of saturated or unsaturated hydrocarbons. These acids are broadly classified into various forms on the basis of their length, saturation or unsaturation, even or odd content of carbon, and by linear or branched structures. On the basis of length, they are categorized into short-chain, medium-chain, long-chain and very long chain fatty acids. Some of the reactions exhibited by fatty acids are saponification, hydrogenation, rancidity and ozonolysis. Fatty acids have a wide variety of applications. They are used in the production of soaps, and as emulsifiers, anti-foam agents, texturizing agents, etc. This book unravels the recent studies on fatty acids. It will also provide interesting topics for research which interested readers can take up. Students, researchers, experts and all associated in this area of study will benefit alike from this book.

This book unites the global concepts and researches in an organized manner for a comprehensive understanding of the subject. It is a ripe text for all researchers, students, scientists or anyone else who is interested in acquiring a better knowledge of this dynamic field.

I extend my sincere thanks to the contributors for such eloquent research chapters. Finally, I thank my family for being a source of support and help.

Editor

1

Free Fatty Acids Quantification in Dairy Products

Kieran N. Kilcawley and David T. Mannion

Abstract

Quantification of free fatty acids in dairy products is not only important due to their (fatty acids) impact on the flavour and texture of dairy products but also because of their potential impact on nutrition and health, and as anti-microbial agents. This chapter provides an overview of the practical issues associated with existing lipid extraction techniques and quantification procedures using gas chromatography flame-ionization detection. The most widely used methods are compared and recent advancements in the quantification of free fatty acids in dairy products are discussed.

Keywords: free fatty acids, dairy, gas chromatography

1. Introduction

Bovine milk fat has a complex fatty acid composition with over 400 individual fatty acids [1, 2]. However, only 15 or 16 fatty acids are present in milk at concentrations above 1% [3, 4]. The predominant fatty acids have a straight carbon atom chain with an even number of carbons and may be either saturated or unsaturated [5]. The approximate composition of the fatty acids of bovine, ovine and caprine milk fat is given in **Table 1**. The proportion of fat in bovine, ovine and caprine milk is ~3.8, 7.1 and 3.7%, respectively, and the fatty acid composition of milk within cows, goats and sheep milk is influenced by diet, stage of lactation and breed among others [6]. Consequently, the proportions of some fatty acids can show marked variations.

Animal fats are complex and contain short-chain fatty acids (SCFFA) that are both water soluble and highly volatile and are not present in vegetable fats [3]. Vegetable fat consists of primarily non-volatile fat-soluble fatty acids. The main fatty acids in milk are $C_{16:0}$ and $C_{18:1}$

Fatty acids (g 1 100 g^{-1})	Common name	Bovine	Ovine	Caprine
$C_{4:0}$	Butyric acid	2.87	2.57	2.03
$C_{6:0}$	Caproic acid	2.01	1.87	2.78
$C_{8:0}$	Caprylic acid	1.39	1.87	2.92
$C_{10:0}$	Capric acid	3.03	6.63	9.59
$C_{12:0}$	Lauric acid	3.64	3.99	4.52
$C_{14:0}$	Myristic acid	10.92	10.17	9.83
$C_{16:0}$	Palmitic acid	28.7	25.1	24.64
$C_{18:0}$	Stearic acid	11.23	8.85	8.87
$C_{18:1}$ *cis*-9	Oleic acid	22.36	20.18	18.65
$C_{18:2}$ *cis*-9, *cis*-12	Linoleic acid	2.57	2.32	2.25
$C_{18:2}$ *cis*-9, *trans*-11	Conjugated linoleic acid	0.57	0.76	0.45
$C_{18:3}$ *cis*-9, *cis*-12, *cis*-15	α -Linolenic acid	0.5	0.92	0.77

Source: Adapted from Markiewick-Keszykca et al. [6].

Table 1. Distribution of the major fatty acids in bovine, ovine and caprine milk.

comprising between ~22–35 and 20–30% of total lipids, respectively [3]. Fatty acids are present either in their free state, as free fatty acids (FFA) or esterified as bound fatty acids (FA) on the glycerides. Accurate determination of both FFA and FA can be indispensable for legislative and quality control purposes but also for research and development purposes. This chapter focuses on gas chromatographic detection of FFA(s). FFA(s) are important as they influence product quality, flavour, texture, nutrition, and health. The flavour of many dairy products is directly and indirectly influenced by the FFA profile of the product [2, 7]. This is particularly the case for fermented dairy products, as FFA contributes directly as volatile aromatic components or indirectly *via* volatile products of metabolism, oxidation or heat treatments (e.g. aldehydes, ketones, alcohols, lactones and esters). FFAs can also contribute to texture and functionality, as they impact on surface tension and foaming capacity of milk [8, 9], but some FFAs such as conjugated linoleic acid have also been shown to have beneficial health and nutritional effects [6].

Fat extraction techniques need to be able to take into account differences in solubility and volatility of the different carbon chain lengths of FA(s) present in milk fat. Therefore, any method for the accurate quantification of FA(s) must be efficient in extracting both water soluble SCFFA and organic soluble FA, avoid the use of evaporation steps to prevent losses of volatile SCFFA and remove or negate any water that may be present in the sample. So far, the most common approach to quantify FFA in the dairy industry and in research is gas chromatography (GC) coupled to a flame ionization detector (FID). FID is used because it is relatively cheap, simple/robust, reproducible and widely available. This chapter discusses in detail the advantages and disadvantages of some of the methods used to extract and quantify FFA in dairy products by GC-FID.

2. Free fatty acid determination in dairy products

2.1. Lipid extraction

Solvent extraction of the fat from the sample is widely used. However, as previously mentioned, evaporation steps should be avoided to prevent losses of volatile SCFFA. Solvents such as chloroform [10], acidified diethyl ether [11], hexane/diethyl ether [12], and diethyl ether/heptane [13] have been used. High recoveries (>92%) of FFA have been achieved using these organic solvents for cheese but are much less reliable when applied to milk, due to the natural oil in water emulsion of milk and the nature of the milk fat globule membrane (MFGM) [14]. In addition, extraction procedures that employ high temperatures such as refluxing or distillation are also prone to increased risk of losing volatile SCFFA(s). In many procedures, anhydrous sodium sulphate is added to absorb moisture present in an attempt to prevent losses of water-soluble SCFFA in the subsequent solvent extraction process. When solvent mixtures are employed, recoveries of SCFFA can decrease when the non-polar component of the solvent is increased [11]. Solvents capable of extracting the complete range of FFA(s) will also extract the remaining lipid portion of the sample and depending upon the application, it may be necessary or prudent to isolate the FFA component prior to analysis. A range of different techniques have been employed such as silicic acid/potassium hydroxide (KOH) columns [10, 15, 16], ion exchange resins [17–19], deactivated alumina columns [12, 13] and aminopropyl solid phase extraction columns [13, 20].

Due to the strongly alkaline nature of silicic acid columns/KOH or ion exchange resins, hydrolysis of glycerides can occur [21, 22] resulting in an overestimation of FFA content. Woo and Lindsay [16] implemented the use of two different columns, a pre-column to remove lactic acid followed by a silicic acid-KOH arrestant column to isolate FFA(s) in Cheddar cheese to overcome this issue. The FFA(s) were eluted using 2% formic acid in ethyl ether. Needs et al. [17] described another method to isolate FFA in milk, using a pre-treated amberlite resin. The lipid extract was mixed with the resin followed by solvent removal and washing to isolate the FFA. Deeth et al. [12] utilised deactivated alumina columns to isolate FFA and reported high recoveries of $C_{4:0}$–$C_{18:1}$ from milk, cheese and butter. The acidic nature of the final extract (6% formic acid in di-isopropyl ether) was reported to have a detrimental impact on column performance as the column phase deteriorated and this leads to a modification of the procedure using a lower concentration of formic acid (3%) in diethyl ether [13]. De Jong and Badings [13] using a reference mix compared the performance of aminopropyl columns and deactivated alumina in isolating FFA(s) and reported a 96–101% recovery with aminopropyl columns against an 82–89% recovery for the deactivated alumina. The procedure was a modification of that employed by Kaluzny et al. [20] who obtained 101.4% recovery for FFA(s) isolated from lipid using solid phase extraction with aminopropyl columns.

2.2. Derivitization

In order for an analyte to be analysed by GC, it must be volatile. For most lipids, this is not the case, and techniques using chemical derivitization are widely employed to volatilise the fatty acids. The most established approach is to convert the FA(s) into a more volatile form, such as fatty acid methyl esters, commonly known as FAME(s). Thus the FAME(s) are injected onto a GC column as a liquid volatilised into a gaseous mixture at a specified flow or pressure. Separation occurs through differences in the interaction of the individual FAME with the GC column phase and the use of a temperature gradient in the GC oven. The separated individual FAME passes from the column into the FID and is then quantified. A FAME approach is used in the current international standard for the analysis of FA(s) in milk fat [23]. FFA(s) can be converted into FAME(s) using methanol in the presence of a suitable acid catalyst. The first step is the protonation of the acid with methanol forming an intermediate, which loses a proton to yield the FAME. An excess of methanol is required to drive the reaction to completion as it is a reversible reaction. It is also necessary to exclude water from the reaction as it is a stronger electron donor than aliphatic alcohols and will inhibit the formation of the intermediate [24].

Several methods have been developed for FFA analysis of dairy products that derivitized at room temperature and do not employ aqueous solvents or evaporation steps. Christopherson and Glass [25] outlined the use of 10% hydrochloric acid (HCl) in methanol to a solution of milk fat prior to GC analysis. Luddy et al. [26] used boron trifluoride (BF_3) in methanol to esterify FFA in butter oil. While these methods are useful, determination of FFAs is achieved by derivitizing FFA(s) with glycerides which involves two different derivitizing steps. Tetramethylammonium hydroxide (TMAH) has long been used as an esterification reagent for FA(s) [27]. Robb and Westbrook [28] identified that the TMAH reaction proceeds rapidly at room temperature, and that the salts readily decompose into the heated injection port of a GC to yield methyl esters and trimethylamine (TMA). The yields obtained ranged from 85 to 95%, and it was concluded that these variable yields made the method suitable for qualitative purposes only. A limitation of many methylation procedures is the necessity to extract the acids from an aqueous solution prior to esterification. Downing [29] investigated the preparation of tetramethylammonium salts in an aqueous solution followed by GC analysis and found the method to be quantitative and reproducible. Downing and Greene [30] used TMAH to esterify polyunsaturated fatty acids (PUFAs) and found that the strong alkaline nature of the TMAH solution interfered with the esterification process. The authors overcome this by reducing the pH (7.5–8.0) using a 5% acetic acid solution. An advantage of this approach was that the ammonium salts of FA(s) could be pyrolysed to form pure esters within the GC injection port as existing methods at the time required the saponification of the acids before the addition of esterifying agents. These methods described the saponification of FA mixtures with a strong base such as sodium hydroxide (NaOH) or KOH followed by esterification with a methanolic acid such as HCl, sulphuric acid (H_2SO_4) or BF_3 to determine FFA content. If the glyceride bound FA also required determination, a methanolic base such as sodium methoxide (CH_3NaO) could be used to form the methyl esters. Since quaternary ammonium hydroxides are inherently strong bases, a methanolic solution can form methyl esters of the glyceride bound FA(s), and as the ammonium salts of FFA(s) degrade to form methyl esters, TMAH was considered to be a more suitable reagent in FAME analysis. Other quaternary ammonium hydroxides were also used in FAME analysis. McCreary et al. [31] used trimethyl (a,a,a-trifluoro-m-tolyl) ammonium hydroxide in methanol to determine the FA content of vegetable oils prepared in benzene. They compared this approach to the more common CH_3NaO and achieved comparable results. They were also able to simultaneously trans-esterify glycerides while forming ammonium salts of the FFA(s), which upon injection in the GC inlet were esterified in a single preparation step. Again, this provided a significant advantage over existing procedures, which required FFA(s) and glycerides to be derivitized and analysed separately. Later, Metcalffe and Wang [32] also used TMAH in methanol in a single step process on different lipid mixtures in diethyl ether. This resulted in different phases where the transesterified methyl esters of the glycerides (FA) were contained in the organic phase, and the ammonium salts of the FFA(s) were in the aqueous phase. Conveniently, each phase was suitable for direct injection to a GC for FFA or glyceride (FA) characterization. This approach was later applied for the quantification of FFA in milk and cheese [33–36]. Martínez-Castro et al. [33] also investigated the effect of making the reaction mixture neutral, prior to the analysis which is usually recommended for GC analysis to protect the column and any PUFA(s) that may be present. They discovered that neutralizing the basic TMAH solution resulted in losses of SCFFA(s) and increased standard deviations in the analytical data. They attributed this to the dissociation of the ammonium salts at the selected pH (7.5–8.0). An advantage of TMAH is that when pyrolysed it degrades to TMA and methanol [31], which are highly volatile and thus suitable for GC analysis. However, TMA was also reported to interfere with peak determination [34]. This approach has obvious benefits in that both the FFA and the triglyceride components can be analysed from a sample in a single extraction; however, a limitation of the procedure using the TMAH reaction was highlighted by Martínez-Castro et al. [33]. These authors noted that some FA(s) from the organic glyceride layer were detected in the aqueous FFA layer, resulting in an overestimation of the FFA content in cheese. To overcome this, the authors separated each layer and washed with an appropriate solvent before analysis. Chavarri et al. [36] identified that if there is a large triglyceride to FFA ratio (which is the case in most dairy products), the issue with FFA dissociation is even more pronounced. The authors concluded that it was necessary to isolate FFA from the lipid mixture before employing the TMAH extraction/esterification method to alleviate this error.

2.3. Direct on-column addition FFA

The isolation of FFA(s) by aminopropyl solid phase extraction (SPE) columns followed by GC-FID analysis is widely used to quantify FFA [37–41], because the isolation process can be automated to a degree and is relatively simple to perform. Overall, it is a convenient alternative to derivatizition and as mentioned was developed by Kaluzny et al. [20] and subsequently improved [13, 37]. The approach works without the need for derivatizition because FFA(s) are volatile and thus can be vapourised in a heated injection port. A cold on-column injection is employed followed by a programmed temperature ramp of the injector, as this allows for the increased separation of FFA(s) based on their volatility within the injector. There are also commercially available columns with specific free fatty acid phases (FFAP) that achieve complete separation of FFA(s) of chain lengths from $C_{2:0}$ to $C_{22:0.}$ However, the acidic extracts reduce column performance and the high affinity of the FFA to the column phase can lead to irreversible adsorption, peak tailing, ghost peaks, or double peak formation [42, 43].

2.4. Comparison of direct on-column addition FFA and TMAH FAME methods

Chavarri et al. [36] compared the TMAH extraction esterification procedure to an on-column chromatographic procedure described by De Jong and Badings [13], where FFAs were isolated using aminopropyl columns before direct injection to GC-FID. Substantial discrepancies were evident between both methods in the analysis of FFA in Cheddar cheese. Levels of FFA were much higher using the on-column direct injection method (4007 ppm) than the TMAH method (1683 ppm). Typically, levels of FFA in Cheddar cheese are below 2000 ppm, thus a considerable error existed with the on-column direct injection method. The higher FFA levels were subsequently shown to be due to the dissociation of glycerides into the FFA layer. This can be rectified by isolating the FFA from the lipid mixture before employing the TMAH extraction/esterification. Mannion et al. [44] also compared and validated both methods. They employed an identical fat extraction and FFA isolation technique for both methods, using the diethyl ether/heptane procedure described by De Jong and Badings [13]. As well as, investigating analytical robustness accuracy, precision, limits of detection (LOD) and limits of quantification (LOQ) were also assessed. A wide range of dairy samples were analysed to assess method suitability to quantify FFAs in dairy products that have different sample matrices, lipid composition and FFA concentration. The products investigated ranged from cheeses (Cheddar, Brie and Blue Stilton), whole milk powder, infant formula, milk, yogurt, ice cream and enzyme-modified cheeses. Repeatability was expressed as percent (%) relative standard deviation (RSD). Recoveries were assessed by spiking a known amount of FFA(s) into each sample with calculations based on recovery of the FFA between spiked and unspiked samples. The FFA(s) levels measured ranged from 173 ppm in infant formula to 126,615 ppm in enzyme-modified cheese. Both methods provided similar results for each sample type, and repeatability was excellent (0.8–13.8% RSD) except for milk, ice cream and yogurt (up to 46.2% RSD). Diethyl ether/heptane was used as the extracting solvent for both methods and is not suitable to reliably extract FFA(s) where the MFGM remains intact. This is easily remedied using an alternate ethanol-based extraction as described by De Jong and Badings [13] or De Jong et al. [37]. Analysis of FFA(s) directly as acids using the direct on-column addition method resulted in issues with column degradation due to the acidic nature of the FFA extract. This resulted in retention time peak shifts, peak broadening, and loss of resolution overtime. Additional steps were also required to prevent and monitor carryover due to the high affinity of FFA with the column phase. Analysis of the FFA with the TMAH FAME method was also not without issues. The most volatile SCFFA eluted with the solvent peak which impacted on sensitivity, and artefact formation periodically interfered with the quantification of other SCFFA. Overall, the direct on-column method proved to be the most sensitive, with an LOD of 0.7 ppm and an LOQ of 3 ppm and required less sample preparation as the FFAs were injected directly after isolation without the need for derivitization. Although the TMAH FAME method was the less sensitive of the two methods (LOD of 5 ppm; LOQ of 20 ppm), it proved to be a much more robust method where column integrity was not affected during analysis and retention times, and peak chromatography remained stable. In addition, the authors described an automated procedure for the dervitization which was significantly faster, used reduced solvent volumes than traditional procedures.

2.5. Future developments in GC-FID methods

The use of higher molecular weight alcohols may be considered an alternative to overcome many issues experienced with methyl esters such as co-elution of solvent peaks with the most volatile FAME(s) and artefact formation. Studies by Parodi [45] demonstrated increased recoveries of SCFFA(s) when butyl esterification was carried out instead of methyl esterification. Parodi [45] evaluated several different methylation methods such as using BF_3 in methanol [46], CH_3NaO [47] and butylation methods using H_2SO_4 in butanol [48], di-n-butyl carbonate [49] and BF_3 in butanol [45] to determine the fatty acid composition of butter fat. Parodi [45] expressed the data as ratio amounts of each fatty acid ester relative to the corresponding myristate ester and obtained better recovery when butyl esters were employed in comparison to BF_3 in methanol or CH_3NaO methylation methods. Parodi [50] utilised KOH in butanol to form butyl esters and was a modification of the methods described by Christopherson and Glass [25], and Kim Ha and Lindsay [51] where they utilised BF_3 in butanol to quantify FFA in milk and cheese. Thus, employing butyl esters may be a more suitable alternative to methyl esters in the analysis of FFA in dairy products.

Most publications on the determination of FFA in dairy products reference long established extraction methods. Little or no development or validation seems to have taken place or has been published in recent years. Some efforts to improve and validate existing fat extraction methods have been undertaken. The widely used extraction method of Folch et al. [52] was modified by Firl et al. [53] to extract lipids from milk samples, which were subsequently converted to methyl esters using trimethylsulfonium hydroxide (TMSH) and analysed by GC-FID. They validated their approach by spiking milk samples with triglycerides and reported LODs, LOQs, accuracy and precision, which was particularly useful. However, quantification of FFA in dairy samples was not undertaken. Reis et al. [54] described a new method using thermal desorption to isolate FA in milk, using TMSH to convert the triglycerides into FAME(s). The reagent and milk samples were simply mixed into auto sampler vial and a heat-assisted reaction (a process they referred to as thermochemolysis) took place on initiation of the instrument sequence, with the FA quantified by GC mass spectrometry (MS). They compared this to the Rose-Gotlieb extraction method [55], where conventional transesterification using KOH in methanol had achieved comparable results in relation to recoveries and repeatability with the exception of $C_{4:0}$. These authors also reported limitations when dealing with raw nonhomogenised milk with poor repeatability between analyses. This was attributed to the volume of milk that was employed for the analysis not being representative of the entire milk sample. The evaluation of the THM technique was based on linearity, repeatability as a comparison against a conventional extraction method, FFA determination was not incorporated. Yurchenko et al. [56] performed a validated approach for the determination of FA(s) in bovine colostrum. They performed the extraction and preparation of FAME(s) as per the AOAC procedure [57], where methanolic NaOH followed by methanolic BF_3 was added to the sample to form the FAME(s). These were subsequently extracted from the sample by phase separation using heptane and a saturated sodium chloride (NaCl) solution. Linearity, accuracy, precision, LOD and LOQ were reported. There is scope for this method to be applied to milk and other dairy samples; however, individual determination of FFA(s) was not included, and only FA of carbon chain lengths $C_{8:0}$–$C_{18:0}$ were investigated.

Amer et al. [58] described a new approach for the quantification of FFA(s) in milk by GC-MS. They used ethyl chloroformate to form ethyl esters insolution with pyridine added as a catalyst in chloroform. The recovery of deuterated FA internal standards was used to quantify each FFA. The authors also validated the method and reported repeatability, linearity, recoveries, LOD and LOQ. The method was applied to different bovine milk samples, raw, full fat (3.55%), semi-skimmed (1.5%) and skimmed milk (0.1%). The stability of the method appears excellent with an RSD of <4% for all FFA(s) and >99% recoveries. Some of the issues experienced with volatility and water solubility of SCFFA(s) associated with methyl esters are overcome by the use of ethyl esters. In addition, the elimination of the requirement of prior extraction before the addition of the derivitizing agent is of some benefit. However, this method appears limited to aqueous samples only but appears a much more suitable alternative to many existing methods.

3. Conclusion

Despite the importance of FFA determination in dairy products for research, legislative, process development and quality control purposes, very little method development has been under-taken. In addition, little information on validation, analytical robustness, linearity, accuracy, LOD and LOQ was reported. Recent developments to improve extraction methods, where validation was carried out and reported, are a positive step forward [44, 52, 54, 58]. However, scope exists for further development to create methods that can rapidly, accurately and pre-cisely quantify FFA(s) in a wide variety of dairy products in an efficient and robust manner.

Author details

Kieran N. Kilcawley* and David T. Mannion

*Address all correspondence to: kieran.kilcawley@teagasc.ie

Teagasc Food Research Centre, Co. Cork, Ireland

References

[1] Jensen RG, Ferris AM, Lammi-Keefe CJ. The composition of milk fat. Journal of Dairy Science. 1991;**74**(9):3228-3243

[2] Collins YF, McSweeney PLH, Wilkinson MG. Lipolysis and free fatty acid catabolism in cheese: A review of current knowledge. International Dairy Journal. 2003;**13**(11):841-866

[3] Jensen RG. The composition of bovine milk lipids: January 1995 to december 2000. Journal of Dairy Science. 2001;**85**(2):295-350

[4] Taylor MW, MacGibbon AKH. Lipids | fatty acids a2 – Roginski, Hubert. Encyclopedia of Dairy Sciences. Oxford: Elsevier; 2002. pp. 1550-1554

[5] Christie WW. Composition and structure of milk lipids. In: Fox PF, editor. Advanced Dairy Chemistry Volume 2 Lipids. London, UK: Chapman and Hall; 1995. pp. 1-36

[6] Markiewicz-Keszycka M, Czyzak-Runowska G, Lipinska P, Wojtowski J. Fatty acid profile-a review. Bulletin of the Vetinary Institute in Pulawy. 2013;**57**:135-139

[7] Fox PF, Wallace JM. Formation of flavour compounds in cheese. Advances in Applied Microbiology. 1997;**45**:17-85

[8] Kamath S, Wulandewi A, Deeth H. Relationship between surface tension, free fatty acid concentration and foaming properties of milk. Food Research International. 2008;**41**(6): 623-629

[9] Kuzdzal-Savoie S. Determination of free fatty acids in milk and milk products. IDF Bulletin. 1980;**118**:53-66

[10] Gray IK. The quantitative analysis of free fatty acids in dairy products. New Zealand Journal of Dairy Science and Technology. 1975;**10**:158-162

[11] Salih AMA, Anderson M, Tuckley B. The determination of short- and long-chain free fatty acids in milk. Journal of Dairy Research. 1977;**44**(03):601-605

[12] Deeth HC, Fitzgerald CH, Snow AJ. A gas chromatographic method for the quantitative determination fo free fatty acids in milk and milk products. New Zealand Journal of Dairy Science and Technology. 1983;**18**:13-20

[13] De Jong C, Badings HT. Determination of free fatty acids in milk and cheese procedures for extraction, clean up, and capillary gas chromatographic analysis. Journal of High Resolution Chromatography. 1990;**13**(2):94-98

[14] IDF. Determination of indiviual free fatty acids—reference method. IDF Bulletin no 265. Brussels, Belgium: International Dairy Federation. 1991

[15] Harper WJ. Direct determination of acetic, propionic, and butyric acids in cheese Journal of Dairy Science. 1953;**36**(8):808-816

[16] Woo AH, Lindsay RC. Rapid method for quantitative analysis of individual free fatty acids in cheddar cheese. Journal of Dairy Science. 1982;**65**(7):1102-1109

[17] Needs EC, Ford GD, Owen AJ, Tuckley B, Anderson M. A method for the quantitative determination of individual free fatty acids in milk by ion exchange resin adsorption and gas-liquid chromatography. Journal of Dairy Research. 1983;**50**(03):321-329

[18] McNeill GP, O'Donoghue A, Connolly JF. Quantification and identification of flavour components leading to lipolytic rancidity in stored butter. Irish Journal of Food Science and Technology. 1986;**10**(1):1-10

[19] McNeill GP, Connolly JF. A method for the quantification of individual free fatty acids in cheese: Application to ripening of cheddar-type cheeses. Irish Journal of Food Science and Technology. 1989;**13**(2):119-128

[20] Kaluzny MA, Duncan LA, Merritt MV, Epps DE. Rapid separation of lipid classes in high yield and purity using bonded phase columns. Journal of Lipid Research. 1985;**26**(1):135-140

[21] Woo AH, Lindsay RC. Development and characterization of an improved silicic acid-koh arrestant column for routine quantitative isolation of free fatty acids. Journal of the American Oil Chemists Society. 1980;**57**(12):414-416

[22] Stark W, Urbach G, Hamilton JS. Volatile compounds in butter oil: Iv. Quantitative estimation of free fatty acids and free δ-lactones in butter oil by cold-finger molecular distillation. Journal of Dairy Research. 1976;**43**(03):469-477

[23] ISO 15884/IDF 182. Milk fat preparation of fatty acid methyl esters. Genova, Switzerland: International Standard Organisation and Brussels, Belgium: International Dairy Federation; 2002.

[24] Christie WW. Preparation of ester deriviataves of fatty acids for chromatographic analysis In: Christie WW, editor. Advances in lipid methodology 2. Dundee: Oily Press; 1993. pp. 69-111

[25] Christopherson SW, Glass RL. Preparation of milk fat methyl esters by alcoholysis in an essentially nonalcoholic solution. Journal of dairy science. 1969;**52**(8):1289-1290

[26] Luddy F, Barford R, Herb S, Magidman P. A rapid and quantitative procedure for the preparation of methyl esters of butteroil and other fats. Journal of the American Oil Chemists Society. 1968;**45**(8):549-552

[27] Prelog V, Plantanida M. Eine methode zur darstellung von methylestern schwer veresterbarer säuren. Hoppe-Seyler′ s Zeitschrift für physiologische Chemie. 1936;**244**(1-2):56-58

[28] Robb EW, Westbrook JJ. Preparation of methyl esters for gas liquid chromatography of acids by pyrolysis of tetramethylammonium salts. Analytical Chemistry. 1963;**35**(11):1644-1647

[29] Downing DT. Analysis of aqueous solutions of organic acids by pyrolysis of their tetramethylammonium salts in the gas chromatograph. Analytical Chemistry. 1967;**39**(2): 218-221

[30] Downing DT, Greene RS. Methylation of fatty acids by pyrolysis of their tetramethylammonium salts in the gas chromatograph. Analytical Chemistry. 1968;**40**(4):827-828

[31] McCreary DK, Kossa WC, Ramachandran S, Kurtz RR. A novel and rapid method for the preparation of methyl esters for gas chromatography: Application to the determination of the fatty acids of edible fats and oils. Journal of Chromatographic Science. 1978;**16**(8):329-331

[32] Metcalffe L, Wang C. Rapid preparation of fatty acid methyl esters using organic base-catalyzed transesterification. Journal of Chromatographic Science. 1981;**19**(10):530-535

[33] Martínez-Castro I, Alonso L, Juárez M. Gas chromatographic analysis of free fatty acids and glycerides of milk fat using tetramethylammonium hydroxide as catalyste. Chromatographia. 1986;**21**(1):37-40

[34] Martin-Hernández MC, Alonso L, Juárez M, Fontecha J. Gas chromatographic method for determining free fatty acids in cheese. Chromatographia. 1988;**25**(2):87-90

[35] Juárez M, de la Fuente MA, Fontecha J. Improved gas chromatographic method for the determination of the individual free fatty acids in cheese using a capillary column and a ptv injector. Chromatographia. 1992;**33**(7-8):351-355

[36] Chavarri F, Virto M, Martin C, Najera AI, Santisteban A, Barron LJR, et al. Determination of free fatty acids in cheese: Comparison of two analytical methods. Journal of Dairy Research. 1997;**64**(03):445-452

[37] De Jong C, Palma K, Neeter R. Sample preparation before capillary gas-chromatographic estimation of free fatty acids in fermented dairy products. Netherlands Milk and & Dairy Journal 1994;**48**:151-156

[38] Kilcawley KN, Nongonierma AB, Hannon JA, Doolan IA, Wilkinson MG. Evaluation of commercial enzyme systems to accelerate cheddar cheese ripening. International Dairy Journal. 2012;**26**(1):50-57

[39] Hickey DK, Kilcawley KN, Beresford TP, Sheehan EM, Wilkinson MG. Starter strain related effects on the biochemical and sensory properties of cheddar cheese. Journal of Dairy Research. 2007;**74**(1):9-17

[40] Hickey DK, Kilcawley KN, Beresford TP, Sheehan EM, Wilkinson MG. The influence of a seasonal milk supply on the biochemical and sensory properties of cheddar cheese. International Dairy Journal. 2006;**16**(6):679-690

[41] Calzada J, del Olmo A, Picon A, Nuñez M. Effect of high-pressure-processing on lipolysis and volatile compounds of brie cheese during ripening and refrigerated storage. International Dairy Journal. 2014;**39**(2):232-239

[42] Emery EM, Koerner WE. Double-column programmed temperature gas chromatography. Volatile polar column packings and quantitative aspects. Analytical Chemistry. 1962;**34**(10):1196-1198

[43] Ackman RG, Burgher RD. Quantitative gas liquid chromatographic estimation of volatile fatty acids in aqueous media. Analytical Chemistry. 1963;**35**(6):647-652

[44] Mannion DT, Furey A, Kilcawley KN. Comparison and validation of 2 analytical methods for the determination of free fatty acids in dairy products by gas chromatography with flame ionization detection. Journal of Dairy Science. 2016;**69**(7):5047-5063

[45] Parodi PW. A study of ester preparation for the gas chromatographic analysis of the fatty acids composition of butterfat. Australian Journal of Dairy Technology. 1967;**22**:144-145

[46] Metcalfe LD, Schmitz AA, Pelka JR. Rapid preparation of fatty acid esters from lipids for gas chromatographic analysis. Analytical Chemistry. 1966;**38**(3):514-515

[47] DeMan JM. Determination of the fatty acid composition of milk fat by dual column temperature programmed gas-liquid chromatography. Journal of Dairy Science. 1964;**47**(5):546-547

[48] Gander GW, Jensen RG, Sampugna J. Analysis of milk fatty acids by gas-liquid chromatography. Journal of Dairy Science. 1962;**45**(3):323-328

[49] Sampugna J, Pitas RE, Jensen RG. Gas-liquid chromatographic analysis of milk fat: Rapid preparation of butyl esters. Journal of Dairy Science. 1966;**49**(11):1462-1463

[50] Parodi P. Fatty acid composition of australian butter and milk fats. Australian Journal of Dairy Technology. 1970;**25**:200-205

[51] Kim Ha J, Lindsay RC. Method for the quantitative analysis of volatile free and total branched-chain fatty acids in cheese and milk fat. Journal of Dairy Science. 1990;**73**(8): 1988-1999

[52] Folch J, Lees M, Stanley GHS. A simple method for the isolation and purification of total lipides from animal tissues. Journal of Biological Chemistry. 1957;**226**(1):497-509.

[53] Firl N, Kienberger H, Rychlik M. Validation of the sensitive and accurate quantitation of the fatty acid distribution in bovine milk. International Dairy Journal. 2014;**35**(2):139-144

[54] Reis MG, dos Reis MM, Leath S, Stelwagen K. Direct analysis of fatty acid profile from milk by thermochemolysis–gas chromatography–mass spectrometry. Journal of Chromatography A. 2011;**1218**(2):316-323

[55] Walstra P, Mulder H. Gravimetric methods for the determination of fat content of milk and milk products Netherlands Milk and & Dairy Journal. 1962;**16**:172-184

[56] Yurchenko S, Sats A, Poikalainen V, Karus A. Method for determination of fatty acids in bovine colostrum using GC-FID. Food Chemistry. 2016;**212**:117-122

[57] AOAC. Methods of analysis. 96933 Fatty acids in oils and fats (17th edition). Virginia, USA. 2000

[58] Amer B, Nebel C, Bertram HC, Mortensen G, Hermansen K, Dalsgaard TK. Novel method for quantification of individual free fatty acids in milk using an in-solution derivatisation approach and gas chromatography-mass spectrometry. International Dairy Journal. 2013;**32**(2):199-203

2

The Synthesis of Milk Medium-Chain Fatty Acids in Mammary Gland

Jiangjiang Zhu and Jun Luo

Abstract

The fatty acid de novo synthesized in mammary gland is mainly catalyzed by fatty acid synthase (FASN) and acetyl CoA carboxylase (ACC), including all the short- and medium-chain fatty acid and half part of the palmitate in ruminants. However, the synthesis mechanism of medium-chain fatty acid among different species is different. In non-ruminants, a tissue-specific enzyme thioesterase II (TE II) can interact with TE I, which is a part of FASN, and terminate the elongation of fatty acids at about 10 carbons. However, in ruminants' mammary-gland acetyl/malonyl-CoA transferase (MAT) is predicted to be involved in the termination of medium-chain fatty acid without the presence of (TE II). A more exact understanding about the mechanism of synthesis of medium-chain fatty acid in different species is still unclear. This review gives the research development of synthesis mechanism of medium-chain fatty acid in mammary gland among different species.

Keywords: mammary gland, medium-chain fatty acid, FASN, ruminants

1. FASN and fatty acid synthesis

The composition of milk fatty acid varies greatly among different species. Long-chain fatty acid is the most abundant fatty acid (C14-C18) in guinea pigs, medium-chain fatty acid takes a large part of medium-chain fatty acid (C8-C10), and long-chain and short-chain (C4-C6) fatty acids are the most abundant in cow's milk [1]; however, in goat milk all carbon-chain fatty acids are observed including short-chain, medium-chain, and long-chain fatty acids [2]. The short-chain and medium-chain fatty acids are mainly de novo synthesized in mammary gland [3–5].

The synthesis of fatty acid in mammary gland is mainly catalyzed by fatty acid synthase (FASN) and acetyl CoA carboxylase (ACC). In mammalian cells, the functional form of FASN is a homodimer (MW ~540,000 Da) [6]. FASN is composed of seven domains, including b-ketoacyl synthase (KS), acetyl/malonyl-CoA transferase (MAT), b-hydroxyacyl dehydratase (DH), enoyl reductase (ER), b-ketoacyl reductase (KR), acyl-carrier protein (ACP), and thioesterase I (TE I) (**Figure 1**). The core region between the DH and ER domains has no catalytic activity. At first, the acyl moiety of acetyl-CoA (initiation substrate) is transferred to the ACP catalyzed by MAT to generate malonyl-CoA (elongation substrate). And then, the acyl moiety is momentarily transferred to KS, and transacylation of malonyl-CoA is catalyzed by MAT to ACP. Acetoacetyl-ACP is then generated with decarboxylative condensation catalyzed by KS. This process follows catalyzation by KR and DH, which are responsible for the NADPH-dependent reduction of the b-carbon, and the dehydration of b-hydroxyacyl-ACP to a,b-enoyl, respectively. And then, by the catalyzation of ER, a four-carbon acyl chain is produced from NADPH-dependent reduction of the enoyl. The subsequent elongation cycles are performed with the malonyl-CoA as two-carbon units. Lastly, TE I is responsible for the release of fatty acid, with a length of 16 carbons, from ACP [6].

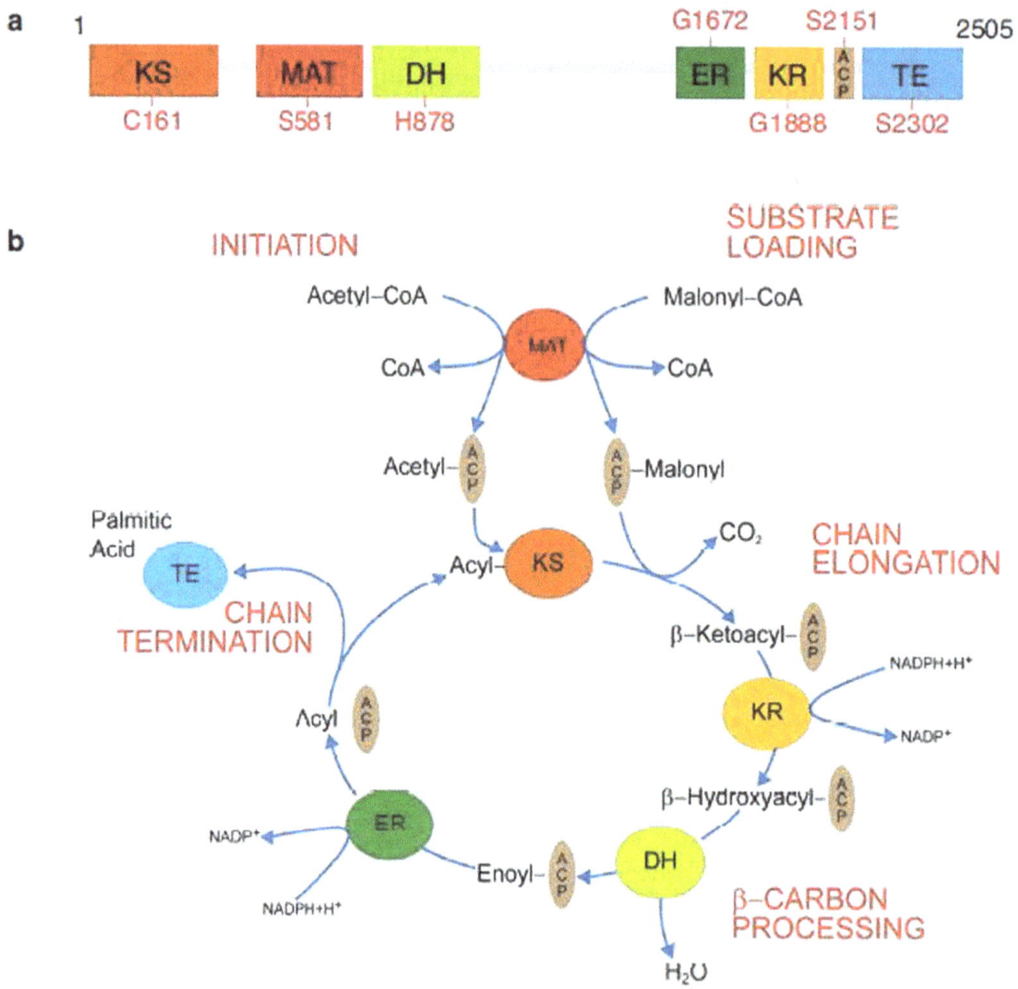

Figure 1. The structure of FASN gene and the fatty acid synthesis cycle. (a) The domain structure of FASN and the active site residues. (b) De novo fatty acid synthesis cycle catalyzed by domains of FASN.

FASN is the crucial enzyme for milk fat synthesis [7]. It plays an important role in the regulation of energy metabolism, cell membrane formation, signaling pathway regulation, and epigenetics [8]. By comparing the sequences of FASN in different species, including goat, sheep, cattle, human, chicken, and rats, FASN genes contain 42 exons in cattle and 43 exons in human and mice, and so on [9] (**Figure 2**).

FASN is also important for the regulation of the genes related to fatty acid synthesis in mammary gland. The inhibition of FASN with orlistat, a natural inhibitor of FASN, suppressed the expression of ACC, lipoprotein lipase (LPL), and heart-type fatty acid-binding protein (H-FABP) [10]. The inhibition of FASN in mammary cells by C75-mediated interference, a synthetic inhibitor of FASN activity, and short hairpin RNA-mediated interference significantly suppressed the deposition of triglyceride, decreased the expression of glycerol-3-phosphate acyltransferase (GPAT), 1-acylglycerol-3-phosphate O-acyltransferase (AGPAT6), and diacylglycerol acyltransferase (DGAT2), which are important for cellular triglyceride synthesis. The inhibition of FASN also enhanced the expression of adipose triglyceride lipase (ATGL) and hormone-sensitive lipase (HSL), both of which are crucial for lipolysis. This is consistent with the markedly lower expression of genes related to lipid-droplet formation and secretion (hormone-sensitive lipase, TIP47; adipose differentiation-related protein, ADFP; xanthine oxidoreductase, XDH; butyrophilin 1A1, BTN1A1) [10].

Despite the effect of FASN inhibition on gene expression and lipid metabolism, the exact mechanism underlying the effect remains unclear. As a terminal enzyme for de novo fatty acid synthesis, previous studies indicated that FASN could help control lipid metabolism through indirect, hence, secondary effects. For instance, FASN could be involved in the control of fatty acid synthesis and oxidation through a direct effect on the concentration of malonyl-CoA, which was shown to help control lipid metabolism through the inhibition (allosteric) of β-oxidation by carnitine palmitoyltransferase 1 (CPT1) [11, 12]. In addition, FASN also helps generate ligands for transcription regulators, including peroxisome proliferator-activated receptors (PPARs) [13, 14], sterol-regulatory element-binding protein 1 (SREBP1), hepatocyte nuclear factor 4a (HNF4α), NF-E2-related factor-2 (NRF2), and Toll-like receptor 4 (TLR4) [14], all of which are important for lipid metabolism regulation. FASN also may affect protein activity indirectly, for example, endothelial nitric-oxide synthase, through palmitoylation [15]. Lastly, FASN may interact directly with caveolin-1 and lipid raft, which was involved in lipid secretion [16, 17]. However, the exact mechanism of how FASN affects the expression of genes related to milk fat synthesis remains unclear.

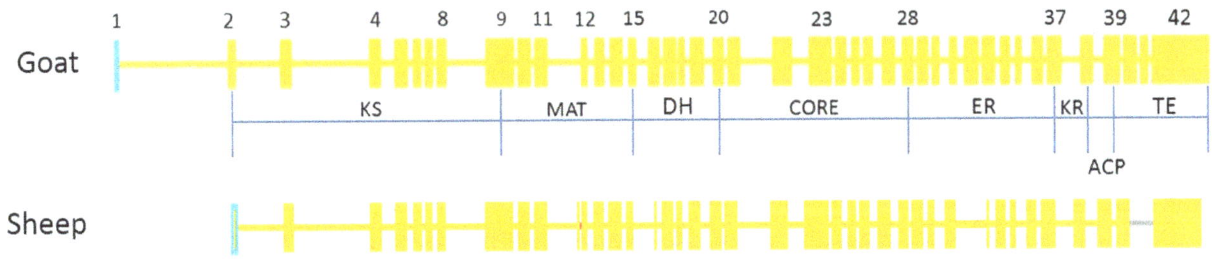

Figure 2. Prediction of exons and functional domains of FASN cloned from the goat mammary gland. Goat FASN gene includes 42 exons represented by seven function domains, in which three catalytic domains in the N-terminal section (KS, MAT, and DH) were separated by a core region from four C-terminal domains (ER, KR, ACP, and TE I). The sheep FASN gene contains 41 exons (XM 004013447).

2. The synthesis of medium-chain fatty acids in non-ruminants

The activity of hydrolase was necessary for the control of the length of the fatty acids produced by FASN with the main product of FASN being C16 palmitate. The thioesterase domain, TE I, catalyzes the termination step by hydrolyzing the thioester bond between palmitate and the 4'-phosphopantetheine moiety of the acyl-carrier protein (ACP) domain. Orlistat is a natural inhibitor of FASN, which was captured in the active sites of two thioesterase molecules as a stable acyl-enzyme intermediate and as the hydrolyzed product. The release of the thioesterase from FASN by limited proteolysis, however, results in the production of fatty acids containing 20–22 carbons [18]. Thus, the thioesterase domain is essential in regulating the length of the fatty acid chain.

Carey et al. [4, 19] purified the FASN protein from rabbit, mouse, and pig; the results showed that they just synthesized long-chain and short-chain fatty acids but not medium-chain fatty acids. However, with the help of TE II, FASN can produce all kinds of fatty acids, including C8, C10, C12, and so on [20, 21]. The medium-chain fatty acids synthesized de novo can be incorporated directly into triacylglycerol without the need of an activation step [22].

TE II is a tissue-specific enzyme independent of FASN. TE II can interact with TE I, which is a part of FASN, and terminate the elongation of fatty acids at about 10 carbons [23]. Decanoyl-CoA and decanoyl-pantetheine are the best substrates for TE II [24]. Ser101 and His237 of TE II are critical for the interaction. Asp236 of TE II enhances but is not essential for the reactivity of Ser101 and His237. For Leu262, it is involved in the interaction of TE II with TE I [23]. The interaction between TE II and FASN is stimulated by polyethylene glycol and suppressed by high concentrates of salt. Orlistat is a natural inhibitor and has a significant inhibition effect on FASN activity by curbing the binding of TE domain between FASN subunits [25]. TE II also can interact with P53 and is involved in the cancer development [25].

3. MAT and medium-chain fatty acid synthesis in ruminants

As a tissue-specific enzyme, TE II is only observed in non-ruminants. A similar enzyme is not present in goat mammary gland, although octanoic acid, decanoic acid, and dodecanoic acid amount to 20 mol% of the fatty acids synthesized in this tissue [26]. By contrast, goat mammary-gland fatty acid synthetase is by itself able to synthesize medium-chain fatty acids in the presence of the microsomal fraction and substrates for triacylglycerol synthesis [27]. Goat mammary-gland fatty acid synthetase exhibits both medium-chain thioesterase [21] and transacylase activity [28]. It seems that there are some differences about the activity of FASN between goat and other species. By treating the FASN protein of goat FASN with PMSF, an inhibitor of TE activity results in the termination of long-chain fatty acid synthesis, but not the medium-chain fatty acids. These results showed that there must be another hydrolase center in FASN protein. Hansen and Knudsen found that the supplementary of malonyl-CoA changed the proportion of medium-chain fatty acids [29]. Considering that the malonyl-CoA is transferred into the reaction by MAT activity, it was predicted that MAT may be involved in the synthesis of medium-chain fatty acid. Engese et al. first found that MAT can transfer not only C2 but also C4, C6, and C10 fatty acids from CoA [28, 30], although the malonyl-CoA showed the strongest affinity with MAT. Supporting the results, the interference of FASN reduced the synthesis of C10:0 and C12:0, increased the content of C14:0, but without an effect on C16:0 and C18:0, in goat mammary epithelial cells [31]. These literatures indicate that the MAT of goat FASN may be responsible for the termination of short- and medium-chain fatty acids in mammary gland.

Rangan and Smith cloned the MAT protein from rat, and showed that MAT expressed in *Escherichia coli*, and refolded in vitro as a catalytically active malonyl-/acetyltransferase [32], similar as our previous study in goat expressed for goat mammary epithelial cells that the MAT domain is capable of folding correctly as an independent protein (unpublished). Replacement of the highly conserved residue His-683 with Ala reduced the activity by 99.95%, and the residual activity was relatively unaffected by diethyl pyrocarbonate. The rate of acylation of the active site serine residue was also reduced by several orders of magnitude in the His-683 to Ala mutant, indicating that His-683 plays an essential role in catalysis, likely by accepting a proton from the active site serine, thus increasing its nucleophilicity. In addition, Ser581 is also important for the activity of MAT. The Ser-581 to Ala mutant was completely inactive with either substrate. The Ser-581 to Cys mutant, however, retained approximately 1% of the activity of the "wild-type" enzyme compared with about 0.05% retained by His-683 to Ala mutation.

Actually, many other stimulants altered the relative content of medium-chain fatty acids. By treating the goat FASN with 5.2 mg/mL, albumin increases the content of C10 fatty acids relative to C12, while 10.4 mg/mL results in four times of C10 than C12 [33]. The supplementary of Malonyl-CoA enhances the synthesis of C12, similar as the effect of globin treatment. These treatments can only change the relative content among different kinds of medium-chain fatty acids, but not alter the total proportion of medium-chain fatty acids relative to long-chain fatty acids [33, 34].

The role of MAT is predicted to be critical for the goat medium-chain fatty acid synthesis; however, there is no direct evidence for the effect of MAT on fatty acid compositions. Although several genes have been proved to be involved in the regulation of fatty acid metabolism in mammary gland, how the activity of MAT, not for the whole FASN, on medium-chain fatty acid synthesis is controlled remains unclear. Recently, our study showed that malonyl-CoA is not only the elongation subtract for de novo fatty acid synthesis but also a regulatory for fatty acid deposition as triglyceride and fatty acid oxidation by inhibiting the activity of carnitine palmitoyl transferase 1 (CPT1), which is important for the transportation of long-chain fatty acids from cytoplasm to mitochondria [35]. Thus, it is the controller that directly interacts with MAT domain indicating that malonyl-CoA may be the balancer of lipid metabolism. The concentration of malonyl-CoA is mainly regulated by the activity of ACC and FASN, both of which are the terminates of lipid metabolism regulation. The hypothesis is that malonyl-CoA may be the indicator of fatty acid metabolism in mammary gland and may be the main mediator for the length of fatty acid responses to the environment around the mammary gland. This may be a good point for revealing the mechanism of fatty acid chain length determination in the future study.

Author details

Jiangjiang Zhu[1,2,3*] and Jun Luo[3*]

Address all correspondence to: zhujiang4656@hotmail.com and luojun@nwsuaf.edu.cn

1 Key Laboratory of Sichuan Province for Qinghai-Tibetan Plateau, Animal Genetic Resources Reservation and Exploitation, Chengdu, P.R. China

2 Key Laboratory of State Ethnic Affairs Commission and Ministry of Education for Animal Genetics & Breeding, Chengdu, Sichuan, P.R. China

3 College of Animal Science and Technology, Northwest A&F University, Yangling, Shaanxi, P.R. China

References

[1] Breckenridge WC, Kuksis A. Molecular weight distributions of milk fat triglycerides from seven species. Journal of Lipid Research. 1967;**8**(5):473-478.

[2] Marai L, Breckenridge W, Kuksis A. Specific distribution of fatty acids in the milk fat triglycerides of goat and sheep. Lipids. 1969;**4**(6):562-570.

[3] Annison EF, Linzell JL, Fazakerley S, Nichols BW. The oxidation and utilization of palmitate, stearate, oleate and acetate by the mammary gland of the fed goat in relation to their overall metabolism, and the role of plasma phospholipids and neutral lipids in milk-fat synthesis. Biochemical Journal. 1967; **102**(3):637-647.

[4] Carey EM, Dils R. The pattern of fatty acid synthesis in lactating rabbit mammary gland studied in vivo. Biochemical Journal. 1972;**126**(4):1005-1007.

[5] Cowie AT, Duncombe WG, Folley SJ, French TH, Glascock RF. Synthesis of milk fat from acetic acid ($CH_3{}^{14}COOH$) by the perfused isolated bovine udder. Biochemical Journal. 1951;**49**(5):610-615.

[6] Asturias FJ. Mammalian fatty acid synthase: x-ray structure of a molecular assembly line. ACS Chemical Biology. 2006;**1**(3):135-138.

[7] Smith S. The animal fatty acid synthase: one gene, one polypeptide, seven enzymes. FASEB Journal. 1994;**8**(15):1248-1259.

[8] Olsen CP. Why Make Fat?: The Impact of de novo Fatty Acid Synthesis on Fat Storage, Reproduction, and Longevity in C. e/egans [thesis]. Washington: University of Washington; 2011. 148 p.

[9] Roy R, Taourit S, Zaragoza P, Eggen A, Rodellar C. Genomic structure and alternative transcript of bovine fatty acid synthase gene (FASN): comparative analysis of the FASN

gene between monogastric and ruminant species. Cytogenetic and Genome Research. 2005; **111**(1):65-73.

[10] Zhu JJ, Luo J, Sun YT, Shi HB, Li J, Wu M, Yu K, Haile AB, Loor JJ. Short communication: Effect of FASN inhibition on triglyceride accumulation and impact on lipid metabolism genes in goat mammary epithelial cells. Journal of Dairy Science. 2015; **94**(5):3485-3491.

[11] Saggerson D. Malonyl-CoA, a key signalling molecule in mammalian cells. Annual Review of Nutrition. 2008; **28**:253-272.

[12] Foster DW. Malonyl-CoA: the regulator of fatty acid synthesis and oxidation. Journal of Clinical Investigation. 2012; **122**(6):1958-1959.

[13] Chakravarthy MV, Lodhi IJ, Yin L, Malapaka RR, Xu HE, Turk J, Semenkovich CF. Identification of a physiologically relevant endogenous ligand for PPARalpha in liver. Cell. 2009; **138**(3):476-488.

[14] Georgiadi A, Kersten S. Mechanisms of gene regulation by fatty acids. Advances in Nutrition. 2012; **3**(2):127-134.

[15] Wei X, Schneider JG, Shenouda SM, Lee A, Towler DA, Chakravarthy MV, Vita JA, Semenkovich CF. De novo lipogenesis maintains vascular homeostasis through endothelial nitric-oxide synthase (eNOS) palmitoylation. Journal of Biological Chemistry. 2011; **286**(4):2933-2945.

[16] Witkiewicz AK, Nguyen KH, Dasgupta A. Co-expression of fatty acid synthase and caveolin-1 in pancreatic ductal adenocarcinoma: Implications for tumor progression and clinical outcome. Cell Cycle. 2008; **7**(19):3021-3025.

[17] Vizio DD, Adam RM, Kim J, Kim R, Sotgia F, Williams T, Demichelis F, Keith R. Solomon, Loda M, Rubin MA et al. Caveolin-1 interacts with a lipid raft-associated population of fatty acid synthase. Cell Cycle. 2008; **7**(15):2257-2267.

[18] Pemble CW, Johnson LC, Kridel SJ, Lowther WT. Crystal structure of the thioesterase domain of human fatty acid synthase inhibited by Orlistat. Nature Structural & Molecular Biology. 2007; **14**(8):704-709.

[19] Carey EM, Dils R. Fatty acid biosynthesis V. Purification and characterisation of fatty acid synthetase from lactating-rabbit mammary gland. Biochimica et Biophysica Acta (BBA) – Lipids and Lipid Metabolism. 1970; **210**(3):371-387.

[20] Knudsen J, Clark S, Dils R. Purification and some properties of a medium-chain acylthioester hydrolase from lactating-rabbit mammary gland which terminates chain elongation in fatty acid synthesis. Biochemical Journal. 1976; **160**(3):631-691.

[21] Grunnet I, Knudsen J. Medium chain acyl-thioester hydrolase activity in goat and rabbit mammary gland fatty acid synthetase complexes. Biochemical and Biophysical Research Communications. 1978; **80**(4):745-749.

[22] Knudsen J, Hansen JK, Grunnet I. Quantitative extraction and separation from [1-^{14}C] acetate and preparation for radio gas chromatography of microgram amounts of

a ^{14}C-labeled mixture of butyrate and longer-chain-length fatty acids. Analytical Biochemistry. 1981; **112**(1):190-193.

[23] Tai MH, Chirala SS, Wakil SJ. Roles of Ser101, Asp236, and His237 in catalysis of thioesterase II and of the C-terminal region of the enzyme in its interaction with fatty acid synthase. Proceedings of the National Academy of Sciences of the United States of America. 1993; **90**(5):1852-1856.

[24] Smith S. Mechanism of chain length determination in biosynthesis of milk fatty acids. Journal of Mammary Gland Biology and Neoplasia. 2009; **14**(3):245-260.

[25] Chuang HY, Chang YF, Hwang JJ. Antitumor effect of orlistat, a fatty acid synthase inhibitor, is via activation of caspase-3 on human colorectal carcinoma-bearing animal. Biomedicine & Pharmacotherapy. 2011; **65**(4):286-292.

[26] Grunnet I, Knudsen J. Fatty-acid synthesis in lactating-goat mammary gland 1. Medium-chain fatty-acid synthesis. European Journal of Biochemistry. 1979; **95**(3):497-502.

[27] Grunnet I, Knudsen J. Fatty-acid synthesis in lactating-goat mammary gland 2. Medium-chain fatty acid synthesis. European Journal of Biochemistry. 1979; **95**(3):503-507.

[28] Knudsen J, Grunnet I. Primer specificity of mammalian mammary gland fatty acid synthetases. Biochemical and Biophysical Research Communications. 1980; **95**(4):1808-1814.

[29] Hansen K, Knudsen J. Transacylation as a chain-termination mechanism in fatty acid synthesis by mammalian fatty acid synthetase. Synthesis of butyrate and hexanoate by lactating cow mammary gland fatty acid synthetase. Biochemical Journal. 1980; **186**(1):287-294.

[30] Mikkelsen J, Højrup P, Hansen HF, Hansen JK, Knudsen J. Evidence that the medium-chain acyltransferase of lactating-goat mammary-gland fatty acid synthetase is identical with the acetyl/malonyltransferase. Biochemical Journal. 1985; **227**(3):981-985.

[31] Zhu JJ, Luo J, Wang W, Yu K, Wang HB, Shi HB, Sun YT, Lin XZ, Li J. Inhibition of FASN reduces the synthesis of medium-chain fatty acids in goat mammary gland. Animal. 2014; **8**(9):1469-1478.

[32] Hoppel CL, Kerner J, Turkaly P, Turkaly J, Tandler B. The malonyl-CoA-sensitive form of carnitine palmitoyltransferase is not localized exclusively in the outer membrane of rat liver mitochondria. Journal of Biological Chemistry. 1998; **273**(36):23495-23503.

[33] Knudsen J, Grunnet I. Transacylation as a chain-termination mechanism in fatty acid synthesis by mammalian fatty acid synthetase. Synthesis of medium-chain-length (C8-C12) acyl-CoA esters by goat mammary-gland fatty acid synthetase. Biochemical Journal. 1982; **202**(1):139-143.

[34] Sumper M, Oesterhelt D, Riepertinger C, Lynen F. Synthesis of various carboxylic acids by the fatty acid synthetase multienzyme complex of yeast and the explanation for their structure. European Journal of Biochemistry. 1969; **10**(2):377-387.

3

Viable Alternatives Study for Reusing Lipids from Microalgae Biomass Present in the Generated Sludge in the Supply Water Treatment Processes

Livia de Oliveira Ruiz Moreti, Rosa Maria Ribeiro,
Letícia Nishi and Rosângela Bergamasco

Abstract

This chapter aims to evaluate the microalgae species' removal efficiency, using *Moringa oleifera* powder seeds as a natural coagulant with subsequent lipid profile characterization. For the tests were used deionized water artificially contaminated with cell cultures of *Anabaena flos-aquae* and *Chlorella vulgaris*, with a cell density in the order of 10^4 and 10^6 cells mL^{-1}, respectively. Coagulation/flocculation/dissolved air flotation (C/F/DAF) tests were conducted using 'Flotest' equipment, using *M. oleifera* powder seeds in the dosage range of 50–1000 mg L^{-1}. For fatty acid profile analyses, a gas chromatograph equipped with a flame ionization detector was used. Variations of the coagulant dosages showed that there was a difference between dosages and that 100 mg L^{-1} provided the best removal efficiency for *A. flos-aquae* (96.5, 80.5 and 78.1%) and 140 mg L^{-1} for *C. vulgaris* (90.5, 78.34 and 70%) of the tested parameters of chlorophyll, color and turbidity, respectively. In relation to the produced sludge, it was observed that the use of this coagulant in the treatment of water contaminated with microalgae produces a biodegradable sludge, rich in lipids, especially oleic acid (>60%). Thus, these results indicate that the sludge's reutilization could be a good alternative to biodiesel production, as it represents an environmentally viable method for reusing residual biomass produced in the water treatment process.

Keywords: reusing lipids, microalgae, water treatment processes

1. Introduction

Several natural coagulants have been studied for water purification; recently, *Moringa oleifera* Lam comes to stand out because it has good color and turbidity removal and promotes large bacteria removal, of above 90% [1]. Currently, there are few studies on the coagulant application in eutrophic waters, but in some works, can be seen an excellent microalgae cells removal efficiency [2–4].

The dissolved air flotation has been considered a viable alternative to the sedimentation step when applied to water treatment in the microalgae's presence, since this process is capable of removing whole cells, besides decreasing the time contact between the generated sludge with water treatment. The waste generated removal is performed by mechanical equipment, which is installed in water and has an easy maintenance [5, 6].

The proper disposal of the generated sludge by water treatment plants (WTPs) is essential, according to NBR 10004 [7], is considered a solid residue, which if released without proper treatment in waterways, can cause direct effects in the aquatic environment, damaging receiving fauna and flora. The nonchalance towards the waste generated, causes impacts, such as increase in the amount of solids in water body, water body siltation, increased of color, turbidity and aluminum concentration in the water, decrease of the water's pH, releasing of odors, decreases the amount of dissolved oxygen in the water body and chronic toxicity on aquatic organisms and their vision.

However, this chapter seeking environmentally viable alternatives to microalgae biomass reuse present in the generated sludge by water treatment plants, for example, to characterize the lipid content produced by these microalgae, to consign them to a further biodiesel production.

Thus, this chapter shows the removal of *Anabaena flos-aquae* and *Chlorella vulgaris* cells by coagulation/flocculation/dissolved air flotation (C/F/DAF) processes, using *M. oleifera* as a natural coagulant, with subsequent lipid characterization of the generated sludge in order to check the potential for reusing this biomass.

2. Microalgae

Phylogenetically, microalgae are prokaryotes or eukaryotes organisms, according to the period when they appeared on the planet, belonging to a very heterogeneous group of microorganisms. According to Andrade et al. [8] the microalgae are photosynthetic microorganisms that combining water and atmospheric carbon dioxide with sunlight to produce biomass (polysaccharides, proteins, lipids and hydrocarbons). Which can be used in biofuel production, feeding supplements and also can be used in atmospheric carbon dioxide capture. Microalgae produce more oxygen than all plants in the world, accounting for at least 60% of the Earth's primary production.

The microalgae biomass presents about 50% carbon in its composition, so the supply of this nutrient to these microorganisms' cultures represents an important component of the production costs [9]. However, it is necessary to take care of microalgae culture systems, considering the peculiarities of each species, adaptation to the environment, as well as the availability of nutrients associated with economic viability [10].

The exact number of microalgae species is not yet known, but many species may already grow in cropping systems. The most difficult task, however, is to grow specific species for oil production [10].

Some microalgae species under adverse environmental conditions, such as nutrient stress (lack of nitrogen or phosphorus), may accumulate lipids. The green algae specie, *C. vulgaris*, is widely used in research on the biofuel production from microalgae [11, 12], and in the present study, was chosen for the comparison of the lipid content with the cyanobacteria *A. flos-aquae*.

2.1. Microalgae lipid content

The microalgae biomass contains three main components: carbohydrates, proteins and lipids [13]. In biological systems, the lipids function as membrane components, reserve products, metabolites and as energy sources, with most of them consists of fatty acids. Thus, the lipids are classified in storage lipids (neutral lipids), triacylglycerols and membrane lipids (polar lipids), phospholipids, glycolipids and sterols [14].

Fatty acids are fundamental units most of the lipids. They are short-chain and long-chain organic acids having 4–24 carbon atoms, and short-chain fatty acids are ideal for the biodiesel production [15].

Some fatty acids synthesized by microalgae, such as omega 3 and 6 (ω-3 and ω-6), which are the main precursors of some hormones such as prostaglandins, prostacyclins, leukotrienes and thromboxanes, have a high economic value in the food and pharmaceutical industry and are fundamental for the development and physiological regulators [16].

Fatty acids in microalgae correspond to the largest lipid fraction, and, in some species, polyunsaturates represent between 25 and 60% of total lipids [17].

The polyunsaturated fatty acids from microalgae have a very promising market in biotechnol ogy, especially in the functional food industry [18]. Studies presented by Favaro-Trindade et al.(2008) [48] show that lipids, especially unsaturated fatty acids, have been encapsulated to reduce susceptibility to oxidation.

According to Nelson e Cox [14], fatty acids have a unique carboxyl group and a non-polar hydrocarbon tail, which give lipids their oily and fatty nature, insoluble in water. They occur in cells or tissues in forms covalently attached to different lipids' classes. Different fatty acids have been isolated from lipids of various species.

They differ by the chain extension and its presence, number and double bonds position, and some fatty acids also have methyl-branched groups.

The glycolipids that are composed of glycerol have been found in many organisms, being observed as the main lipid component of microalgae photosynthetic membranes, including cyanobacteria (blue microalgae). Its structure is analogous to that of glycerophospholipids with a sugar molecule, glycosidically attached to glycerol three position and fatty acids esterified in the other two positions.

Among the main glycosylacylglycerols of microalgae and plant photosynthetic membranes is monogalactosyl-diacylglycerol (MGDG), which occurs abundantly in plants and algae, especially in chloroplasts. Contains high proportions of polyunsaturated fatty acids. For *C. vulgaris*, the MGDG presents mainly oleic acid (C18:1) and linoleic acid (C18:2) when cultivated in the dark and 20% of linolenic acid (C18:3) when cultivated in the light [19].

The lipid production estimation by microalgae ranges from 15,000 to 30,000 L.km^2, and its extraction is simple and can be applied to traditional methods used in the chemical industry, including solvent extraction [20].

It is known that among the nutrients that can influence the lipids and fatty acids production are the sources of nitrogen and sulfur, which are used by microalgae in the synthesis of amino acids and fatty acids [20].

The main applications of fatty acids microalgae occur in the enrichment of fish feed, biodiesel production and sources of essential unsaturated fatty acids in the human diet [21].

Although there are many microorganism types capable to accumulating lipids, not all of them have favorable characteristics for the application in the biodiesel production. The microalgae stand out because they present, in some cases, compatibility in the ratio of their oil produced to the vegetable oil used in the transesterification process [22, 23].

According to Schimitz et al. [10], the presence of polyunsaturated compounds produced by microalgae causes a decrease in the stability of produced biodiesel. However, due to the presence of these fatty acids, biodiesel from microalgae presents a high yield at low temperatures, a characteristic that is not presented by conventional oilseed biodiesel, which have little yield at relatively low temperatures.

2.2. *Moringa oleifera* Lam

M. oleifera Lam is a tropical tree that grows naturally in India, South Saharan Africa and South America [24]. The leaves, flowers, seeds, roots and bark may be used as food or for medicinal and therapeutic purposes [25], especially in developing countries [24, 26]. In addition, other applications were pointed out for cosmetics preparation, mechanical lubricants and even for potential biodiesel elaboration [27].

According to Ndabigengesere and Narasiah e Talbot [28], *M. oleifera* seeds contain about 37% protein, 35% lipids and 5% carbohydrates (oligosaccharides). The carbohydrate content is very low whereas the high lipid content explains why the seeds can be used as a source of vegetable oil. This oil resembles olive oil in its composition, being rich in oleic acid, which makes it suitable for edible purposes [29].

The *M. oleifera* seeds are also very useful as a coagulant in drinking water clarification and effluent treatment since 1979, due to the presence of a water soluble cationic coagulant protein capable of reducing the turbidity of the treated water. The seeds can be used in the form of powder, such as aqueous or saline extracts [24, 30, 31].

2.3. Removal of microalgae from water using *Moringa oleifera*, as natural coagulant

Coagulation/flocculation (C/F) followed by dissolved air flotation (DAF) is suitable for the treatment of naturals and synthetic eutrophic waters [5, 32]. When it comes to the removal of cyanobacterial cells, DAF is an effective alternative, as shown by some studies in the literature [5, 6]. However, to achieve good efficiency, water treatment plants (WTPs) use a series of auxiliary products in the process, especially the use of inorganic coagulants, usually based on metals such as aluminum, as well as pH control. However, these coagulants do not generate biodegradable sludge, causing problems in terms of disposal and treatment; this may be also related to some diseases, such as Alzheimer's disease, due to residual aluminum in treated water [33, 34]. Thus, the search becomes necessary for alternative natural coagulants that are biodegradable and safe to human health [35].

The *M. oleifera* (MO) seeds can be used for efficient clarification of water [26, 36]. This efficiency has been shown by high values of color, turbidity and bacteria removal [2, 37] and even cyanobacteria cells in the water treatment process [38], as well as some economic and environmental advantages related to decreasing the costs of synthetic products to correct the pH of water and produce a sludge without metals.

The water treatment processes in WTPs produce residues, mostly water used for washing the filters and sludge from sedimentation tanks/floaters [39]. Particularly, in WTPs with cyanobacteria problems, the sludge generated is composed of microalgae biomass. Knowing that such biomass has a relatively high amount of lipids in their composition [40], which could be used for biodiesel production.

Firstly, by evaluating the results obtained (**Figure 1**), one can observe the percentages of the removal of the parameters color, turbidity and chlorophyll-a and compounds with absorp-

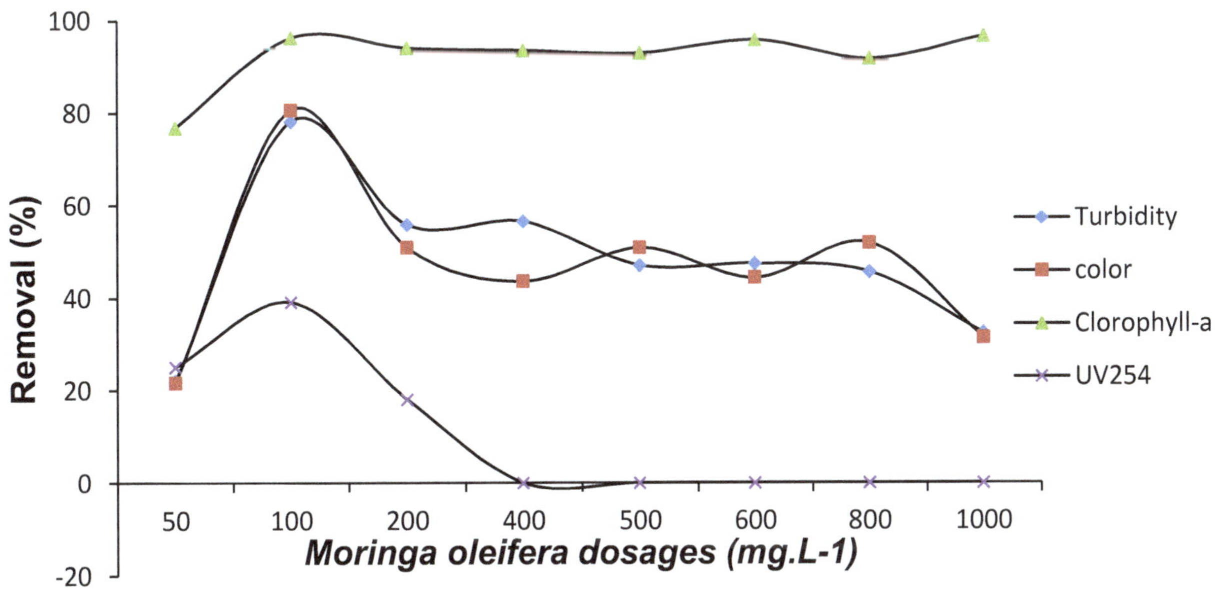

Figure 1. Color, turbidity, compounds with absorption in $UV_{254\,nm}$ and chlorophyll-a removal in relation to the dosage of *Moringa oleifera* powder seeds for *Anabaena flos-aquae*.

tion in $UV_{254\,nm}$, indicating the optimum dosage of the *M. oleifera* powder seeds used in water contaminated with cyanobacteria species (*A. flos-aquae*).

The results indicate that the *M. oleifera* powder seed added directly to cell suspensions was effective in removing cells, color and turbidity, reaching up to 96.4, 80.5 and 78.1%, respectively, for the dosage of 100 mg L^{-1}, which was considered ideal for this study.

Regarding the removal of $UV_{254\,nm}$, it was observed that *M. oleifera* didn't obtain very satisfactory results, reaching 39.1% removal in the dosage of 100 mg L^{-1}. There was a drop in the removal efficiency as the *M. oleifera* dosage was increased. This result can be attributed to the fact that *M. oleifera* is an organic coagulant, basically composed of proteins, lipids and carbohydrates, responsible for the organic residual in the treated water.

The optimum *M. oleifera* dosage for the *C. vulgaris* species, which is a unicellular microalga, was also evaluated in order to verify if the different morphology of the microalgae interferes in the parameters of removal efficiency. In this way, it can be observed that the optimum coagulant dosage was different among the species.

For *C. vulgaris*, the optimum coagulant dosage was 400 mg L^{-1}, verified by the removal efficiency for color (78.34%), turbidity (70%), chlorophyll-a (90.5%) and $UV_{254\,nm}$ (16%) absorption compounds as shown in **Figure 2**.

Thus, it was observed that the C/F/DAF processes used together with *M. oleifera* as a coagulant had an excellent efficiency for both species that were tested.

In relation to the microalgae lipid profile analysis, the fatty acids and esters microalgae were first identified without the *M. oleifera* treatment.

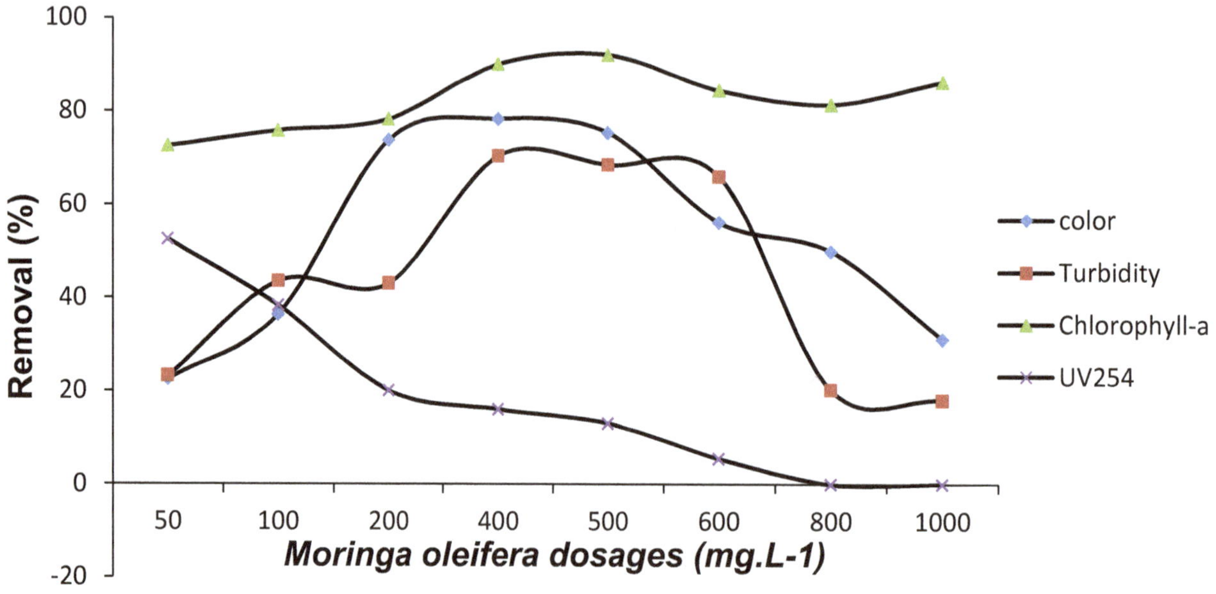

Figure 2. Color, turbidity, compounds with absorption in $UV_{254\,nm}$ and chlorophyll-a removal in relation to the dosage of *Moringa oleifera* powder seeds for *Chlorella vulgaris*.

It can be verified that the saturated fatty acids corresponded to 40.4% composition of *C. vulgaris* and 35.85% of *A. flos-aquae,* whereas the unsaturated ones presented values of 39.58 and 40.1%, respectively.

Among the acids with the highest values in *C. Vulgaris* were C20:0 (arachidic acid) with 21.15%, C18:1n9 (oleic acid) with 18.85% followed by C16:0 (palmitic acid) and C18:2n6 (linoleic acid) with about 15% each. Already *A. flos-aquae,* the highest percentages are due first to C16:0 (palmitic acid) with 30.55%, then C18:2n6 (linoleic acid) presented 17% and, finally, C18: 1n9 (oleic acid) presented 7.4% of its composition, shown in **Table 1**.

		% Fatty acids means present in microalgae without treatment	
Fatty acids		*Chlorella vulgaris*	*Anabaena flos-aquae*
Saturated	**10:00**	0.2	0.25
	12:00	0.35	0.35
	14:00	0.55	0.6
	15:00	0.55	-
	16:00	15.35	30.55
	17:00	0.2	-
	18:00	1.1	2.3
	20:00	21.15	0.1
	21:00	0.2	0.25
	22:00	0.6	1.45
	24:0	0.15	-
Subtotal		40.4	35.85
Monounsaturated	**16:01**	0.8	1.2
	20:01	1.45	2.15
	24:1	0.3	0.55
Subtotal		2.55	3.9
polyunsaturated	**15:1n5**	1	0.95
	18:1n9	18.85	7.4
	18:2n6	15.13	17
	18:3n6	0.95	2.35
	20:3n3	0.8	1.9
	20:3n6	0.3	-
	20:5n3	0.2	4
	22:02	-	2.6

		% Fatty acids means present in microalgae without treatment	
Fatty acids		*Chlorella vulgaris*	*Anabaena flos-aquae*
Subtotal		37.23	36.2
Not identified		19.8	23.95
	Total	100	100
Total lipids		5%	3.05%

Table 1. Microalgae chromatographic profile without *Moringa oleifera* treatment.

The species *C. vulgaris* presented a lipid content of 5% of its dry weight, a value higher than the cyanobacteria studied (*A. flos-aquae*), which was 3.05%, but this result can be reversed, since the medium and conditions as potentiated as lipid productions by microalgae, as light [41], carbon dioxide concentration, temperature [42, 43], nitrogen source concentration [43], among other nutrients. The value obtained for *C. vulgaris* is in agreement with the results obtained by Radman and Costa [20], which presented concentrations of approximately 5.97% of lipid content for the same microalgae under the same culture conditions.

Embora haja poucos trabalhos relatando o perfil lipídico de *Anabaena flos-aquae*, os resultados obtidos neste estudo apresentam valores próximos aos apresentados por Nichols e Wood (1967) em que apresentam C16:0 (39,5%), C16:1 (5,5%), C18:1 (5,2%) C18:2 (36,5%) como os principais ácidos graxos pertencentes da maioria da composição desta alga.

Although there are few studies reporting the *A. flos-aquae* lipid profile, the results obtained in this chapter present values close to those presented by Nichols and Wood [44] in which they present C16:0 (39.5%), C16:1 (5.5%), C18:1 (5.2%) and C18:2 (36.5%) as the main fatty acids belonging to the majority of this algae composition.

After treatment with the *M. oleifera* optimal dosages, the analysis was repeated, and it was observed that results presented in **Table 2** demonstrate that after the treatment with the coagulant optimal dosages, the total lipid percentages of each sample suffered an increase, seen by the values of 16.4% of the total lipids for *C. vulgaris* and 6.2% for *A. flos-aquae*.

This increase is probably related to the residual *M. oleifera* coagulant in the sludge because, according to the *M. oleifera* seeds' physicochemical characterization from Aracaju–SE, they present approximately 37% of lipids in their composition. Therefore, most of the fatty acids present in the samples for both microalgae species were unsaturated fatty acids.

The acid responsible for this increase was C18:1n9 (oleic acid), presenting 69.5% in the *C. vulgaris* sample against 61.7% in the *A. flos-aquae* sample. These results are in agreement with those reported by Silva et al. [45] in which the oil extracted from *M. oleifera* is characterized and with the presence of 78% of oleic acid. Rashid et al. [46] also presented more than 70% of oleic acid in its *Moringa* samples. According to him, some oscillations in the fatty acids' values can occur related to the conditional variations used by the farmers such as fertilizers, soil and the seed variety.

		% Fatty acids means presented in microalgae with MO treatment	
Fatty acids		*Chlorella vulgaris*	*Anabaena flos-aquae*
Saturated	**10:00**	0.02	0.15
	12:00	0.03	0.4
	14:00	0.11	0.2
	15:00	0.03	-
	16:00	5.83	10.8
	17:00	0.1	-
	18:00	1.2	5.9
	20:00	5.4	0.2
	21:00	0.2	3.85
	22:00	0.1	0.2
	24:0	0.8	0.95
Subtotal		13.82	22.65
Mono-unsaturated	**16:01**	0.1	1.4
	20:01	0.05	1.75
	24:1	0.03	0.1
Subtotal		0.18	1.85
poli-unsaturated	**15:1n5**	1.01	1
	18:1n9	69.5	61.7
	18:2n6	0.04	0.2
	18:3n6	0.1	0.2
	20:3n3	0.1	0.25
	20:3n6	0.1	-
	20:5n3	0.25	3.4
	22:02	-	0.15
Subtotal		71.1	66.9
Not identified		14.4	8.6
	total	100	100
Total lipids		16.4%	6.20%

Table 2. Microalgae chromatographic profile after treatment with *Moringa oleifera*.

According to Qu et al. [47], oils with high oleic acid values (>70%) improve the biodiesel properties, such as cold flow, cloud point and pour point, in this way, the sludge produced after treatment with the *M. oleifera* optimum dosage presented high oleic acid values in its composition, data that make interesting the use of this residue to a future production of biodiesel.

Author details

Livia de Oliveira Ruiz Moreti[1], Rosa Maria Ribeiro[2], Letícia Nishi[1]* and Rosângela Bergamasco[1]

*Address all correspondence to: leticianishi@hotmail.com

1 State University of Maringa, Maringá-PR, Brasil

2 Cesumar Institute of Science, Technology and Innovation (ICETI), Maringá-PR, Brazil

References

[1] T. Nkurunziza, J. B. Nduwayezu, E. N. Banadda, I. Nhapi. The effect of turbidity levels and *Moringa oleifera* concentration on the effectiveness of coagulation in water treatment. Water Science & Technology. 2009;**59**:1551-1558. doi:10.2166/wst.2009.155

[2] F. P. Camacho, M. Bongiovani, F. S. Arakawa, A. M. S. Vieira, R. Bergamasco. Advanced processes of cyanobacteria and cyanotoxins removal in supply water treatment. Chemical Engineering Transactions. 2013;**32**:421-426. doi:10.3003/CET1332071

[3] L. O. R. Moreti, P. F. Coldebella, F. P. Camacho, M. C. Bongiovani, A. H. P de Souza, A. K. Gohara, M. Matsushita, M. F. Silva, L. Nishi, R. Bergamasco. Removal of *Anabaena flos-aquae* in water treatment process using *Moringa oleifera* and assessment of fatty acid profile of generated sludge. Environmental Technology. 2015;**37**(11):1408-1417. doi:10.1080/09593330.2015.1117144

[4] M. S. Carvalho, B. R. R. Alves, M. F. Silva, R. Bergamasco, L. A. Coral, F. J. Bassetti. $CaCl_2$ applied to the extraction of *Moringa oleifera* seeds and the use for *Microcystis aeruginosa* removal. Chemical Engineering Journal. 2016;**304**:469-475. doi:10.1016/j.cej.2016.06.101

[5] M. R. Teixeira, M. J. Rosa. Comparing dissolved air flotation and conventional sedimentation to remove cyanobacterial cells of *Microcystis aeruginosa*: Part II. The effect of water background organics. Separation and Purification Technology. 2007;**53**(1):126-134. doi:10.1016/j.seppur.2006.07.001

[6] P. A. P. do Amaral, L. A. Coral, M. E. Nagel-Hassemer, T. J. Belli, F. R. Lapolli. Association of dissolved air flotation (DAF) with microfiltration for cyanobacteria removal in water supply. Desalination and Water Treatment. 2012;**51**(7-9):1664-1671. doi:10.1080/19443994.2012.715128

[7] Brazilian Association of Technical Standards. NBR 10004: Solid wastes-classification. Rio de Janeiro. 2004:71.

[8] M. R. Andrade, J. A. V. Costa. Cultivation of Spirulina platensis microalgae in alternative sources of nutrients. Science and Agrotechnology. 2008;**32**(5):1551-1556. doi:10.1590/S1413-70542008000500029

[9] A. Vonshak, editor. Spirulina Platensis (arthrospira) Physiology, Cell-biology and Biotechnology. London: Taylor & Francis; 1997. 252 p.

[10] R. Schimitz, C. D. Magro, L. M. Colla. Aplicações ambientais das microalgas. Revista CIATEC. 2012;4(1):48-60. doi:10.5335/ciatec.v4i1.2393

[11] A. Scragg, J. Morrison, S. Shales. The use of a fuel containing *Chlorella vulgaris* in a diesel engine. Enzyme and Microbial Technology. 2003;**33**(7):884-889. doi:10.1016/j.enzmictec.2003.01.001

[12] C. M. L. L. Teixeira, F. V. Kirsten, P. C. N. Teixeira. Evaluation of *Moringa oleifera* seed flour as a flocculating agent for potential biodiesel producer microalgae. Journal of Applied Phycology. 2012;**24**(3):557-563. doi:10.1007/s10811-011-9773-1

[13] V. Gressler, N. S. Yokoya, M. T. Fujii, P. Colepicoloc, J. M. Filhod, R. P. Torresd, E. Pinto. The microalgae biomass contains three main components: Carbohydrates, proteins and lipids. Food Chemistry. 2010;**120**(2):585-590. doi:10.1016/j.foodchem.2009.10.028

[14] David L. Nelson, Michael M. Cox, editors. Princípios de Bioquímica de Lehninger. 5th ed. Porto Alegre: Editora Artmed; 2011. 1304 p.

[15] Y. Christi. Biodiesel from microalgae. Biotechnology Advances. 2007;**25**:294-306. doi:10.1016/j.biotechadv.2007.02.001

[16] C. M. P. Pereira, C. B. Hobuss, J. V. Maciel, L. R. Ferreira, F. B. Del Pino, M. F. Mesko, E. Jacob-Lopes, P. C. Neto. Renewable biodiesel derived from microalgae: advances and technological perspectives. New Chemistry. 2012;**35**(10):2013-2018. doi:10.1590/S0100-40422012001000022

[17] A. Richmond. Handbook of Microalgal Culture: Biotechnology and Applied Phycology. Oxford: Blackwell Publishing; 2003. 566 p.

[18] F. C. Bertoldi, E. Sant'Anna, J. L. B. Oliveira. Review: Microalgae biotechnology. Food Processing Research Center Bulletin. 2008;**26**:9-20.

[19] M. I. Gurr, J. L. Harwood, K. N. Frayn. Lipid Biochemistry: An introduction. 5th ed. Wiley-Blackwell; 2002. 340 p.

[20] E. M. Radmann, J. A. V. Costa. Lipid content and fatty acid composition of Microalgae exposed to the gases CO_2, SO_2 and NO. New Chemistry. 2008;**31**(7):1609-1612.

[21] M. G. Morais, J. A. V. Costa. Profile of fatty acids from microalgae grown with dioxide of carbon. Science and Agrotechnology. 2008;**32**(4):1245-1251.

[22] X. Meng, J. Yang, X. Xu, L. Zhang, Q. Nie, M. Xian. Biodiesel production from oleaginous microorganisms. Renewable Energy. 2009;**34**(1):1-5. doi:10.1016/j.renene.2008.04.014

[23] G. Huang, F. Chen, D. Wei, X. Zhang, G. Chen. Biodiesel production by microalgal biotechnology. Applied Energy. 2010;**87**(1):38-46. doi:10.1016/j.apenergy.2009.06.016

[24] S. A. A. Jahn. Using Moringa seeds as coagulants in developing countries. Journal of American Water Works Association. 1988;**80**(6):43-50.

[25] F. Anwar, S. Latif, M. Ashraf, A. H. Gilani. *Moringa oleifera*: A food plant with multiple bio-chemical and medicinal uses- a review. Phytotherapy Research. 2007;**21**(1):17-25. doi:10.1002/ptr.2023

[26] A. Ndabigengesere, K. S. Narasiah. Quality of water treated by coagulation using *Moringa oleifera* seeds. Water Research. 1998;**32**(3):781-791. doi:10.1016/S0043-1354(97)00295-9

[27] K. A. Yongabi, D. M. Lewis, P. L. Harris. A *Moringa oleifera* disinfectant-sand filter integration: A review of an alternative sustainable technology for household treatment. Journal of Environmental Science and Engineering. 2011;**5**(9):1100-1108.

[28] A. Ndabigengesere, K. S. Narasiah, B. G. Talbot. Active agents and mechanism of coagulation of turbid waters using *Moringa oleifera*. Water Research. 1995;**29**(2):703-710. doi:10.1016/0043-1354(94)00161-Y

[29] D. K. Dubey, D. Jyotsna, A. Kumar, R. K. Gulsan. A multipurpose tree: *Moringa oleifera*. International Journal of Pharmaceutical and Chemical Sciences. 2013;**2**(1):415-423.

[30] Bina, B., Mehdinejad, M. H., Dalhammer, G., Rajarao, G., Nikaeen, M., Attar, H. M. Effectiveness of *Moringa oleifera* coagulant protein as natural coagulant aid in removal of turbidity and bacteria from turbid waters. World Academy of Science Engineering and Technology. 2010;**43**:618-620.

[31] G. S. Madrona, G. B. Serpelloni, A. M. Salcedo, L. Nishi, K. Cardoso, R. Bergamasco. Study of the effect of saline solution on the extraction of the *Moringa oleifera* seeds active component for water treatment. Water, Air & Soil Pollution. 2010;**211**(1):409-415. doi:10.1007/s11270-009-0309-0

[32] M. R. Teixeiraa, V. Sousa, M. J. Rosa. Investigating dissolved air flotation performance with cyanobacterial cells and filaments. Water Research. 2010;**44**(11):3337-3344. doi:10.1016/j.watres.2010.03.012

[33] V. Rondeau, D. Commenges, H. Jacqmin-Gadda, J. F. Dartigues. Relation between aluminum concentrations in drinking water and Alzheimer's disease: An 8-year follow-up study. American Journal of Epidemiology. 2000;**152**(1):59-66.

[34] M. C. Bongiovani, F. P. Camacho, L. Nishi, P. F. Coldebella, K. C. Valverde, A. M. Vieira, R. Bergamasco. Improvement of coagulation/flocculation process using a combination of *Moringa oleifera* Lam with anionic polymer in water treatment. Environmental Technology. 2014;**35**(17-20):2227-2236. doi:10.1080/09593330.2014.899398

[35] T. Okuda, A. U. Baes, W. Nishijima, M. Okada. Isolation and characterization of coagulant extracted from *Moringa oleifera* seed by salt solution. Water Research. 2001;**35**(2):405-410. doi:10.1016/S0043-1354(00)00290-6

[36] N. Poumayea, J. Mabinguia, P. Lutgenb, M. Bigan. Contribution to the clarification of surface water from the *Moringa oleifera*: Case M'Poko River to Bangui, Central

African Republic. Chemical Engineering Research and Design. 2012;**90**(12):2346-2352. doi:10.1016/j.cherd.2012.05.017

[37] N. Nwaiwu, B. Lingmu. Studies on the effect of settling time on coliform reduction using *Moringa oleifera* seed powder. Journal of Applied Sciences in Environmental Sanitation. 2011;**6**(3):279-286.

[38] L. Nish, G. S. Madrona, A. L. F. Guilherme, A. M. S. Vieira, A. A. Araújo. Cyanobacteria removal by coagulation/flocculation with seeds of the natural coagulant *Moringa oleifera* Lam. Chemical Engineering Transactions. 2011;**14**:1129-1134.

[39] C. L. Achon, M. M. Barroso, J. S. Cordeiro. Draining beds: Natural system for sludge volume reduction in the water treatment plant. Engenharia Sanitária e Ambiental. 2008;**13**(1):54-62. doi:10.1590/S1413-41522008000100008

[40] B. D. Wahlen, M. W. Robert, C. S. Lance. Biodiesel production by simultaneous extraction and conversion of total lipids from microalgae, cyanobacteria, and wild mixed-cultures. Bioresource Technology. 2011;**102**(3):2724-2730. doi:10.1016/j.biortech.2010.11.026

[41] F. G. A. Fernandez, J. A. S. Perez, J. M. F. Sevilla, F. G. Camacho, E. M. Grima. Modeling of eicosapentaenoic acid (EPA) production from *Phaeodactylum tricornutum* cultures in tubular photobioreactors: Effects of dilution rate, tube diameter, and solar irradiance. Biotechnology and Bioengineering. 2000;**68**(2):173-183. doi:10.1002/(SICI)1097-0290(20000420)

[42] S. M. Renaud, L. Thinh, G. Lambrinidis, D. L. Parry. Effect of temperature on growth, chemical composition and fatty acid composition of tropical Australian microalgae grown in batch cultures. Aquaculture. 2002;**211**(1-4):195-214. doi:10.1016/S0044-8486(01)00875-4

[43] L. M. Colla, T. E. Bertolin, J. A. V. Costa. Fatty acids profile of Spirulina platensis grown under different temperatures and nitrogen concentrations. Zeitschrift für Naturforschung. 2004;**59**(1-2):55-59. doi:10.1515/znc-2004-1-212

[44] B. W. Nichols, B. J. B. Wood. The occurrence and biosynthesis of gamma-linolenic acid in a blue-green alga, Spirulina platensis. Lipids. 1968;**3**(1):46-50. doi:10.1007/BF02530968

[45] J. P.V. da Silva, T. M. Serra, M. Gossmann, C. R. Wolf, M. R. Meneghetti, S. M.P. Meneghetti. *Moringa oleífera* oil: Studies of characterization and biodiesel production. Biomass and Bioenergy. 2010;**34**(10):1527-1530. doi:10.1016/j.biombioe.2010.04.002

[46] U. Rashida, F. Anwara, B. R. Moserb, G. Knothe. *Moringa oleifera* oil: A possible source of biodiesel. Bioresource Technology. 2008;**99**(17):8175-8179. doi:10.1016/j.biortech.2008.03.066

[47] J. Qu, H. Mao, W. C., S. Gao, Y. Bai, Y. Sun, Y. Geng, J. Ye. Development of marker-free transgenic Jatropha plants with increased levels of seed oleic acid. Biotechnology for Biofuels. 2012;**5**(10). doi:10.1186/1754-6834-5-10

[48] FavARO-Trindade, Carmem Silva; Pinho, SCd; Rocha, G. A. Revisão: Microencapsulação de ingredientes alimentícios. Brazilian Journal of Food Technology. 2008;**11**(2)

4

Genetic Factors that Determine the Meat Fatty Acids Composition

Marcos Vinicius Antunes de Lemos,

Angelica S.C. Pereira, Inaê Cristina Regatieri,

Fabieli Louise Braga Feitosa and Fernando Baldi

Abstract

In relation the nutritional attributes of beef meat quality, the composition of fatty acid is important not only because it affects the meat palatability, but also it can affect the human health. The fatty acids harmful to human health have received attenuating attention in recent years. Some studies, with taurine breed, have shown that there is a genetic variation for the trait fatty acid profile of the meat and, therefore, the possibility of genetic improvement of this trait in beef cattle. Meantime, in zebu cattle, the genetic parameter estimates for fatty acid profile are scarce. Furthermore, the trait meat fatty acid profile is something difficult and costly to measure and for this kind of trait is indicated the use of genomic selection, which is a type of marker-assisted selection. The objective of this chapter is showing the genetic variability of meat fatty acid profile different cattle breeds and makes an approach of the implement models and methods that use genomic information to improve the fatty acid composition of beef meat.

Keywords: human health, meat quality, genomic selection, GWAS, SNP, genomic regions, *Bos taurus indicus*

1. Introduction

In response to the constant bombardment of health-related stories, there is a continuing and growing concern on the part of the population and public health institutions about excessive consumption of fats, especially fats of animal origin, as well as the type of fat or fatty acid profile in the meat and their impact on consumer health. The fatty acid profile of intramuscular fat is important for human health, since intramuscular fat cannot be extracted or removed before meat consumption [1]. The composition of fatty acids of intramuscular fat has been widely studied, as it is also related to the succulence, aroma, and tenderness of the meat. For international meat quality standards, the amount of intramuscular fat or marbling deposited on the *longissimus* muscle is the main determinant of the carcass value and predictor of palatability [2].

Although beef is considered a highly nutritious food, being an important source of proteins, micronutrients, and B-complex vitamins, it has a high fat content with undesirable composition, such as high percentage of saturated fatty acids (SFAs). A high intake of SFA is associated with an increase in serum cholesterol and low-density lipoprotein levels (LDLs), which are risk factors for cardiovascular disease [3]. The predominant SFAs (Saturated Fatty Acids) in bovine fat are myristic (C14:0), palmitic (C16:0), and stearic (C18:0) acids [4]. It is noteworthy that C14:0 has a potential to raise serum cholesterol concentrations four to sixfold higher than C16:0 [5].

The fatty ruminant tissue is a natural source of isomers of conjugated linoleic acid (CLA), such as cis–9, trans–11 [6], which is synthesized in the rumen as a consequence of the biohydrogenation process of acids by the microorganisms [7]. CLA has favorable effects on human health, increasing immunostimulatory, antimutagenic, and antioxidant activity [8]. In addition, polyunsaturated fatty acids (PUFAs) present in bovine fat such as linoleic (C18: 2n–6) and linolenic (C18: 3n–3) and monounsaturated fatty acids (AGMI), such as oleic acid (C18: 1, n–9), which offer protection to the cardiovascular system, since balanced consumption of these drugs is associated with a reduction in serum cholesterol levels and an increase in high-density lipoprotein (HDL) [9].

For many years, the composition of fatty acids in meat-producing animals has received considerable interest in view of its implications for human health and meat quality traits [10–12]. Like most traits of economic interest in animal production, the composition of fatty acids is influenced by environmental and genetic factors. A number of studies have demonstrated large changes in fatty acid composition due to alterations in feeding strategies, especially in monogastric animals [13] and in ruminants [11]. However, genetic factors that affect fatty acid composition in cattle have been less investigated, although several studies report differences between breeds for the composition of fatty acids [14, 15]. Despite the differences between breeds for fatty acid composition, they are often confounded by differences in fat deposition or differences in precocity between breeds [1].

2. Fatty acid composition influencing human health and meat quality

The fatty acids composition in beef cattle production system has been studied because of its implications for human health and the traits associated with meat quality. There has been interest in to manipulate the fatty acid composition of meat because it has high nutritional value from children to seniors and is a rich source of protein, iron, zinc, complex B vitamins, and essential polyunsaturated fatty acids such as linoleic (C18:2), linolenic acid (C18:3), and arachidonic (C20:4) [16]. However, meat also is source of fat in the diet, and the presence of cholesterol, low concentration of polyunsaturated fatty acids, and high concentration of saturated fatty acids has been associated with coronary heart disease, diabetes, obesity, and cancer, as well as the ratio of n−6:n−3 polyunsaturated fatty acids, especially in the formation of blood clots leading to a heart attack [13, 17].

The nutritional properties of meat are largely related to its fat content and its fatty acid composition [11]. Different muscles differ in fat content and may also differ in fatty acid composition, which differs between various tissues, including intramuscular and intermuscular, as well as abdominal and subcutaneous adipose tissue [18, 19]. Moreover, genetic and environmental factors can influence the fatty acid composition of the meat [1, 20]. Differences due to the crossing of breeds and between animals within breeds, species, breeds, or lines can change the fatty acid composition of the meat [21]. But generally, the nature and level of deposit of fatty acids in the muscle depends on the diet, ingestion, intestinal absorption, hepatic metabolism, and lipid transportation [22]. Fatty acids composition can influence the meat quality in the fat tissue firmness (hardness), due to the different melting points of the fatty acids; shelf life (lipid and pigment oxidation) due to the propensity of unsaturated fatty acids to oxidize, leading to the development of rancidity and changing the color, flavor due to the production of volatile, odorous, lipid oxidation products during cooking and the involvement of these with Maillard reaction products and aromas [12, 23].

Wood et al. [20] showed that beef has, on average, 50% of saturated (SFA), 40% of monounsaturated fatty acids (MUFA), and 10% of polyunsaturated fatty acids (PUFA). However, in ruminants, linoleic acid (C18:2 n−6) and α-linolenic acid (C18:3 n−3), which are present in many concentrate feed ingredients, are degraded into monounsaturated and saturated fatty acids in the rumen by microbial biohydrogenation, and only a small proportion (around 10% of dietary consumption) is available for incorporation into tissue lipids. In addition to this, their important role is to work together to regulate immune responses and anti-inflammatory processes. Linolenic acid is also associated with the reduction of coronary diseases and plasma cholesterol and also has anticancer properties. The consumption of saturated fatty acids is associated with an increase in serum cholesterol levels and the risk of coronary heart disease. Especially lauric, myristic, and palmitic fatty acids are responsible for increasing plasma total and LDL cholesterol concentrations, and palmitic acid (C16:0) has the most impact on cholesterol levels, because it raises the levels of LDL.

Long-chain polyunsaturated fatty acids of the omega 3 family also are present in the meat, such as eicosapentaenoic, docosapentaenoic, and docosahexaenoic acids (C22: 6n−3). Eicosapentaenoic acid acts by relaxing the blood vessels and preventing the formation of blood clots. Arachidonic acid, resulting from omega 6 metabolism, leads to constriction of the vessels and formation of blood clots. Although they perform opposite functions, both are necessary for the maintenance of the balance of the organism. Moreover, prostaglandins are lipid autacoids derived from arachidonic acid. They both sustain homeostatic functions and mediate pathogenic mechanisms, including the inflammatory response [24]. Therefore, an omega 6/omega 3 ratio of less than four is recommended. The bovine meat analysis has verified values of the omega 6/omega 3 ratio between 1.5 and 10.4, and the lowest values were found in the meat of cattle raised in pasture. Disorders that have been suggested to be linked with lack of omega-3 PUFA include hypertension, inflammatory and immune disorders, depression, and neurological dysfunction. Repeatedly, there are a lot of dietary recommendations to reduce the consumption of saturated fatty acids, such as prevention of cardiovascular diseases. On the other hand, some studies have shown beneficial effects of polyunsaturated fatty acids, mainly the n−3 family, CLA, docosahexaenoic acid, and docosapentaenoic acid on the level of serum lipids and their antithrombotic action on platelets and protection against some diseases [25]. Studies also indicate that stearic acid (C18:0) has been shown not to increase total cholesterol or LDL-cholesterol concentrations and slightly changes serum cholesterol levels in humans; however, it is poorly stored in tissues [26].

To attend the need of good human health, it is necessary to produce meat with a higher ratio of polyunsaturated to saturated fatty acids and a more favorable balance between n−6 and n−3 PUFA. The ratio of n−6:n−3 PUFA is particularly beneficial in cattle, especially from animals that have consumed grass which contains high levels of 18:3 acid. Dietary intake of PUFA from the n−3 series and especially from the n−6 series by the animals favors the production of conjugated isomers of linoleic acid (CLA c9 t11), such as C18:2 cis-9 trans-11 [5], which are synthesized in the rumen as a result of biohydrogenation of fatty acids, performed by microorganisms [7]. Some of these fats, such as CLA (conjugated linoleic acid), could be beneficial to human health. CLA is important in the prevention of specific cancers and in the treatment of obesity, immune functions, and potential beneficial effects on coronary heart disease [8].

In relation to diet, fatty acid composition of concentrate and forage diets is different and leads to different fatty acid compositions in tissues. The presence of the rumen makes fatty acid composition in beef more difficult to manipulate by changing diet, but studies showed that the C18:3 acid, n−3 PUFA concentrations, lipid oxidation, color, and aromas were affected by feeding treatments [23, 24]. Some data demonstrated the feasibility of reducing population cholesterol levels through strategies involving alteration of fat quality within the agricultural and food manufacturing chains. Ruminants consuming fresh pasture, in general, have higher content of unsaturated fatty acid in their meat that those receiving a grain-based concentrate diet. Grass lipids contain high proportions of the unsaturated linolenic acid (C18:3 n−3), and the only way to improve the ratio of PUFA in ruminant meats is by preventing ruminal biohydrogenation or by feeding protected PUFA supplements [27]. For all these reasons, there is an increase interest in research intended to modify the fatty acid composition in meat, especially reducing the concentration of SFA and increasing PUFA.

3. Meat fatty acids profile variation between and within beef cattle breeds

The genetic variability is characterized as the differences between animals within breeds, differences between breeds or lines, and due to the crossing of breeds. The heritabilities and genetic correlations estimate the latter source of variation. The major genes segregation may influence the breed effects, of which the double-muscled gene in cattle is a well-known example [1], and major factors that influence the fatty acid composition of beef are age of animal, diet, and breed type [20]

Several studies have demonstrated that adipose tissues from *Bos indicus* cattle breeds are less saturated when compared to *Bos taurus* [14, 28, 29]. In this sense, Rossato et al. [29] pointed out that Nelore beef is nutritionally healthier than Angus breed, once it has lower percentages of cholesterol and higher amounts of n–3 fatty acids, CLA precursor (C18:1 *trans*). Bressan et al. [30] showed that the production system has an important influence on beef fatty acid profile when compared animals from *Bos taurus* and *Bos indicus* breeds. The *Bos taurus* animals showed the lower percentage of saturated fatty acids (SFA) and higher percentage for monounsaturated fatty acids (MUFA) in relationship to indicine animals finished at the feedlot system. According to these authors, taurine cattle that was finished under feedlot conditions showed higher ability to desaturate SFA than indicine cattle.

Recently, Lemos et al. [31] realized a study to identify regions associated with saturated, mono, and polyunsaturated and n–6 to n–3 ratios, in the *longissimus* thoracis muscle from confined Nelore, using the single-step method. The individual fatty acids with the highest concentration in the intramuscular fat of *longissimus thoracis* found by these authors were C16:0, C18:1 cis-9, C18:1 trans-11, and C18:0, representing 67.3% of its fat composition. These results are in agreement with those reported by some authors [32–34] who observed high levels of palmitic, stearic, and oleic fatty acids (FAs). Some authors [4, 35] also reported that palmitic fatty acid was the predominant FA in beef fat. In Nelore finished in feedlot [34], oleic acid (37.46%) displayed the highest concentration in intramuscular fat. The myristic and palmitic FAs are associated with an increase in circulating LDL cholesterol due to interference with hepatic LDL receptors. The saturated fatty acids were predominant, followed by the MUFAs and PUFAs. Similar results [36] were reported for Nelore cattle, 43.93% (SFA), 42.33% (MUFA), and 12.8% (PUFA). However, studies using taurine [15] and Nelore [34] breeds found similar concentrations for SFA and MUFA, 47 and 47.5%, and 47.23 and 48.34%, respectively.

Information on genetic parameters for carcass and meat traits, fatty acid composition, and genetic-quantitative relationships between these traits is essential to improve meat tenderness and the proportion of fat in the carcass, without harming the fat composition in livestock production. On this concern, some studies have been done to estimate these parameters. In these sense, Feitosa et al. [37] studied the genetic-quantitative relationships between the beef fatty acid profile with the carcass and meat traits of Nelore cattle used a total of 1826 bulls finished in feedlot conditions to analyze the following carcass and meat traits: subcutaneous fat thickness (BF), shear force (SF), and total intramuscular fat (IMF). The fatty acid (FA) profile of the *longissimus thoracis* samples was determined. These authors estimated the heritability, which varied from 0.06 to 0.65 for individual saturated fatty acids (SFA), 0.02 to 0.14 for monounsaturated fatty acids (MUFA), and it ranged from 0.05 to 0.68 for polyunsaturated fatty acids (PUFA). Some traits showed the heritability estimates low to moderate, varying from 0.09 to 0.20, how was the case of Omega 3, Omega 6, SFA, MUFA, and PUFA. For the carcass and meat traits, the heritability estimates for the authors were low (SF (0.06) and IMF (0.07)) unless for BF (0.17) which presented a moderate value.

Aboujaoude et al., [38] in a study to determine genetic parameters for fatty acids in intramuscular fat from feedlot-finished Nelore carcasses, estimated heritability for individual FAs ranged from 0.01 to 0.35. The heritability estimates for myristic (0.25 ± 0.09), palmitic (0.18 ± 0.07), oleic (0.28 ± 0.09), linoleic (0.16 ± 0.06), and α-linolenic (0.35 ± 0.10) FAs were moderate. Stearic, elaidic, palmitoleic, vaccenic, conjugated linoleic acid, docosahexanoic, eicosatrienoic, and arachidonic FAs had heritability estimates below 0.15. Heritability estimates for beef fatty acids were also estimated by Cesar et al. [34] in a study with Nelore breed. The estimates varied from low (<0.10 for lauric, palmitc acid, cis-vaccenic acid, cis-12 octadecenoic, vaccenic acid, eicosanoic acid, aicosatrienoic acid, arachidonic acid, eicosapentaenoic acid, and atherogenic index, respectively) to moderate (up to 0.29 for intramuscular fat, myristic acid, myristoleic acid, palmitoleic acid, margaric acid, heptadecenoic acid, stearic acid, oleic acid, trans-6,7,8 octadecenoic, trans-10,11,12 octadecenoic, linoleic acid, octadecenoic acid, α-linolenic acid, γ-linolenic acid, eicosapentaenoic acid, docosahexaenoic acid, SFA, MUFA, PUFA, Sn–3, Sn–6, and n–6:n–3). For pentadecylic acid, cis-13 octadecenoic, cis-15 octadecenoic, trans-16 octadecenoic, and eicosadienoic acid, the heritability estimates were zero. Differently of these author and working with taurine breed, Tait et al. [39] (Angus) and Nogi et al. [40] (Japanese Black cattle) found the estimates of heritability for IMF fat deposition and composition traits are higher when compared with the results above. The lower values of heritability reported for the populations of some studies could be explained by the reduced sample size [41] or lower amount of genetic variation in the population [42].

Comparing these values with another study that was accomplished with a great number of animals [31], the estimate of linolenic FA heritability, for example, was similar to that found by Cesar et al. (2014) (0.13) and lower than that reported by Nogi et al. [40] (0.58). However, higher estimates have been reported for linolenic acid in other studies (0.21) [39] and also for palmitoleic acid (0.15) [34] and (0.49) [43]. Higher heritability estimates were reported for linoleic FA, 0.34 and 0.58, respectively, in the intramuscular fat of Japanese Black cattle, suggesting that genetic influence on linoleic acid varies among breeds [40]. Recently, authors also estimated high heritability for SFA (0.54) and MUFA (0.54) and, therefore, concluded that there is sufficient genetic variation in the fatty acid profile of cattle subcutaneous fat to respond to selection [33]. Therefore, these results suggest that it is possible to change the beef lipid composition of intramuscular fat of different cattle breeds' through selection. This information is important for breeding programs that aim at improving the beef fatty acid composition.

4. Genetic markers and metabolic pathways associated with meat fatty acids profile

The fatty acid metabolism is a complex process, which includes lipolysis of dietary fat and biohydrogenation in the rumen, *de novo* synthesis of fatty acids by rumen bacteria, absorption and transport of fatty acids by the host animal, *de novo* synthesis in tissues host, elongation and desaturation in animal tissues, hydrolysis of triglycerides and esterification, oxidation of fatty acids, or metabolism to other components [44–46].

In ruminants, the fatty acid synthesis occurs mainly in the adipose tissue, except during the lactation, when the mammary gland becomes the predominant organ [47]. The main point about control of the fatty acids synthesis is the acetyl-CoA carboxylase, and it seems that the endocrine control is very similar in, at least, adipose tissue (insulin activation, inhibition of catecholamine) of ruminants and nonruminants [48].

The principal precursor of fatty acid synthesis in ruminants is the acetate, which should be converted into acetyl-CoA by the action of acetyl CoA synthetase and subsequently incorporated into fatty acids. The conversion of acetate to acetyl-CoA is performed in adipose tissue, which is the largest synthesizer fatty acids in ruminants [49]. Some studies have been carried out to evaluate gene expression pattern in cattle for fatty acid composition and also identified genomic regions and metabolic pathways involved in those process, aiming to improve the beef fatty acid profile. In this sense, Berton et al. [50] studied the gene expression profile in Nelore cattle with extreme phenotypes for intramuscular fatty acid composition, found the *ACSM3* (acyl-CoA synthetase medium-chain family member 3) gene as differentially expressed for linoleic, monounsaturated, polyunsaturated, saturated, and omega-3 acids, participates in the metabolism of lipids and in metabolic pathways that involves the precursor acetyl-CoA metabolism. Also, the *ACSS1* (acyl-CoA synthetase short-chain family member 1) gene acts in the transformation of acetyl-CoA into fatty acids, through chemical reactions and metabolic pathways involving acetyl-CoA, being differentially expressed ($q < 0.05$), for saturated fatty acids such as palmitic, stearic, oleic, and total saturated acids.

Some studies has been realized in attempt do identify and describe the genes which play this important role on the beef fatty acids metabolic pathways. In a previously study, Lemos et al. [31] found several regions close to QTL (Quantitative Trait Loci) associated with saturated, polyunsaturated, and monounsaturated fatty acid groups in the meat of Nelore cattle. These regions have found interesting PCGs (pyruvate candidate genes) that are involved in lipid metabolism, such as receptors for reproductive hormones, transport and use of fatty acids and cholesterol, elongation factors and synthesis of long-chain fatty acids in different species, constituents of cell membranes, biosynthesis and hydrolysis of phospholipids and membrane constituents, energy metabolism, and protein kinase synthesis. The ELOVL5 (ELOVL fatty acid elongase 5) gene, among the many identified, is the most prominent. It is located on chromosome BTA23 at 25 Mb and associated with arachidonic acid (C20: 4 n−6). ELOVL genes are responsible for the elongation of long-chain fatty acids, which encode enzymes that play an important role in the elongation of long-chain fatty acids. The FASN (fatty acid synthetase) enzyme responsible for fatty acid synthesis is located on the BTA19 chromosome between 51,384,922 and 51,403,614 bp, variations of this enzyme were related to the fatty acid composition of Angus beef [51]. In mammals, *FASN* synthesizes the fatty acids that contain up to 16 carbon atoms, and the genes of the *ELOVLs* group produce long-chain fatty acids with 18 carbon atoms or more [13].

In additional, we employed the ingenuity pathway analysis (IPA) online software to detect the canonical pathways involving the genes of the above study. No canonical pathway was significant (p-value < 0.05). A large proportion of the pathways acted on fucose and cholesterol biosynthesis, and peroxisome proliferator-activated receptors alpha (PPARa) activation, which would provide valuable insights into explaining the molecular mechanism of lipid metabolism. As one of the pathways showed on canonical pathway, the PPAR alpha has a great role in the regulation in the fatty acids metabolism. Peroxisome proliferator-activated receptors (PPARs) are nuclear hormone receptors that are activated by fatty acids and their derivatives and play an essential physiological role in the regulation of adipocyte tissue development lipogenesis and skeletal muscle lipid metabolism [52]. Doran et al. [53] performing the study genome-wide association studies (GWAS) in Holstein-Friesian cattle identified PPAR signaling pathway as the biological pathway more significantly involved in the performance of carcass traits, suggesting that PPARs play a key role in the control of carcass weight, carcass fat, and carcass conformation traits.

Li et al. [54] sampled spleen tissues from grain-fed and grass-fed Angus steers and performed a comparative study of gene expression using RNA-Seq method. Based on the DEGs (differentially expressed genes), they identified potential mechanisms, by implemented a functional analysis, that could contribute to the difference observed between both groups. The authors have detected 123 DEGs between grass-fed and grain-fed spleen of Angus cattle. In the grass-finished group, 87 were up-regulated, while the other 36 decreased their gene activity. Based on these genes, they identified 9 significant molecular networks and 13-enriched biological pathways through performed an IPA (ingenuity pathway analysis). The pathways, Nur77 signaling in T lymphocytes and calcium-induced T lymphocyte apoptosis, which are immune related, contain a pair of genes HLA-DRA and NR4A1 with dramatically altered expression level.

In a recent study, Berton et al. [50] analyzed the gene expression profile of intramuscular muscle in Nelore cattle with extreme values of fatty acid and identified several genes associated with fatty acid metabolism, such as those involved in intra- and extra-cellular transport of fatty acid synthesis precursors in intramuscular fat of *longissimus thoracis* muscle. The authors found some genes that play important traits on the metabolic pathways of fatty acids, such as precursors in the synthesis of fatty acids (CSM3 (Chromosome segregation in meiosis protein 3) and ACSS1); deposition of saturated fat in adipose tissue (DGAT2); support in insulin synthesis, stimulating both glucose synthesis and the entry of amino acids into cells (GPP and LPL); and synthesis and degradation of ketone bodies used in the synthesis of ATP (BDH1).

5. Genomic selection and genome-wide association studies for beef fatty acid composition

Marker-assisted selection (MAS) is recommended to increase the annual genetic gain for traits of economic importance in several animal species [55]. In this kind of selection, molecular information from markers is used together with phenotypic data of production and pedigree to select the animals. This way, MAS provides possibility to improve difficult and/or high cost measurement traits, such as the meat fatty acids composition. Some studies in several countries have mainly used microsatellite as genetic marker to study the fatty acids composition in taurine breeds [56]. However, genotype using microsatellite markers is expensive and just a small proportion of the total genetic variance can be show for the markers, limiting the progress or genetic gain [57]. Sequencing of the bovine genome has led to the discovery of thousands of single-nucleotide polymorphism (SNP) markers and subsets of SNPs that can characterize the bovine genome with a wider range and lower cost [58]. In bovine, genome-wide association studies (GWAS) and genomic selection have been done using high-density SNP chips, with thousands of genetic markers for traits related to milk or meat quality, as the fatty acid composition [59–61].

In dairy cattle, Bouwman et al. [61] performed a genome-wide association analysis using 50,000 SNP markers for the contente of saturated fatty acids (C4:0–C18:0), monounsaturated (C10:1–C18:1), and polyunsaturated (C18:2cis9trans11-CLA), to identify genomic regions associated with individual fatty acids in bovine milk. The authors found 54 regions on 29 chromosomes that were significantly associated with one or more fatty acids. In beef cattle, studies involving genomic association or selection are scarce. Uemoto et al. [60] found 32 SNPs located on the chromosome 19 associated with the amount of oleic acid (C18:1) in the intramuscular fat of the trapezius muscles in Japanese black cattle. The content of oleic acid is positively correlated with the sensorial quality of the meat [62]. In the study of Uemoto et al. [60], the authors used the Illumina BovineSNP50 BeadChip and genotyped only 160 bovines (80 animals with higher values and 80 animals with lower values of oleic acid) selected from 3.356 animals based on corrected phenotype.

Another study with a significantly higher number of animals has been shown for Reecy et al. [59]. They used the BovineSNP50 beadchip (54 k) to genotype 2.285 Angus bulls to analyze the fatty acid concentration of the meat. Effects of SNPs on each trait were estimated using the Bayes C module, considering the probability of a SNP not influencing the trait (π) = 0.90. Depending on the fatty acid considered in the analysis, 2.3–48.5% of observed variance could be explained by an animal's 54 K genotype. According to the authors, long-chain fatty acids appear to be lowly heritable traits with a low proportion of variance accounted by markers, in relation to short-chain fatty acids. They concluded that a large proportion of variation in fatty acid composition is associated with a relatively low number of SNPs. Therefore, genetic progress can be achieved by implementation of whole genome selection to improve fatty acid composition in meat. Similarly, Saatchi et al. [43] found in other GWAS with 2.177 Angus cattle, using a 54-K genotyping panel, 57 genomic regions associated with the fatty acids profile trait in meat. The authors concluded that this large number of genomic regions might indicate the presence of an elaborate molecular mechanism that control fatty acid content in skeletal muscle.

The first genome-wide association study involving intramuscular fat deposition and fatty acid composition in Nelore cattle (*Bos indicus*) was carried out by Cesar et al. [34]. The authors genotyped 386 Nelore steers using a BovineHD BeadChip (770 k) and used Bayesian methods (Bayes B) to identify genomic regions and putative candidate genes that could be involved with fatty acid composition in Nelore. The authors found eight genomic regions (1 Mb windows) for saturated fatty acids that explained more than 1% of genotypic variation for C12:0, C14:0, C16:0, and C18:0. Ten genomic regions for monounsaturated fatty acids, which relates C14:1 cis-9, C16:1 cis-9, C18:1 cis-9, and C18:1 trans-16. For polyunsaturated fatty acids, nine genomic regions which relates C18:2 cis-9 cis12 n−6, C18:2 trans-11 cis-15, C18:3 n−3, C18:3 n−6, C20:3 n−6, C20:5 n−3, and C22:5 n−3. They concluded that intramuscular fat composition is affected by many loci with small effects, and the identification of genomic regions associated to fatty acid composition can lead to selection to improve human nutrition and health.

Trying to identify regions of the genome associated with saturated, mono, and polyunsaturated fatty acids, Lemos et al. [31] genotyped 1616 Nelore using the high-density Bovine BeadChip (770 k) and the single-step method to perform the GWAS. The authors used the single-step method because it allows to combine the information of genotyped and nongenotyped animals in the genetic evaluation process, expanding the scope and identification of potential regions associated with loci responsible for variations in the studied traits [63]. Interestingly, the results showed that a total of 31 genomic regions that explain more than 1% of genotypic were found for total saturated fatty acids, C12:0, C14:0, C16:0, and C18:0; 37 genomic regions for monounsaturated fatty acids, which relates to total monounsaturated fatty acids, C14:1, C16:1, C18:1 trans11, C18:1 cis9, and C18:1 trans9 and 40 genomic regions for the polyunsaturated fatty acids group as C20:4 n−6, C18:2 cis-9 cis12 n−6, C18:3 n−3, C22:6 n−3, and C20:3 n−6 cis-8 cis-11 cis-14. Additionally, a total 21 genomic regions accounted for more than 1% of the genetic variance for n−3 and n−6 fatty acids and the n−6:n−3 ratio. The authors could conclude that the identification of such regions and the respective candidate genes should contribute to improve the genetic knowledge regarding the fatty acids profile of Nelore cattle and help to improve the selection of such traits to favor human health.

Some authors have been testing different methodologies to predict the direct genomic value for many traits in livestock production, such as SNP-BLUP (single-nucleotide polymorphisms best linear unbiased predictor), which assumes a normal distribution for SNP effect and com-mon variance for all markers [64]; the LASSO (least absolute shrinkage and selection oper-ator) assumes a double exponential distribution for the SNPs effect [65, 66] and Bayesian methods—BayesC and BayesCπ, which considered a variable with binomial distribution that reports whether a marker (SNP) has (1) or not (0) effect on the trait under study, with π vari-able probability to be zero and a normal distribution with probability $1-\pi$, assuming that part of the effect markers follows a normal distribution. These methods differ in the assumptions about the genetic model associated with quantitative traits, and the best method is the one that reflects the biological nature of polygenic traits, in terms of genic effects [67].

In this sense, studying an Angus population, Saatchi et al. [43] concluded that genomic selection for beef FA profile using Bayesian models is feasible. Moreover, Onogi et al. [68] evaluated the predictive ability of genomic selection in FA composition of Japanese Black cattle, using single-step genomic-best linear unbiased method. Recently, Chiaia et al. [69] evaluated the genomic predictability for beef fatty profile in Nelore breed and concluded that genomic information can assist in improving FA profile in Zebu animals, since the use of genomic information yielded genomic values for FA profile with accuracies ranging from low to moderate. The authors also concluded that none of the evaluated methods (SNP-BLUP, Bayesian Lasso, BayesC, and BayesCπ) excelled in terms of accuracy; however, the SNP-BLUP method allows obtaining less-biased genomic evaluations; thereby, this method is more feasible when taking account the computational cost. The genomic selection has the potential to increase the genetic gain for hard measure traits, like the FA profile, however, the most suitable model to evaluate those traits are still being studied. The divergence between studies suggests that the difference within the methods is due to the genetic architecture of the trait that is the accuracy tends to increase as the model adjusts itself to the genetic architecture of the trait [70] (Lund et al., 2009). For traits that are affected by moderate to major genes effect, higher accuracies can be reached through Bayesian methods [71]. Traits that are controlled by many genes with small effects, polygenic trait, and the SNP-BLUP method showed better prediction ability [72].

6. Final considerations

The review was to give a comprehensive approach of current knowledge about the genetic influence on the beef fatty acid profile. Several studies have reported genomic regions and genes that are involving witch the lipid metabolic pathways in cattle and other livestock species. With this information, the elucidation of the genetic basis for the improvement of meat quality traits, with an emphasis on human health, becomes closer to reality. Another contribution is the improvement of the knowledge about the biosynthesis of fatty acids and the selection of animals with better nutritional quality. However, more research with focus on genomic and fatty acid composition needed to improve meat is required since the use of genomic information can produce genomic values for FA profile more accurate. Together, this information can be implanted in future breeding programs for cattle, in order to select animals according to the fatty acid profile of the meat.

Acknowledgements

MVA Lemos (FAPESP, Fundação de Amparo à Pesquisa do Estado de São Paulo). F Baldi (FAPESP, Fundação de Amparo à Pesquisa do Estado de São Paulo grant no. 2011/21241-0).

Author details

Marcos Vinicius Antunes de Lemos[1*], Angelica S.C. Pereira[2], Inaê Cristina Regatieri[1], Fabieli Louise Braga Feitosa[1] and Fernando Baldi[1]

*Address all correspondence to: marcoslemoszootec@gmail.com

1 São Paulo State University, Jaboticabal, Brazil

2 University of São Paulo, Pirassununga, Brazil

References

[1] De Smet S, Raes K. & Demeyer D.: Meat fatty acid composition as affected by fatness and genetic factors: a review, Animal Research, 53, 81-98, 2004.

[2] Ferraz J.B.S., Felício P.E.: Production systems – An example from Brazil, Meat Science, 84, 238-243, 2010.

[3] Katan M.B., Zoock P.M., Mensink, R.P.: Effects of fats and fatty acids on blood lipids in humans: an overview, American Journal of Clinical Nutrition, Bethesda, 60, 1017-1022, 1994.

[4] Lawrie, R.A., Ciência da carne, 6th ed., Porto Alegre: Artmed, 384p., 2005.

[5] Mensink, R.P. & Katan M.B.: Effect of dietary fatty acids on serum lipids and lipoproteins: A meta-analysis of 27 trials, Arteriosclerosis and Thrombosis, 12, 911-919, 1992.

[6] French P., Stanton C., Lawless F.: Fatty acid composition, including conjugated linoleic acid, of intramuscular fat from steers offered grazed grass, grass silage or concentrate based diets, Journal of Animal Science, 78, 2849-2855, 2000.

[7] Tamminga S., Doreau M.: Lipids and rumen digestion, In: Jouany J.P., editor. Rumen microbial metabolism and ruminant digestion, INRA: Paris, 151-164, 1991.

[8] Ip C., Scimeca J.A., Thompson H., Effect of timing and duration of dietary conjugated linoleic acid on mammary cancer prevention, Journal of Nutrition and Cancer, Volume 24, Issue 3, 1995.

[9] Pensel N.: The future of red meat in human diets, Nutrition Abstracts and Reviews (Series A), Farnham Royal, 68, 1-4, 1998.

[10] Xie Y.R., Busboom J.R., Gaskins C.T., Johnson K.A., Reeves J.J., Wright R.W., Cronrath J.D.: Effects of breed and sire on carcass characteristics and fatty acid profiles of crossbred Wagyu and Angus steers, Meat Science, 43, 167-177, 1996.

[11] Wood J.D., Richardson R.I., Nute G.R., Fisher A.V., Campo M.M., Kasapidou E., Sheard P.R., Enser M.: Effects of fatty acids on meat quality: A review, Meat Science, 66, 21-32, 2004.

[12] Gandemer G.: Lipids and meat quality: lipolysis, oxidation, Maillard reaction and flavor. Science Aliments, 19, 439-458, 1999.

[13] Jakobsen M.U., Overvad K., Dyerberg J.: Heitmann Bl: Intake of ruminant trans fatty acids and risk of coronary heart disease, International Journal of Epidemiology, 37, 173-182, 2008.

[14] Huerta-Leidenz N.O., Cross H.R., Savell J.W., Lunt D.K., Baker L.S., Smith B.: Fatty acid composition of subcutaneous adipose tissue from male calves at different stages of growth, Journal of Animal Science, 74, 1256-1264, 1996.

[15] Pitchford W.S., Deland M.P.B., Siebert B.D., Malau-Aduli A.E.O., Bottema C.D.K.: Genetic variation in fatness and fatty acid composition of crossbred cattle, Journal of Animal Science, 80, 2825-2832, 2002.

[16] Mcneill S., Van Elswyk M.E.: Red meat in global nutrition, In Meat Science Volume 92, 2012 Elsevier Ltd: England, 166-173,2012.

[17] Enser M.: The role of fats in human nutrition, In Rossell B. (Ed.), Oils and fats, Vol. 2, Animal carcass fats, 77-122, Leatherhead Publishing: Leatherhead, Surrey, UK, 2001.

[18] Marmer W.N., Maxwell R.J., Williams J.E.: Effects of dietary regimen and tissue site on bovine fatty acid profiles, Journal of Animal Science, 59, 109-121, 1984.

[19] Webb E.C., De Smet S., Van Nevel C., Martens B., Demeyer D.I.: Effect of anatomical location on the composition of fatty acids in double-muscled Belgian Blue cows, Meat Science, 50, 45-53, 1998.

[20] Wood J.V., Enser M., Fisher A.V., Nute G.R., Sheard P.R., Richardson R.I. Hughes, S.I. & Whittington, F.M.: Fat deposition, fatty acid composition and meat quality: A review, Meat Science, 78, 343-358, 2008.

[21] Fisher A.V., Enser M., Richardson R.I., Wood J.D., Nute G.R., Kurt E., Sinclair L.A., Wilkinson R.G.: Fatty acid composition and eating quality of lamb types derived from four diverse breed production systems. Meat Science, 55, 141-147, 2000.

[22] Geay Y., Bauchart D., Hocquette J., Culioli J.: Effect of nutritional factors on biochemical, structural and metabolic characteristics of muscles in ruminants, consequences on dietetic value and sensorial qualities of meat, Reproduction, Nutrition, Development, v.41, 1-26, 2001.

[23] Campo M.M., Nute G.R., Wood J.D., Elmore S.J., Mottram D.S., Enser M.: Modelling the effect of fatty acids in odour development of cooked meat in vitro: Part I-sensory perception, Meat Science, 63, 367-375, 2003.

[24] Simopoulos A.P.: Omega-3 fatty acids in health and disease and in growth and development, The American Journal of Clinical Nutrition, 54, 438-463, 1991.

[25] Bonanome A., Grundy S.M.: Effect of dietary stearic acid on plasma cholesterol and lipoprotein, New England Journal of Medicine, 318, 1244-1248, 1988.

[26] Vatansever L., Kurt E., Enser M., Nute G.R., Scollan N.D., Wood J.D., Richardson R.I.: Shelf life and eating quality of beef from cattle of different breeds given diets differing in n–3 polyunsaturated fatty acid composition, Animal Science, 71, 471-482, 2000.

[27] Scollan N.D., Enser M., Gulati S.K., Richardson I., Wood J.D.: Effects of including a ruminally protected lipid supplement in the diet on the fatty acid composition of beef muscle, British Jornal of Nutrition, 90, 709-716, 2003.

[28] Perry D.; Nicholls, P.J., Thompson, J.M.: The effect of sire breed on the melting point and fatty acid composition of subcutaneous fat in steers, Journal of Animal Science, 76, 87-952, 1998, PRADO, I.N.; ITO, R.H.; PRADO, J.M. et al. The influence of dietary soyabean and linseed on the chemical composition and fatty acid profile of the *Longissimus* muscle of feedlot-finished bulls. Journal of Animal and Feed Science, v.17, p.307-317, 2008.

[29] Rossato, L.V., Bressan, M.C., Rodrigues, E.C., Gama, L.T., Bessa, R.J.B., Alves, S.P.A.: Physicochemical parameters and fatty acid profiles in Angus and Nellore cattle finished on pasture, Revista Brasileira de Zootecnia, 39, 1127-1134, 2010.

[30] Bressan M.C., Rossato L.V., Rodrigues E.C., Alves S. P., Bessa R.J.B., Ramos E.M.: Genotype × environment interactions for fatty acid profiles in *Bos indicus* and *Bos taurus* finished on pasture or grain, Journal of Animal Science, 89, 221-232, 2011.

[31] Lemos M.V.A., Chiaia H.L.J., Berton M.P., Feitosa F.L.B., Aboujaoud C., Camargo G.M.F., Pereira A.S.C., Albuquerque L.G., Ferrinho A.M., Mueller L.F., Mazalli M.R., Furlan J.J.M., Carvalheiro R., Gordo D.M., Tonussi R., Espigolan R., Silva R.M.O., Oliveira H.N., Duckett S., Aguilar I., Baldi F.: Genome-wide association between single nucleotide polymorphisms with beef fatty acid profile in Nellore cattle using the single step procedure, BMC Genomics, 17, 213, 2016.

[32] Prado, I.N., Moreira, F.B., Matsushita, M. & Souza, N.E.: Longissimusdorsi fatty acids composition of *Bosindicus* and *Bosindicus* × *Bostaurus* crossbred steers finished in pasture, Brazillian Archives of Biology and Technology, 46, 599-606, 2003.

[33] Kelly, M.J., Tume, R.K., Newman, S. & Thompson, J.M.: Genetic variation in fatty acid composition of subcutaneous fat in cattle, Journal compilation, Animal Production Science, 53, 129-133, 2013.

[34] Cesar A.S.M., Regitano L.C.A., Tullio R.R., Lanna D.P.D., Nassu R.T., Mudado M.A., Oliveira P.S.N., Do Nascimento M.L., Chaves A.S., Alencar M.M., Sonstegard T.S., Garrick D.J., Reecy J.M. and Coutinho L.L.: Genome-wide association study for intramuscular fat deposition and composition in Nellore cattle, BMC Genetics, 15, 2014.

[35] Rossato L.V., Bressan M.C., Rodrigues E.C., Carolino M.D.C., Bessa, R.J.B., Alves, S.P.P.: Lipid composition of meat from zebu and taurine cattle finished in confinement, Revista Brasileira de Zootecnia, 38, 1841-1846, 2009.

[36] Prado J.M., Prado I.N., Visentainer, J.V., et al: The effect of breed on the chemical composition and fatty acid profile of the *Longissimus dorsi* muscle of Brazilian beef cattle, Journal of Animal and Feed Sciences, v.18, pp.231-240, 2009.

[37] Feitosa F.L.B., Olivieri B.F., Aboujaoude C., Pereira A.S.C., Lemos M.V.A., Chiaia H.L.J., Berton M.P., Peripolli E., Ferrinho A.M., Mueller L.F., Mazalli M.R., De Albuquerque L.G., De Oliveira H.N., Tonhati H., Espigolan R., Tonussi R.L., De Oliveira Silva R.M., Gordo D.G.M., Magalhães, A.F.B., Aguilar, I., Baldi F.: Genetic correlation estimates between beef fatty acid profile with meat and carcass traits in Nellore cattle finished in feedlot, Journal of Applied Genetics, v. 1, p. 1, 2016.

[38] Aboujaoude C., Pereira A.S.C., Feitosa F.L.B., Lemos, M.V.A., Chiaia H.J.L., Berton, M.P., Peripolli E., Silva R.M.L., Ferrinho A.M., Mueller L.F., Olivieri B.F., Galvão L.G., Oliveira H.N., Tonhati H., Espigolan R., Tonussi R., Gordo D.M., Magalhaes A.F.B., Baldi F.: Genetic parameters for fatty acids in intramuscular fat from feedlot-finished Nelore carcasses, Animal Production Science (Print) JCR, v. 1, p. 1, 2016

[39] Tait R. Jr, Zhang S., Knight T., Minick Bormann J., Beitz D., Rj M.: Heritability estimates for fatty acid quantity in Angus beef, Journal of Animal Science, 85(Suppl. 2), 58. Abstr, 2007.

[40] Nogi T., Honda T., Mukai F., Okagaki T., Oyama K.: Heritabilities and genetic correlations of fatty acid compositions in longissimus muscle lipid with carcass traits in Japanese Black cattle, Journal of Animal Science, 89(3), 615-621, 2011.

[41] Casas E., Stone R.T., Keele J.W., Shackelford S.D., Kappes S.M., Koohmaraie M.: A comprehensive search for quantitative trait loci affecting growth and carcass composition of cattle segregating alternative forms of the myostatin gene, Journal of Animal Science, 79(4), 854-860, 2001.

[42] Casas E., Shackelford S.D., Keele J.W., Koohmaraie M., Smith T.P., Stone R.T.: Detection of quantitative trait loci for growth and carcass composition in cattle, Journal of Animal Science, 81(12), 2976-2983, 2003.

[43] Saatchi M., Garrick D.J., Tait Jr R.G., Mayes M.S., Drewnoski M., Schoonmaker J., Diaz C., Beitz D.C., Reecy J.M.: Genome-wide association and prediction of direct genomic breeding values for composition of fatty acids in Angus beef cattle, BMC Genomics, 14, 730, 2013.

[44] Bauchart D.: Lipid absorption and transport in ruminants, Journal of Dairy Science, 76, 3864-3881, 1993.

[45] Chilliard Y.: Dietary fat and adipose tissue metabolism in ruminants, pigs, and rodents: A review, Journal of Dairy Science, 76, 3897-3931, Cloonan, N. et al. Stem cell transcriptome profiling via massive-scale mRNA sequencing. Nat. Methods 5, 613-619(2008), 1993.

[46] Jenkins T.C.: Lipid metabolism in the rumen, review, Journal of Dairy Science, 76, 3851-3863, 1993.

[47] Vernon R.G. & Flint D.J.: Roles of insulin and growth hormone in the adaptations of fatty acid synthesis in White adipose tissue during the lactation cycle in shee, The Biochemical Journal, 256, 873-878, 1998.

[48] Vernon R.G. & Flint D.J., Proceedings of the Nutrition Society, 42, 315-331,1983.

[49] Polizel Neto A., Branco R.H., Bonilha S.F.M., Gomes H.F.B., Corvino T.L.S., Functions of Volatiles Fatty Acids on Intramuscular Fat Deposition – Review, 2008, http://www.infobibos.com/Artigos/2008_3/AcidosGraxos/index.htm, Accessed 10 Dec 2016.

[50] Berton M.P., Fonseca L., Gimenez D.F.J., Utembergue B.L., Cesar A.L., Coutinho L.L., Lemos M.V.A. Aboujaoude C., Pereira A.S.C., Silva R.M., Stafuzza N.B., Feitosa F.L.B., Chiaia H.L.J., Olivieri B., Peripolli E., Tonussi R.L., Gordo D.G.M., Espigolan R., Ferrinho A., Mueller L.F., Albuquerque L.G., Oliveira H.N., Duckett S., Baldi F.: Gene expression profile of intramuscular muscle in Nellore cattle with extreme values of fatty acid, BMC Genomics, 17, 972, 2016.

[51] Zhang S., Knight T.J., Reecy J.M., Beitz D.C.: DNA polymorphisms in bovine fatty acid synthase are associated with beef fatty acid composition, Animal Genetics, 39(1), 62-70, 2008.

[52] Melissa J. & Chastity B.: Omega-3, omega-6 and omega-9 fatty acids: Implications for cardiovascular and other diseases, Journal of Glycomics Lipidomics, 4(4), pp.1-8, 2014.

[53] Doran A.G., Berry D.P., Creevey C.J.: Whole genome association study identifies regions of the bovine genome and biological pathways involved in carcass trait performance in Holstein-Friesian cattle, BMC Genomics, 15, 837, 2014.

[54] Wang H., Misztal I., Aguilar I., Legarra A., Muir W.M.: Genome-wide association mapping including phenotypes from relatives without genotypes, Genetical Research, 94, 73-83, 2012.

[55] Dekkers J.C.M.: Commercial application of marker and gene assisted selection in livestock: Strategies and lessons, Journal of Animal Science, 82, 313-328, 2004.

[56] Abe T., Saburi J., Hasebe H., Nakagawa T., Kawamura T., Saito K., Nade T., Misumi S., Okumura T., Kuchida K., Hayashi T., Nakane S., Mitsuhasi T., Nirasawa K., Sugimoto Y., Kobayashi E.: Bovine quantitative trait loci analysis for growth, carcass, and meat quality traits in an F2 population from a cross between Japanese Black and Limousin, Journal of Animal Science, 86, 2821-2832, 2008.

[57] Solberg T.R., Sonesson A.K., Woolliams J.A., Meuwissen T.H.E.: Genomic selection using different marker types and densities, Journal of Animal Science, 86, 2447-2454, 2008.

[58] Matukumalli L.K., Lawley C.T., Schnabel R.D., Taylor J.F., Allan M.F., Heaton M.P., C'connell J., Moore S.S., Smith T.P.L., Sonstegard T.S., Van Tassell C.P.: Development and characterization of a high density SNP genotyping assay for cattle, PloS One, 4, e5350, 2009.

[59] Reecy J.M., Tait R.G., Vanoverbeke D.L., Garmyn A.J., Mateescu R.G., Van Eenennaam A.L., Duan Q., Liu Q., Schoonmaker J.P., Drewnoski M.E., Beitz D.C., Kizilkaya K., Fernando R.L., Garrick D.J.: Use of genomics to improve healthfulness and quality of meat, In: 9th World Congress on Genetics Applied to Livestock Production, 1 to 6 of august of 2010, Leipzig, Proceedings of 9th World Congress on Genetics Applied to Livestock Production, CD ROM, 2010.

[60] Uemoto Y., Abe T., Tameoka N., Hasebe H., Inoue K., Nakajima H., Shoji N., Kobayashi M., Kobayashi E.: Whole-genome association study for fatty acid composition of oleic acid in Japanese Black cattle, Animal Genetics, 42, 141-148, 2010.

[61] Bouwman A.C., Bovenhuis H., Visker M.H., Van Arendonk J.A.: Genome-wide association of milk fatty acids in Dutch dairy cattle, BMC Genetics, 43: 2-12, 2011.

[62] Melton S.L., Amiri M., Davis G.W., Backus W.R.: Flavor and chemical characteristics of ground beef form grass-, forage-grain- and grain-finished steers, Journal of Animal Science, 55, 77-87, 1982.

[63] Kemper K.E., Goddard M.E.: Understanding and predicting complex traits: Knowledge from cattle, Human Molecular Genetics, 21, R45–R51, 2012.

[64] Meuwissen, T.H., Hayes, B.J., & Goddard, M.E.: Prediction of total genetic value using genome-wide dense marker map, 2012Genetics, 157, 1819-1829, 2001.

[65] Park, T., & Casella, G.: The bayesian lasso. Journal of the American Statistical Association, 103, 681-686, 2008.

[66] De Los Campos, G., Hickey, J.M., Pong-Wong, R., Daetwyler, H.D., & Calus, M.P.L.: Whole-genome regression and prediction methods applied to plant and animal breeding, Genetics, 193, 327-345, 2013.

[67] Resende, M.D.V., Lopes, P.S., da Silva, R.L., & Pires, I.E.: Genome wide selection (GWS) and maximization of the genetic improvement efficiency, Pesquisa Florestal Brasileira, 56, 63-77, 2008.

[68] Onogi A., Ogino A., Komatsu T., Shoji N., Shimizu K., Kurogi K., Yasumori T., Togashi K., Iwata H.: Whole-genome prediction of fatty acid composition in meat of Japanese Black cattle, Animal Genetics, 46, 557-559, 2015.

[69] Chiaia, H.J.J., Peripoli E., Silva R.M.O., Aboujaoude C., Feitosa F.L.B., Lemos M.V.A., Berton M.P., Olivieri B.F., Espigolan R., Tonussi R., Gordo D.G.M., Bresolin T., Magalhães A.F.B., Fernandes Júnior G.A., Albuquerque L.G., Oliveira H.N., Furlan J.J.M., Ferrinho A.M., Mueller L.F., Tonhati H., Pereira A.S.C., Baldi F.: Genomic prediction for beef fatty acid profile in Nellore cattle, Meat Science, 128, 60-67, 2017.

[70] Neves, H.H., Carvalheiro, R., O'brien, A.M.P., Utsunomiya, Y.T., Do Carmo, A.S., Schenkel, F.S., Sölkner, J., Mcewan, J.C., Van Tassell, C.P., Cole, J.B., Silva, M.V.G.B., Queiroz, S.A., Sonstegard, T.S., & Garcia, J.F.: Accuracy of genomic predictions in *Bos indicus* (Nellore) cattle, Genetics, Selection, Evolution, 46, 17, 2014.

[71] Ekine-Dzivenu, C., Chen, L., Vinsky, M., Aldai, N., Dugan, M.E.R., Mcallister, T.A., Wang, Z., Okine, E.& Li, C.: Estimates of genetic parameters for fatty acids in brisket adipose tissue of Canadian commercial crossbred beef steers, Meat Science, 96, 1517-1526, 2014.

[72] Clark, S.A., Hickey, J.M., & Van Der Werf, J.H.: Different models of genetic variation and their effect on genomic evaluation, Genetics, Selection, Evolution, 43, 18, 2011.

5

Fatty Acids in Fish

Oğuz Taşbozan and Mahmut Ali Gökçe

Abstract

The human body cannot synthesize certain fatty acids: these essential fatty acids must be consumed in the diet. Fish and other aquatic foods are known to be the main sources of polyunsaturated fatty acids (PUFA); therefore, humans obtain most of their eicosapentaenoic acid (EPA) and docosahexaenoic acid (DHA) by consuming fish, aquatic invertebrates, and algae. The increasing demand for fish and the stabilization of marine fish and freshwater landings have contributed to a widening gap between demand and supply for fish and fish products. This leads to a necessity to improve aquaculture production. Fish are the main contributors of n-3 PUFA in the human diet, although there are some interspecific and intraspecific differences in fatty acid profiles. The fatty acid composition of fish differs depending on a variety of factors, including species, diet, as well as environmental factors such as salinity, temperature, season, geographical location, and whether the fish are farmed or wild. In this chapter, information will be provided on fish fatty acids based on their ecology, feeding habits, lipid contents, and environmental conditions where they are harvested.

Keywords: marine fish, freshwater fish, EPA, DHA, PUFA, HUFA, n3/n6

1. Introduction

Many studies have investigated the effects of lipids and fatty acids in human nutrition on health. This has resulted in an increasing consumer interest and a tendency to consume healthy foods.

Among the fatty acids, highly unsaturated n-3 fatty acids (n-3 HUFA) or long-chain n-3 poly unsaturated fatty acids (LC n-3 PUFA), particularly 20:5 n-3 (eicosapentaenoic acid [EPA]) and 22:6 n-3 (docosahexaenoic acid [DHA]) affect human health, early development, and the prevention of some diseases; therefore, dieticians increasingly recommend consuming foods containing these fatty acids [1]. The recommended n-6/n-3 fatty acid ratio in human nutrition is 5:1, but this ratio (n-6/n-3) varies between 7:1 and 20:1 in the diets of most West Europeans and North Americans [1, 2]. The n-3/n-6 fatty acid ratio recommended by the World Health Organization is 1:1 or above [3]; hence, fish consumption should be increased or foods rich in n-3 fatty acids should be consumed for proper nutrition and disease prevention.

Fish are the most important sources of these fatty acids; fatty fish, such as sardines, mackerel, anchovies, and some salmon species, are rich in EPA and DHA. In these fish, the ratio of n-3 fatty acid to n-6 fatty acid approaches 7. Fish cannot synthesize these fatty acids; they obtain them from food they consume (algae and planktons) [4].

However, lipid composition and thus fatty acid composition in fish differ depending on various factors: usually, their aquatic environment (marine water, freshwater, and cold or warm water) and the biological, physical, and chemical properties of that environment. Also, seasonal changes, migration, sexual maturity and spawning period, species, feeding habits, and whether reared in aquaculture or grown in natural habitats affect the lipid/fatty acid composition [5].

Therefore, detailed information on the changes in lipid and fatty acid compositions among fish and the importance of fish consumption in human health are provided in this chapter by examining these subjects.

2. The importance of fatty acids and fish consume

The interest in fat, which holds an important place in human nutrition, has increased with the recently increasing interest in, and awareness of, human health. Fats are important components of hormones, cell membranes, and signaling molecules, as well as being important energy sources. Fat ingested into the body is first stored in the liver, hypodermic connective tissues, mesentery, and muscles and used when necessary [6].

Fatty acids have a methyl group on one end and have long hydrocarbon chains carrying a carboxyl group on the other end. Fatty acid molecules are classified based on the presence and number of double bonds: saturated fatty acids have no double bond, and monounsaturated fatty acids have a single double bond; polyunsaturated fatty acids (PUFA) have two or more double bonds. The number and position of the double bonds determine the physical properties and functional characteristics of fatty acids. The human body can produce some of these fatty acids, but others, some of which also contain n-3 and n-6 PUFA, cannot be produced by the body. These essential fatty acids (EFAs) need to be obtained through food intake. In current human diets, n-6 fatty acid and, especially, linoleic acid (LA) sources are soya and maize oil, and arachidonic acid (AA); the main n-3 alpha-linolenic acid (ALA) source is meat. Linoleic acid (LA), an n-6 fatty acid, can be converted to fatty acids with longer chains, and n-3 ALA can be converted to eicosapentaenoic acid (EPA) and docosahexaenoic acid (DHA); these conversion rates vary between 1 and 10%. Even though EPA and DHA, with 20–22 long chain n-3 fatty acids, have a critical role in human health, their consumption is relatively low, stemming from the deficiency in consumption of fish and fish products in developed countries [7–9].

Many recent studies have shown the importance of and necessity for n-3 fatty acids in human development and health. Some studies show that they have a positive effect on maternal and fetal health during pregnancy and on newborn and childhood health. These studies emphasize an important role for fatty acids in prevention of hormone-related cancers and important functions in the prevention of cardiovascular diseases. These fatty acids also are purported to relieve dementia, hyperactivity, and some psychiatric disorders [10].

Many studies have carefully evaluated the effects of the lack of fatty acids in the diet following the onset of pregnancy on the prenatal and on postnatal development of newborns and children [6, 10–12]. One study investigated the effect of the lack of n-6 and n-3 long chain PUFA on children with attention-deficit hyperactivity [13]. Moreover, several studies have evaluated the relationship between n-3 PUFA deficiency and depression and mood disorders [14–16]. Various studies have reported the cardio-protective effect of n-3 PUFA (EPA and DHA) supplementation and recommend 1 g EPA intake per day to prevent coronary heart disease [17]. Although their anti-cancer roles have yet to be proven, many studies have shown that n-3 and n-6 PUFA positively affected the prevention of development of different types of tumors. The n-3 PUFA has been reported to alter cell growth by modulating cell replication, interfering with the components of the cell cycle, or by increasing cell death through necrosis or apoptosis [18, 19]. Other studies on the same subject have focused on breast cancer, colon and colorectal cancers, prostate cancer, lung cancer, and neuroblastoma [20].

Fish consumption has an important role in n-3 PUFA (EPA and DHA) intake. Although fatty acids of the n-3 group that cannot be synthesized in the body vary in different species and individuals, fish contain significant amounts of n-3 fatty acids. The nutritional contents and fatty acid compositions of fish differ, depending on various factors. For example, fatty fish, such as herring and mackerel have 400 mg of PUFA per 15 g; thus, weekly consumption of 300 g fatty fish or a daily 200 mg EPA and DHA intake is reported to be sufficient [21]. Furthermore, the n-3/n-6 ratio is reported to be a good index for comparison of the relative nutritional value of fish oils [22]. Although there is no specific recommendation for the n-3/n-6 ratio, evidence found in wild animals and estimated food intake during human evolution indicates a dietary ratio of 1:1 [3].

3. How the fatty acid profiles vary in fish?

Among vertebrates, fish have the highest species diversity and, as stated previously, the nutritional content and fatty acid compositions of fish vary. The most important causes of the variation in the fatty acid profiles of fish are the differences among species. Moreover, living in aquatic environments with different ecological conditions is an important source of variation in nutritional components. The fishing season, size, and reproduction status of the individuals from the same species living in a certain region also affect this variation. Moreover, the aquaculture conditions and feeds used in fish aquaculture also cause variations in the fatty acid compositions of fish that were supplied to the market using aquaculture. These factors are elaborated in the sub-sections that follow.

3.1. Fish bioecology

Fish are divided into two groups based on their habitat: marine fish and freshwater fish. Water temperature and salinity are the most important environmental factors; thus, fishes are first studied based on water temperature and then divided into two groups, the warm-water fish group and cold-water fish group. The optimal temperatures for warm-water species are around 25–30°C, whereas cold-water species prefer temperatures below 20°C· In addition to this classification, both cold-water and warm-water fish are further classified as freshwater and marine fish. Moreover, some fish species migrate from seas to fresh waters or from fresh waters to seas. These fish, such as salmonids and eels, are termed diadromous species [23]. The nutritional compositions of fish (especially lipids and fatty acids) substantially vary due to the differences in their habitats.

In general, freshwater fish require either linoleic acid (18:2 n-6), linolenic acid (18:3 n-3) or both, whereas marine fish require EPA (20:5 n-3) and/or DHA (22:6 n-3) [24].

The studies on the essential fatty acid (EFA) requirement of marine fish showed that their n-3 EFA requirement can be met by EPA (20:5 n-3) and DHA (22:6 n-3). These two fatty acids are termed n-3 HUFA (highly unsaturated fatty acid) or LC (long chain) n-3 PUFA (poly unsaturated fatty acid). The EPA and DHA requirements of fish respond to the n-3 HUFA rich nutrients in marine environments because primary food sources, such as marine algae and planktons, and also other food sources, are known to be rich in EPA and DHA and contain lower levels of linoleic acid and linolenic acid [25, 26].

Studies on freshwater fish have shown that their n-3 EFA requirements are mostly focused on linolenic acid (18:3n-3). The fact that EPA and DHA in vertebrates are biologically active forms of n-3 EFA led to the conclusion that many freshwater fish can convert 18:3 n-3 fatty acids to EPA and DHA. This is less straightforward in marine fish. When compared to marine microalgae, freshwater microalgae contain higher levels of 18:3 n-3 fatty acids than EPA and DHA. Moreover, even though there is not a great amount of 18:2 n-6 fatty acid in marine microalgae, freshwater microalgae contain plenty of this fatty acid. This explains why freshwater fish require higher amounts of linoleic and linoleic acid as compared to marine fish [25, 26]. On the other hand, only one species among the freshwater fish (*Esox lucius*) has the ability to convert 18:3 n-3 fatty acid to EPA and DHA. Because this fish species is an extreme carnivore consuming smaller fish, it does not have high amounts of 18:3 n-3 and 18:2 n-6 fatty acids [27].

The nutritional compositions of the foods in the natural environment of marine and freshwater fish necessitate providing farmed fish with food sources that meet the requirements of their species. The essential fatty acid amount required in the feeds of commercial freshwater and marine fish (preadult or older juvenile) is determined in terms of dry diet%. For example, the 18:3 n-3 fatty acid requirement of trout is 0.7–1.0, and the n-3 HUFA requirement of trout is 0.4–0.5. The 18:2 n-6 fatty acid requirement of common carp is 1.0, whereas their 18:3 n-3 fatty acid requirement ranges from 0.5 to 1.0. The 18:2 n-6 fatty acid requirements of *Tilapia zilli* and *Oreochromis niloticus* are 1.0 and 0.5, respectively. Channel catfish require 18:3 n-3 fatty acid between 1.0 and 2.0, and n-3 HUFA between 0.5 and 0.75. Among the marine fish species, turbot require n-3 HUFA at a ratio of 0.8, and red sea bream (*Pagrus major*) require n-3 HUFA at a ratio of 0.5. The EPA and DHA requirements of red sea bream are 1.0 and 0.5, respectively. The n-3 HUFA requirement of gilthead sea bream (*Sparus aurata*) is between 0.9 and 1.9, whereas the n-3 HUFA requirement of sea bass (*Dicentrarchus labrax*), another important marine species, is 1.0 [25, 26].

Many studies have focused on determining the lipid and fatty acid compositions of marine and freshwater fish (both cold water and warm water). The goal of these studies was both to find the differences among fatty acid compositions of fish from different aquatic environments and to evaluate these fatty acids in terms of human health.

The results obtained in a study from Turkey on the fatty acid compositions of eight different marine fish species that were either farmed or caught in their natural habitats (waker, tub gurnard, whiting, mackerel, blue fish, sea bream, sea bass, and marbled spinefoot) and six different freshwater fish species caught in their natural habitats offer an insight into this subject. The PUFA ratios of the marine species were between 25.2 and 48.2%, whereas they were between 23.2 and 43.8% in freshwater species. The EPA and DHA values of marine species were 4.23–7.02% and 11.7–36.01%, respectively. The EPA and DHA values of freshwater species were between 2.10 and 13.8%, and between 6.72 and 24.8%, respectively. Among marine fish, at 22.6%, the lowest n-3 PUFA ratio was found in waker and, at 44.2%, the highest ratio was found in blue fish. Among the freshwater fish, North African catfish had the lowest n-3 PUFA (11.05%) value, whereas at 28.4%, zander had the highest value. In addition, the n-6 PUFA ratios in the marine fish were between 0.43 and 14.4% and between 5.27 and 16.8% in the freshwater fish [28]. The researchers reported that n-6/n-3 ratios in both the freshwater fish and marine fish were below the ratio recommended by the UK Department of Health (4.0 at maximum) [29].

The results obtained in a study on 34 different marine fish species from the Mediterranean Sea showed that the fatty acid levels of all fish were at the desired levels for human health and quality food consumption. The EPA and DHA values of fish were between 1.94 and 10.0%, and 3.31 and 31.03%, respectively. The n-3 PUFA levels were between 12.66 (annular sea bream) and 36.54% (European hake), whereas oceanic puffer had the lowest n-6 PUFA level at 1.24% and flathead mullet had the highest n-6 PUFA level at 12.76%; the n-6/n-3 ratio varied between 0.04 and 0.91 [30].

A study was carried out with marine and freshwater fish from Malaysia found EPA ratios in freshwater fish between 0.63 and 1.41%, whereas they were between 4.68 and 10.62% in marine fish. The DHA levels were between 0.14 and 0.25%, and 2.50 and 10.05% in freshwater fish and marine fish, respectively. Moreover, that study reported that the n-3/n-6 ratios in all marine species were above 1, whereas the highest level reached was 0.73 in freshwater fish [31]. The World Health Organization recommended that the n-3/n-6 ratio should be at least 1 [3].

Overall, the studies reported that marine species and species that show carnivorous feeding habits and species living in cold water contained high amounts of EPA and DHA and therefore can be used as an important source of food for human health.

3.2. Feeding habits

Aquatic animals (organisms) have environmental and biological characteristics. The most important biological characteristics are feeding habits. Fish are classified as carnivorous, herbivorous, omnivorous, and detritivorous (detritivore, detrivore, or detritus feeder) based on their usual food source preferences in their natural habitats [32]. Moreover, each class is further classified based on their food source preferences as euryphagous (feed on a great variety of foods), stenophagous (feed on a limited variety of foods), or monophagous (feed on only one type of food) [33]. The detritivorous species is *Cirrhinus molitorella*, known as mud carp, and does not have much commercial value.

The most frequently consumed and farmed fish species worldwide have carnivorous, herbivorous, and omnivorous feeding habits; thus, these fish are rich in nutrients and popular among consumers. The fish species that are widely farmed are: euryphagous carnivores, such as salmon, basses, breams, halibut, flounders, and groupers; euryphagous herbivores, such as some carp and tilapia species, milkfish; or euryphagous omnivores, such as common carp, channel catfish, grey mullet, and eels.

The anatomic structures of fish digestion systems differ depending on their feeding habits. Carnivores have shorter intestines and larger stomachs as compared with the other groups, and their stomachs are usually tube-shaped [34]. The digestion ratio of food is higher in carnivorous fish when compared with other groups because semi-digested foods are stored in the chyme, that is, in the stomach, for shorter periods [35].

Common carp, one of the omnivorous species, does not have a stomach because it tends to consume herbal foods; however, some omnivorous species have pouch-shaped stomachs that are smaller than those of the carnivorous species. Moreover, their intestinal structure is more developed and longer. Herbivorous species do not have stomachs and have the longest and most complex intestinal structure because they consume only herbal food sources [34, 36].

The energy requirements of fish differ depending on their feeding habits; therefore, lipid digestion and requirement for lipids, the most important energy source, vary among the fish species. In addition to fish species and feeding habits, some other factors also affect lipid digestion. The age of the fish is the most important factor in lipid digestion [37–39]. The ability of young fish and, especially, fish at the larval stage to digest foods containing high amounts of lipid and lipids in feeds is markedly insufficient [36, 37, 39, 40]. Temperature also affects lipid digestion: warm-water fish species have the greatest ability to digest lipids [41, 42].

In general, carnivorous species can better digest the lipids in high-fat nutrients in their natural habitat—or pellet feeds under farming conditions [43–46]. Their ability to better digest lipids is attributable to their genetic potential to store lipids [47]. In contrast, fish species that have herbivorous and omnivorous feeding habits have a lower capacity to digest nutrients or feeds with high lipid content [48]. Although the ability to digest lipids is affected by many factors, the superior lipid digestion ability of carnivorous species is attributable to their more specific and higher lipase activity when compared with herbivorous and omnivorous species [49].

Based on the results obtained in relevant studies, the fatty acid requirements and compositions of fish were divided into three groups. The first group is the salmon and rainbow trout group, which are freshwater and anadromous carnivore species from the Salmonidae family; the second group is the carnivorous marine fish group which contains sea bass and sea bream; the third group is the temperate-climate fish group, that mostly have herbivorous and omnivorous feeding habits (tilapia, carp, eel, among others) [50].

For the fish in the first group, α-linolenic acid (18:3 n-3) is the main fatty acid that must be in their feeds, especially under farming conditions. Certain levels of EPA and DHA can only be synthesized from linolenic acid by elongation if there is sufficient α-linolenic acid and insufficient of EPA and DHA in feeds. This does not imply that EPA and DHA are unimportant for trout; on the contrary, trout require these two fatty acids in high amounts but can partially met their requirements with linolenic acid when EPA and DHA are insufficient; however, in some cases, the synthesized amount may itself be insufficient [46, 50].

The most important fatty acids for carnivorous marine fish, especially for sea bass and sea bream, are EPA and DHA. Their ability to synthesize EPA and DHA using other fatty acids is inferior to that of the fish of the Salmonidae group; therefore, they tend to feed on nutrients rich in EPA and DHA, either in their natural habitats or under farming conditions [50].

Linoleic acid (18:2n-6) is known to be the most important fatty acid requirement of species that have herbivorous and omnivorous feeding habits. These fish species, along with linoleic acid, require linolenic acid (18:3n-3) and arachidonic acid (20:4n-6) [50].

Table 1 shows the fatty acid requirements of some carnivorous and herbivorous fish species [50].

3.3. Lipid contents

There are three different lipid compositions of fish: lean fish (less than 5% fat), mid-fat fish (5–10% fat), and fatty fish (10–25% fat). Although the lipid contents of fish depend on many factors, they are generally divided into three groups based on their composition: lean, mid-fat or medium fat, and fatty fish. This classification and the lipid levels in fish have been reported somewhat inconsistently in publications and research. For example, in one paper, fish were divided into four groups based on their lipid levels and lean fish were evaluated under two categories. These groups were:

Very low fat (less than 2%): cod, haddock, flounder/sole, and tuna

Low fat (2–5%): tilapia, halibut, ocean perch, and salmon (chum, pink)

Medium fat (5–10%): bluefish, catfish, rainbow trout, and sword fish

High fat (10% or more): herring, mackerel, sardines, and salmon (Atlantic, sockeye, coho, and chinook) [51].

In another study, fish were separated into three different classes. Fish having lipid levels below 2% were regarded as lean fish; fish having lipid levels between 2 and 8% were regarded

Species	Requirements of fatty acids (in dry feed%)
1. Carnivores	
Rainbow trout (*Oncorhynchus mykiss*)	Linolenic 1% Linolenic 0.8% EPA + DHA 0.4–0.5%
Sea bass (*Dicentrarchus labrax*)	EPA + DHA 1%
Sea bream (*Sparus aurata*)	EPA + DHA 1%, EPA:DAHA = 1 EPA + DHA 1.9%, EPA:DAHA = 0.5
2. Omnivores	
Common carp (*Cyprinus carpio*)	Linoleic 1%; linolenic 0.5–1%
Japanese eel (*Anguilla japonicus*)	Linoleic 0.5%; linolenic 0.5%
3. Herbivores	
Grass carp (*Ctenopharyngodon idella*)	Linoleic 1%; linolenic 0.5%
Tilapia (*Tilapia zilli*)	Linoleic 1%; arachidonic 1%
Tilapia (*Oreochromis niloticus*)	Linoleic 0.5%

Table 1. Fatty acid (in dry feed%) requirements of some carnivorous, omnivorous, and herbivorous fish species [50].

as mid-fat fish; fish having lipid levels above 8% were regarded as fatty fish. Cod fish was given as the best example of lean fish, and some salmon species, herring and mackerel, were placed in the fatty fish group. Another important issue, which should not be overlooked, is that the lipid content in fish can vary significantly. In wild fish, seasonal changes, sexual maturity, reproduction period, and the nutrients they consume; in farmed fish, the feed content and quality directly affect the lipid content [52].

The lipid ratio in lean or fatty fish usually depends on how and where the lipids are stored. Cod fish are known to be lean fish; they do not store lipids in their muscle tissues (fillet) but store them only in the liver, whereas salmon and trout species store lipids in their muscle tissues and the surrounding organs and do not store lipids in their liver [53].

Table 2 shows the nutritional composition in lean, mid-fat, and fatty fish [52].

The lipids in fish vary with body composition. In general, the differences in lipid compositions of certain fish species depend on the spawning period and seasonal changes. For instance, seasonal changes were reported to significantly affect the lipid compositions of herring (*Clupea harengus*) and mackerel (*Scomber scombrus*). In herring, the lowest lipid level was around 5% and observed in April; the highest lipid level was above 25% and was observed in July. In fillets of mackerel, the lowest lipid level was 5% and observed between June and July, whereas the lipid level was above 20% during September–January and approached 30% in December [54].

Because these fish are rich in n-3 fatty acids, the consumption of fatty fish is also important for human health, both EPA and DHA-rich species including salmon, herring, mackerel, anchovies, and sardines. In these fish, the n-3 fatty acid contents are sevenfold or more higher than their n-6 fatty acid contents [55].

Fish species	Fat (g)	Protein (g)	Water (g)
Lean fish			
Haddock	0.2	16.6	81
Cod	0.3	18.1	80
Common sole	0.5	14.8	84
Bluefish tuna	1.0	24.0	74
European perch	1.3	18.1	81
Medium fat fish			
Turbot	2.4	15.9	79
Redfish/ocean perch	2.8	17.1	79
Sea trout	3.3	20.0	74
Char	7.1	16.1	73
Fatty fish			
Farmed trout	10.2	17.2	70
Halibut	10.4	16.2	72
Wild salmon	11.5	19.7	66
Farmed salmon	13.4	19.9	67
Mackerel	20.2	18.5	60
Fatty herring	25.0	17.0	56
Eel	32.5	17.3	46

Table 2. Nutritional content per 100 g in different fish species.

The lipid, EPA, and DHA levels in fillets determined in a study that investigated the lipid levels in fish are given in **Table 3** [52].

In India, researchers examined the EPA, DHA, and fatty acid compositions of 34 fish, 3 prawns, and 2 mussels with different lipid contents (lean, mid-fat, and fatty) and, again, found that EPA and DHA levels were high in the fatty fish group. Among these fish, 12 were obtained from marine water, 3 were obtained from brackish water, 14 were obtained from freshwater, and 5 were obtained from cold water. *Sardinella longiceps* (marine water) and *Tenualosa ilisha* (freshwater) had the highest lipid contents (9.2 and 10.5%, respectively). It was observed that the fatty acid compositions of both fatty fish were especially rich in HUFA, EPA, DHA, and n-3 fatty acids. *Sardinella longiceps* had 12.3% EPA, 6.9% DHA, and 21.4% $\sum$ n-3 fatty acid. The fatty acid composition of *Tenualosa ilisha,* a freshwater species, contained 2.9% EPA, 8.9% DHA, and 14.2% $\sum$ n-3 fatty acid [56].

Although the lipid level of *Catla catla* fish, which is a freshwater species having low lipid levels (2–4%) and was obtained from farms, was low in the body composition, it had high levels of EPA, DHA and $\sum$ n-3 fatty acid. Again, although *Rastrelliger kanagurta* and *Stolephorus waitei*

Fish species	Fat (g)	EPA (g)	DHA (g)
Lean fish			
Haddock	1.0	0.07	0.27
Cod	0.6	0.07	0.16
Saithe/coalfish	0.3	–	–
Tusk	0.3	–	–
European plaice	1.5	0.24	0.26
Medium fat fish			
Halibut	3.9	0.28	0.41
Atlantic wolffish	2.7	0.40	0.20
Rainbow trout	6.7	0.32	1.16
Spotted wolffish	4.8	0.70	0.40
Fatty fish			
Greenland Halibut	15.6	1.00	0.90
Salmon	10.0	0.65	1.80
Mackerel	24.4	1.27	3.17
Herring (summer)	14.5	0.57	1.25
Herring (winter)	19.0	2.48	2.24
Eel	31.5	1.27	2.07

Table 3. Fat, EPA, and DHA content in selected fish species.

(lean fish; less than 2% fat) from marine water had low lipid content, they were rich in EPA, DHA, and $\sum$ n-3 fatty acid relative to the other fish in the same group [56].

3.4. Wild or farmed fish

There are significant differences in nutritional compositions of farmed fish and wild fish. Many recent studies have focused on this issue and have tried to determine to what degree the nutritional composition of fish affects human health and has nutritional benefits [57–64]. The nutritional quality of farmed fish has improved in the recent years thanks to environmentally friendly and advanced aquaculture techniques. In addition, the advancing feed sector now can offer the most suitable and best quality feeds.

In its early years, aquaculture was carried out in small areas using artificial feeds and simple techniques, and the only way to handle disease factors was to use antibiotics. However, more recently, aquaculture has improved significantly and has begun to yield quality products that are both environmentally friendly and beneficial to human health. Many studies have shown that the nutritional, and especially, the lipid compositions in farmed fish are more consistent than in wild fish; therefore, they are richer in n-3 fatty acids [59, 63, 65].

A study that compared individual farmed fish with individual wild fish using the sharp snout sea bream (*Diplodus puntazzo*). In the farmed fish, EPA, DHA, ∑PUFA, ∑n-3 fatty acid levels, and the n-3/n-6 ratio were 4.23 (g/100 g total fatty acid), 10.09 (g/100 g total fatty acid), 35.39, 28.65, and 4.25, respectively. In the wild fish, the EPA level was 6.86, DHA level was 9.28, ∑PUFA level was 32.29, ∑n-3 level was 24.75, and n-3/n-6 ratio was 3.53 [57].

Sea bass (*Dicentrarchus labrax*) is frequently farmed, both in Europe and in Turkey; many studies have focused on this species. Alasavar et al. reported the nutritional compositions of farmed and wild sea bass. In the farmed fish, EPA and DHA values were 6 and 18.1%, respectively, whereas they were 10.06 and 19.5% in the wild fish. In that study, the n-3/n-6 ratios were 2.88 and 3.02 in farmed and wild fish, respectively. The researchers attributed the lower levels in farmed fish to SFA and MUFA-rich, but PUFA-poor, feeds. They also stated that the wild fish were living in a nutrient-rich region and fed on n-3 fatty acid-rich food sources [58].

Different results were obtained in a study that compared the nutritional compositions of three different fish species (sea bass, sea bream, and rainbow trout) to each other. For sea bass, the EPA level was 7.32%, the DHA level was 14.8%, and the ∑n-3 level was 26.2%; for sea bream, the levels were 5.48, 12.4, and 23.3%, respectively; for rainbow trout, the levels were 6.16, 19.04, and 31.1%, respectively. The n-3/n-6 ratios for each fish species were 3.84, 2.85, and 4.54, respectively. Considering that, in general, the n-3/n-6 ratio in a healthy aquaculture food should be at least 1:1, and these three species were well above this value, an indicator of the quality of the aquaculture environment and feeds used in aquaculture [59].

In a study carried out in Turkey, the nutritional compositions of wild fish caught in the Atatürk Dam Lake were investigated: the EPA level was 7.18%, the DHA level was 5.39%, the ∑PUFA level was 23.09%, the ∑n-3 level was 15.64%, the ∑n-6 level was 7.45%, and the n-3/n-6 ratio was 2.10 [60].

Males and females (a total of 10 individual fish) from the shabbout (*Barbus grypus*) species obtained from the same region were used. The samples taken from the species, an omnivorous and rapidly growing species, weighed between 1.5 and 2 kg. Evaluating individual fishes separately showed that EPA values to be between 2.7 and 3.7% and DHA values between 5.2 and 10.7%; the highest value was determined in a male fish. Their ∑PUFA values were between 19.2 and 26.1%; ∑n-3 values between 14.7 and 18.2%, and, again, the highest value was determined in a male fish. The lowest ∑n-6 value was 3.9 and determined in a female, whereas the highest ∑n-6 value was 7.6 and determined in a male fish; thus, at 2.4, the lowest n-3/n-6 ratio was determined in male fish and, at 4.8, the highest n-3/n-6 was determined in female fish [61].

A similar study was carried out on spiny eel (*Mastacembelus mastacembelus*) and EPA and DHA values were 1.62 and 8.41%, respectively. The ∑PUFA level was 21.74%; ∑n-3 level, 14.16%, and ∑n-6 level, 7.11%. The researchers found a n-3/n-6 ratio = 2, and asserted that it could be a beneficial species for human health [62]. The researchers asserted that these were the first studies on wild shabbout and spiny eel in the region studied and stated that their results showed that the nutritional and fatty acid compositions of both species were of high quality and can have economic value.

Interesting results were obtained in a study carried out with individual farmed and wild trout. The nutritional compositions of fish obtained from earthen ponds, sea cages, lake (freshwater) cages, and from their natural habitats in lakes were compared. The highest EPA value (8.74%) was found in the wild fish, whereas the lowest EPA value (3.14%) was determined in the lake-caged fish. To the contrary, at 5.66%, the lowest DHA level was determined in the wild fish and, at 18.49%, the highest DHA value was determined in sea-caged fish. The ∑n-3 levels varied between 18.21 and 26.31%; the highest value was determined in sea-caged fish, the ∑n-6 values varied between 7.11% (wild fish) and 13.1% (lake-caged fish). The n-3/n-6 ratios were 1.33, 1.75, 2.58, and 2.71 for lake-caged fish, earthen pond fish, sea-caged fish, and wild fish, respectively. The n-3/n-6 ratios of fish from each different environment were reported to be at acceptable values [63].

4. Conclusion

The nutrients in fish are important for human health, but are easily obtained from fish oils. Fish fatty acids and particularly poly unsaturated fatty acids (PUFA) play an important role in human health, from embryological development to prevention and treatment of some diseases— including arthritis and inflammation, autoimmune disease, type 2 diabetes, hypertension, kidney and skin disorders, and cancer in children and in adults. The human body cannot synthesize certain fatty acids: these essential fatty acids must be consumed in the diet. Therefore, consumption of fish should routinely take place in human nutrition. The fish resources attract consumer interest and have been discussed in detail in the recent years; therefore, many studies have been carried out to investigate the nutritional value of farmed fish. Most of the studies showed that there was no significant difference between farmed and wild fish in terms of nutritional composition. A significant number of these studies mostly focused on the quantity and quality of fish feeds and the edible parts of fish.

Author details

Oğuz Taşbozan* and Mahmut Ali Gökçe

*Address all correspondence to: tasbozan@cu.edu.tr

Cukurova University, Fisheries Faculty, Department of Aquaculture, Adana, Turkey

References

[1] Jobling J, Leknes O: Cod liver oil: feed oil influences on fatty acid composition. Aquaculture International. 2010, **18**:223-230. doi:10.1007/s10499-008-9238-y

[2] Ruxton CHS, Reed SC, Simpson MJA, Millington KJ: The health benefits of omega-3 polyunsaturated fatty acids: a review of the evidence. Journal of Human Nutrition and Dietetics. 2007, **20**:275-285. doi:10.1111/j.1365-277X.2007.00770.x

[3] Simopoulos AP: Omega-3 fatty acids in health and disease and in growth and development, a review. The American Journal of Clinical Nutrition. 1991, **54**:438-463.

[4] Falk-Petersen S, Sargent JR, Henderson J, Hegseth EN, Hop H, Okolodkov YB: Lipids and fatty acids in ice algae and phytoplankton from the Marginal Ice Zone in the Barents Sea. Polar Biology. 1998, **20**(1):41-47. doi:10.1007/s003000050274

[5] Aras NM, Haliloğlu Hİ, Atamanalp M: Balıklarda Yağ Asitlerinin Önemi. Atatürk Üniversitesi Ziraat Fakültesi Dergisi. 2002, **33**(3):331-335.

[6] Lee JH: Polyunsaturated fatty acids in children. Pediatric Gastroenterology, Hepatology & Nutrition. 2013, **16**(3):153-161. doi:10.5223/pghn.2013.16.3.153

[7] Williams CA: Diatery fatty acids in human health. Annales de Zootechnie. 2000, **49:**165-180. doi:10.1051/animres:2000116

[8] Hossu B, Korkut AY, Fırat A: Balık Besleme ve Yem Teknolojisi I. Ege Üniversitesi Basımevi, Bornova İzmir. 2000, p. 276.

[9] Kaya Y, Duyar HA, Erdem ME: Balık yağ asitlerinin insan sağlığı için önemi. E.U. Journal of Fisheries & Aquatic Sciences, 2004, **21**(3-4):365-370

[10] Umhau JC, Dauphinais KM: Omega-3 polyunsaturated fatty acids and health. Low-Cost Approaches to Promote Physical and Mental Health, Springer. New York, USA, 2007, Chapter 4, pp. 87-101

[11] Innis SM: Fatty acids and early human development. Early Human Development 2007, **83**:761-766. doi:10.1016/j.earlhumdev.2007.09.004

[12] Osendarp SJM: The role of omega-3 fatty acids in child development. Oilseeds & fats, Crops and Lipids. 2011, **18**(6):307-313. doi:10.1051/ocl.2011.0417

[13] Burgess JR, Stevens L, Zhang W, Peck L: Long-chain polyunsaturated fatty acids in children with attention-deficit hyperactivity disorder. American Journal of Clinical Nutrition. 2000, **71** (Suppl):327-330.

[14] Deacon G, Kettle C, Hayes D, Dennis C, Tucci J: Omega 3 polyunsaturated fatty acids and the treatment of depression. Critical Reviews in Food and Nutrition. 2017, **57**(1):212-223. doi:10.1080/10408398.2013.876959

[15] McNamara RK: Role of omega-3 fatty acids in the etiology, treatment, and prevention of depression: current status and future directions. Journal of Nutrition & Intermediary Metabolism. 2016, **5:**96-106. doi:10.1016/j.jnim.2016.04.004

[16] Messamore E, Almeida DM, Jandacek RJ, McNamara RK: Polyunsaturated fatty acids and recurrent mood disorders: phenomenology, mechanisms, and clinical application. Progress in Lipid Research. 2017, **66:**1-13. doi:10.1016/j.plipres.2017.01.001

[17] Duda1 MK, O'Shea KM, Stanley WC: n-3 polyunsaturated fatty acid supplementation for the treatment of heart failure: mechanisms and clinical potential. Cardiovascular Research. 2009, **84:**33-41. doi:10.1093/cvr/cvp169

[18] Carayol M, Grosclaude P, Delpierre C. Prospective studies ofdietary alpha-linolenic acid intake and prostate cancer risk: a meta-analysis. Cancer Causes Control. 2010, **21**:347-355. doi:10.1007/s10552-009-9465-1

[19] Burlingame B, Nishida C, Uauy R, Weisell R. Fats and fattyacids in human nutrition: introduction. Annals of Nutrition & Metabolism. 2009, **55**:5-7. doi:10.1159/000228993

[20] Huerta-Yépez S, Tirado-Rodriguez AB, Hankinson O: Role of diets rich in omega-3 and omega-6 in the development of cancer. Boletin Medico del Hospital Infantil Mexico. 2016, **73**(6):446-456. doi:10.1016/j.bmhimx.2016.11.001

[21] Hagstrup IP: Marine n-3 fatty acids, wine intake, and heart rate variability in patients referred for coronary angiography, Circulation. 2001, **103**:651-657. doi:10.1161/01.CIR.103.5.651

[22] Piggott GM, Tucker BW: Seafood: effects of technology on nutrition. New York: Marcel Dekker. 1990, p. 384.

[23] Hertrampf JW, Pascual FP, editors. Handbook on Ingredients for Aquaculture Feeds. 1st ed. Dordrecht, The Netherlands: Kluwer Academic Publishers; 2000. pp. 4-35.

[24] Committee on animal nutrition board on agriculture National Research Council: nutrient requirements of fish. 1st ed. Washington: National Academy Press; 1993. p. 115.

[25] Halver, J E, Hardy, RW: Fish Nutrition. 3th ed. USA: Academic Press, Elsevier Science; 2002. p. 824.

[26] Tocher, DR: Fatty acid requirements in ontogeny of marine and freshwater fish. Aquaculture Research, 2010, **41**:717-732. doi:10.1111/j.1365-2109.2008.02150.x

[27] Henderson, RJ, Park MT, Sargent JR: The desaturation and elongation of ^{14}C-labelled polyunsaturated fatty acids by pike (*Esox lucius* L.) in vivo. Fish Physiology and Biochemistry. 1995, 14:223-235. doi:10.1007/BF00004313

[28] Özoğul Y, Özoğul F, Alagöz S: Fatty acid profiles and fat contents of commercially important seawater and freshwater fish species of Turkey: a comparative study. Food Chemistry. 2007, **103**:217-223. doi:10.1016/j.foodchem.2006.08.009

[29] HMSO, UK: Nutritional aspects of cardiovascular disease (report on health and social subjects No. 46). London: HMSO; 1994.

[30] Özoğul Y, Özoğul F, Çiçek E, Polat A, Küley E: Fat content and fatty acid compositions of 34 marine water fish species from the Mediterranean Sea. International Journal of Food Sciences and Nutrition. 2009, **60(6):**464-475. doi:10.1080/09637480701838175

[31] Muhamad NA, Muhamad J: Fatty acids composition of selected Malaysian fishes. Sains Malaysiana. 2012, 41(1):81-94.

[32] New MB. Feed and fish shrimps – A manual on the preparation and of compounds of feeds for shrimps and fish in aquaculture. In: Hertrampf JW, Pascual FP, editors. Handbook on Ingredients for Aquaculture Feeds. 1st ed. Dordrecht, The Netherlands: Kluwer Academic Publishers; 2000. pp. 4-35.

[33] Moyle P B, Cech J J. Fishes: An Introduction to Ichthyology. In: Halver, J E, Hardy, RW, editors. Fish Nutrition. 3th ed. USA: Academic Press, Elsevier Science; 2002. pp. 368-451.

[34] Çetinkaya O. Balık Besleme. 1st ed. Van: Yüzüncü Yıl Üniversitesi, Yayın No:9; 1995. p. 137.

[35] Friesecke H. Handbuch der praktischen Fütterung von Rind, Schaf, Pferd, Schwein, Geflügel und Süβwasserfischen. Hertrampf JW, Pascual FP, editors. Handbook on Ingredients for Aquaculture Feeds. 1st ed. Dordrecht, The Netherlands: Kluwer Academic Publishers; 2000. pp. 4-35.

[36] Morais S, Koven W, Ronnestad I, Dinis MT, Conceicao LEC: Dietary rotein:lipid ratio and lipid nature affects fatty acid absorption and metabolism in teleost larva. British Journal of Nutrition. 2005, **93**:813-820. doi:10.1079/BJN20051378

[37] Morais S, Conceicao LEC, Ronnestad I, Koven W, Cahu C, Zambonin Infante JL: Dietary neutral lipid level and source in marine fish larvae: effects on digestive physiology and food intake. Aquaculture. 2005, **268**:106-122. doi:10.1016/j.aquaculture.2007.04.033

[38] Bogevik AS, Tocher DR, Waagbo R, Olsen RE: Triacylglycerol-, wax ester- and sterol ester-hydrolases in midgut of Atlantic salmon (*Salmo salar*). Aquaculture Nutrition. 2008, **14**:93-98. doi:10.1111/j.1365-2095.2007.00510.x

[39] Sargent JR, McEvoy LA, Bell JG: Requirements, presentation and sources of polyunsaturated fatty acids in marine fish larval feeds. Aquaculture. 1997, **155**:117-127. doi:10.1016/S0044-8486(97)00122-1

[40] Kolkovski S: Digestive enzymes in fish larvae and juveniles–implications and applications to formulated diets. Aquaculture, 2001, **200**:181-201. doi:10.1016/S0044-8486(01)00700-1

[41] Cho CY, Slinger SJ: Apparent digestibility measurement in feedstuff for rainbow trout. In: Halver JE, Tiews K, editors. Finfish Nutrition and Fish Food Technology. Heenemann GmbH. Et Co KG. Berlin, Germany, 1979, Vol. 2, pp. 239-247.

[42] Hepher B: Nutrition of Pond Fishes. Cambridge: Cambridge University Press; 1988.

[43] Peres H, Oliva-Teles A: Effect of the dietary lipid level on growth performance and feed utilisation by European sea bass juveniles (*Dicentrarchus labrax*). Aquaculture, 1999, **179**:325-334. doi:10.1016/S0044-8486(01)00700-1

[44] Williams KC, Barlow C, Rodgers L, Hocking I, Agcopra C, Ruscoe I: Asian seabass *Lates calcarifer* perform well when fed pelleted diets high in protein and lipid. Aquaculture, 2003, **225**:191-206. doi:10.1016/S0044-8486(03)00278-3

[45] Denstadli V, Vegusdal A, Krogdahl A, Bakke-McKellup AM, Berge GM, Holm H: Lipid absorption in different segments of the gastrointestinal tract of Atlantic salmon (*Salmo salar*). Aquaculture, 2004, **240**:385-398. doi:10.1016/j.aquaculture.2004.06.030

[46] Glencross BD: Exploring the nutritional demand for essential fatty acids by aquaculture species. Reviews in Aquaculture, 2009, **1**:71-124. doi:10.1111/j.1753-5131.2009.01006.x

[47] Lupatsch I, Kissil GW, Sklan D: Comparison of energy and protein efficiency among three fish species *Sparus aurata, Dicentrarchus labrax* and *Epinephelus aeneus*: energy expenditure for protein and lipid deposition. Aquaculture, 2003, **225**:175-189. doi:10.1016/S0044-8486(03)00288-6

[48] Chou BS, Shiau SY: Optimal dietary lipid level for growth of juvenile hybrid tilapia, *Oreochromis niloticus* × *Oreochromis aureus*. Aquaculture, 1996, **143**:185-195. doi:10.1016/S0044-8486(97)89568-3

[49] Patankar VB: Esterases in the stomach of fishes with different food habits. Folia of Histochemistry and Cytochemistry, 1973, **11**:253-282.

[50] Polat A: BalıkBesleme. 1st ed. Adana: Çukurova Üniversitesi, Su Ürünleri Fakültesi, Seri No:10; 2003. P.182.

[51] Seafood health facts [Internet]. 2016. Available from: http://www.seafoodhealthfacts.org/printpdf/seafood-nutrition/patients-and-consumers/seafood-nutrition-overview. [Accessed: 2017.01.25]

[52] Nutritional content and benefit of sea food. Prepared by the Norwegian Food Safety Authority, Norwegian Directorate of Health and the University of Oslo. 3th ed. in English language is designed by Treehouse [Internet]. 2017. Available from: http://www.feap.info/Docdownload.asp?ID=B488944A4D0706030BE2. [Accessed: 2017.01.25].

[53] Fat storage. The National Institute of Nutrition and Seafood Research [Internet]. 2017. Available from: https://www.nifes.no/en/research-topics/aqua-culture/robust-fish/fat-storage/ [Accessed: 2017.01.25].

[54] Çelik M, Küçükgülmez A: Taze balıkta kalite ve kalite değisimleri. 1st ed. Ankara, Nobel Yayın Dağıtım Ltd. Sti.; 2007, p. 195.

[55] Falk-Petersen S, Sargent JR, Henderson J, Hegseth EN, Hop H, Okolodkov YB: Lipids and fatty acids in ice algae and phytoplankton from the Marginal Ice Zone in the Barents Sea. Polar Biology, 1998, **20(1)**:41-7. doi:10.1007/s003000050274

[56] Mohanty BP, Ganguly S, Mahanty A, Sankar TV, Anandan R, Chakraborty K, Paul BN, Sarma D, Dayal JS, Venkateshwarlu G, Mathew S, Asha KK, Karunakaran D, Mitra T, Chanda S, Shahi N, Das P, Akhtar MS, Vijayagopal P, Sridhar N: DHA and EPA Content and Fatty Acid Profile of 39 Food Fishes from India. BioMed Research International, 2016, Article ID 4027437, doi:10.1155/2016/4027437

[57] Rueda FM, Hernandez MD, Egea MA, Aguado F, Garcia B, Martinez FJ: Differences in tissue fatty acid composition between reared and wild sharpsnout sea bream, *Diplodus puntazzo* (Cetti, 1777). British Journal of Nutrition, 2001, **86**:617-622. doi:10.1079/BJN2001438

[58] Alasalvar C, Taylor KDA, Zubcov E, Shahidi F, Alexis M: Differentiation cultured and wild sea bass (*Dicentrarchus labrax*) total lipid content, fatty acid and trace mineral composition. Food Chemistry, 2002, **79 (2)**:145-150. doi:10.1016/S0308-8146(02)00122-X

[59] Testi S, Bonaldo A, Gatta PP, Badiani A: Nutritional traits of dorsal and ventral fillets from three farmed fish species. Food Chemistry, 2006, **98 (1)**:104-111. doi:10.1016/j.foodchem.2005.05.053

[60] Celik M, Gokce MA, Basusta N, Kucukgulmez A, Tasbozan O, Tabakoglu SS: Nutritional quality of rainbow trout (*Onchorhynchus mykiss*) caught from the Ataturk Dam Lake in Turkey. Journal of Muscle Food, 2007, **19**:50-61. doi:10.1111/j.1745-4573.2007.00099.x

[61] Gokce MA, Tasbozan O, Tabakoglu SS, Celik M, Ozcan F, Basusta A: Proximate composition and fatty acid profile of shabbout (*Barbus grypus*Heckel) caught from the Ataturk Dam Lake, Turkey. Journal of Food, Agriculture & Environment, 2011, **9(2):**148-151.

[62] Tasbozan O, Gokce MA, Celik M, Tabakoglu SS, Kucukgulmez A, Basusta A: Nutritional composition of Spiny eel (*Mastacembelus mastacembelus*) caught from the Atatürk Dam Lake in Turkey. Journal of Applied Biological Sciences, 2013, **7(2):**78-82.

[63] Tasbozan O, Gokce MA, Erbas C: The effect of different growing conditions to proximate composition and fatty acid profiles of rainbow trouts (*Oncorhynchus mykiss*). Journal of Applied Animal Research, 2016, **44(1):**442-445. doi:10.1080/09712119.2015.1091323

[64] Gökçe MA, Tasbozan O, Çelik M, Tabakoğlu SS: Seasonal variations in proximate and fatty acid compositions of female common sole (*Solea solea*). Food Chemistry, 2004, **88(3):**419-423. doi:10.1016/j.foodchem.2004.01.051

[65] Cahu C, Salen P, De Lorgeril M: Farmed and wild fish in the prevention of cardiovascular diseases: assessing possible differences in lipid nutritional values. Nutrition Metabolism Cardiovascular Diseases, 2004, **14(1)**:34-41. doi:10.1016/S0939-4753(04)80045-0

6

Fatty Acids on Osteoclastogenesis

Sergio Montserrat-de la Paz, Rocio Abia,
Beatriz Bermudez, Sergio Lopez and
Francisco JG Muriana

Abstract

Excessive bone resorption is a hallmark on the onset and development of bone diseases, including osteoporosis, periodontitis, and rheumatoid arthritis. Osteoclasts are bone-resorbing multinucleated cells that differentiate from hematopoietic progenitors of the myeloid lineage. The regulation of this differentiation process is considered an effective therapeutic intervention to the treatment of pathological bone loss. Dietary fatty acids (FAs), transported in the form of postprandial triglyceride-rich lipoproteins, have been linked with inflammation and oxidative stress associated to the overactivation of circulating leukocytes. Monocyte differentiation by soluble cytokines is known to up-regulate osteoclast maturation via increased expression levels of receptor activator for nuclear factor-κB ligand relative to osteoprotegerin. This review summarizes the effects of dietary omega-3 long-chain polyunsaturated fatty acids, monounsaturated fatty acids, and saturated fatty acids on plasticity during osteoclast formation and function.

Keywords: bone marrow, bone metabolism, fatty acids, osteoclasts, osteoporosis

1. Introduction

The links among bone and nutrition focus on considerable public health and research interests. Over the past 20 years, the fact that there is an inverse relationship between bone mass and marrow adiposity, observed under physiological and pathological conditions, has led to increased recent interest in bone lipids [1, 2]. Under different pathologies, for example, osteoporosis, an increase of bone marrow fat that was associated with osteoclast (OC) overabundance and a low bone mass [3]. Cholesterol, phospholipids, and fatty acids (FAs) either free or in the form of triglycerides, have been demonstrated to act on bone metabolism and bone cell development and functions. Thus, they can be regarded as regulatory molecules important in bone health. A growing body of evidence, including the recognition that specific FA receptors are expressed in bone-related cells, suggests that FAs both circulating and inside bone marrow, could be an active determinant role as messengers on metabolic activity and remodeling rate of bone [4]. This review will provide a current overview on the effects of FAs on OC maturation and function.

2. Osteoclast biology

Bone is a specialized, hard tissue consisting of a soft part (the bone marrow), and the mineralized osseous tissue itself. To ensure bone integrity during childhood and adulthood, bone undergoes a continuous remodeling process that consists of multiple cycles of bone digestion and rebuilding steps [5]. Two cell types mainly determine this remodeling process, the bone-forming osteoblasts (OBs) and the bone-resorbing OCs. A dysregulation of the bone remodeling balance is linked with several skeletal disorders such as osteopetrosis and osteoporosis. Osteopetrosis is characterized by an increase in bone mass due to a lower OC number or activity, whereas osteoporosis is characterized by the loss of bone mass due to an elevated OC activity [6]. Moreover, bone contains interconnected and embedded OBs, called osteocytes, which might respond to the mechanical pressure applied onto bone [7].

During initial bone formation, OBs produce organic bone matrix and promote its mineralization. At the same time, OBs indirectly affect bone resorption by the expression of ligands, including the receptor activator of NF-κB ligand (RANKL), which is important for OC differentiation [5]. In contact with bone, OCs change their plasma membrane to form different domains, including the ruffled border that faces the bone surface. This specialized cell membrane is provided with many lysosomal integral membrane proteins, mainly the V-type H^+-ATPase, ensuring the acidification of the resorption environment that is required to dissolve the bone inorganic matrix. OCs also release lysosomal hydrolases such as cathepsin K to digest the organic bone matrix [8]. Furthermore, the ruffled border is composed by actin-rich podosomes that ensure the attachment of OCs onto the bone. Bone degradation products are endocytosed through the ruffled border, transcytosed, and secreted into the extracellular space [9]. For efficient resorption, OCs undergo several cycles of adhesion, resorption, and migration along bone surfaces.

Bone biology has greatly benefited from studies using animal models. For example, silencing Src tyrosine kinase, receptor–activator of NF-κB (RANK), tartrate-resistant acid phosphatase (TRAP), and cathepsin K in mice result in an osteopetrotic mouse model due to the lack of OC precursor differentiation or a lack of mature OC activity [10, 11]. However, these mutant animal models do not provide an integrated view on the function of a particular gene on OC differentiation and function and its modulation by certain cytokines, nutrients, and drugs, which could provide a better understanding of their effects on OC biology.

3. Osteoclastogenesis in the bone marrow

The bone-resorbing OCs are originated from the differentiation of hematopoietic mononucleated precursors and their subsequent fusion to form multinucleated mature OCs (**Figure 1**). Physiologically, osteoclastogenesis requires two essential hematopoietic factors in the bone marrow: macrophage colony-stimulating factor (M-CSF/CSF-1) and RANKL. M-CSF/CSF-1 is a survival and proliferation factor that induces RANK expression in OC precursor cells [12]. The role of M-CSF/CSF-1 in osteoclastogenesis is highlighted by the osteopetrotic phenotype M-CSF$^{-/-}$ mouse model, in which mutant animals had a deficiency in OCs and circulating monocytes [13]. The second key factor in osteoclastogenesis, RANKL, is a membrane-residing protein found on OBs and their precursors, and is recognized by its cognate receptor RANK expressed in the bone marrow macrophage/OC lineage, promoting their differentiation into OCs [14]. In mice, genetic experiments have shown the importance of RANK/RANKL axis for osteoclastogenesis, as targeted inhibition of RANK or RANKL gene results in a complete absence of OC maturation and osteopetrosis [15]. In humans *in vitro* studies, these two factors are able to generate OCs from circulating monocytes, dendritic cells, and bone marrow–derived macrophages [16]. In addition to RANKL, osteoprotegerin (OPG) is secreted by OBs, which acts as a soluble RANKL decoy receptor; therefore, OPG negatively regulates RANKL activity (**Figure 2**) [17]. From these observations, the RANKL/OPG ratio indicates the rate of osteoclastic bone resorption [18].

The binding of RANK receptor to RANKL triggers signaling cascades that terminally differentiate the hematopoietic precursor cells into OCs. The initial step in RANKL signaling is the binding of RANK receptor to the cytoplasmatic tumor necrosis factor receptor-associated factors (TRAF), mainly to TRAF6 [19]. The Src tyrosine kinase binds to TRAF6, regulating the aspects of OC function such as cytoskeletal reorganization. In addition, RANKL signaling leads to the OC specific gene expression such as β_3 integrins, TRAP, cathepsin K, and calcitonin receptor. It also leads to the morphological conversion of mononucleated cells into large multinucleated cells that are able to efficiently resorb large bone surface areas.

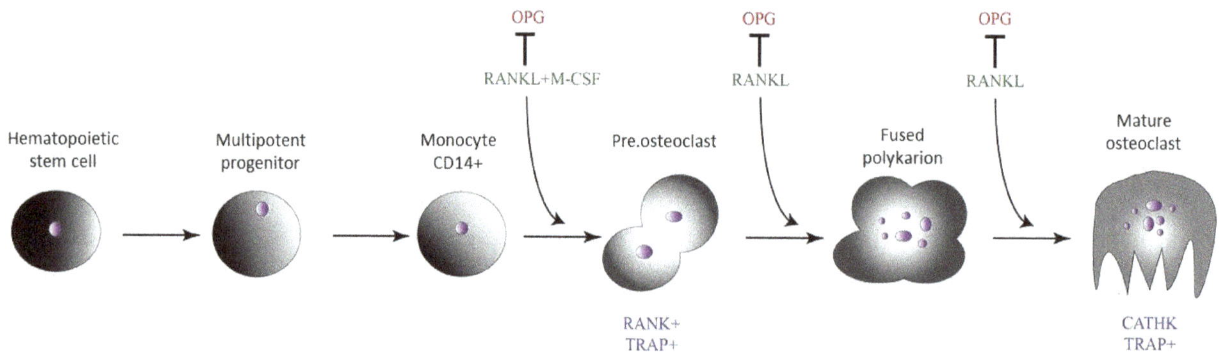

Figure 1. Diagram illustrating the differentiation of hematopoietic mononucleated precursors and their subsequent fusion to form multinucleated mature osteoclasts.

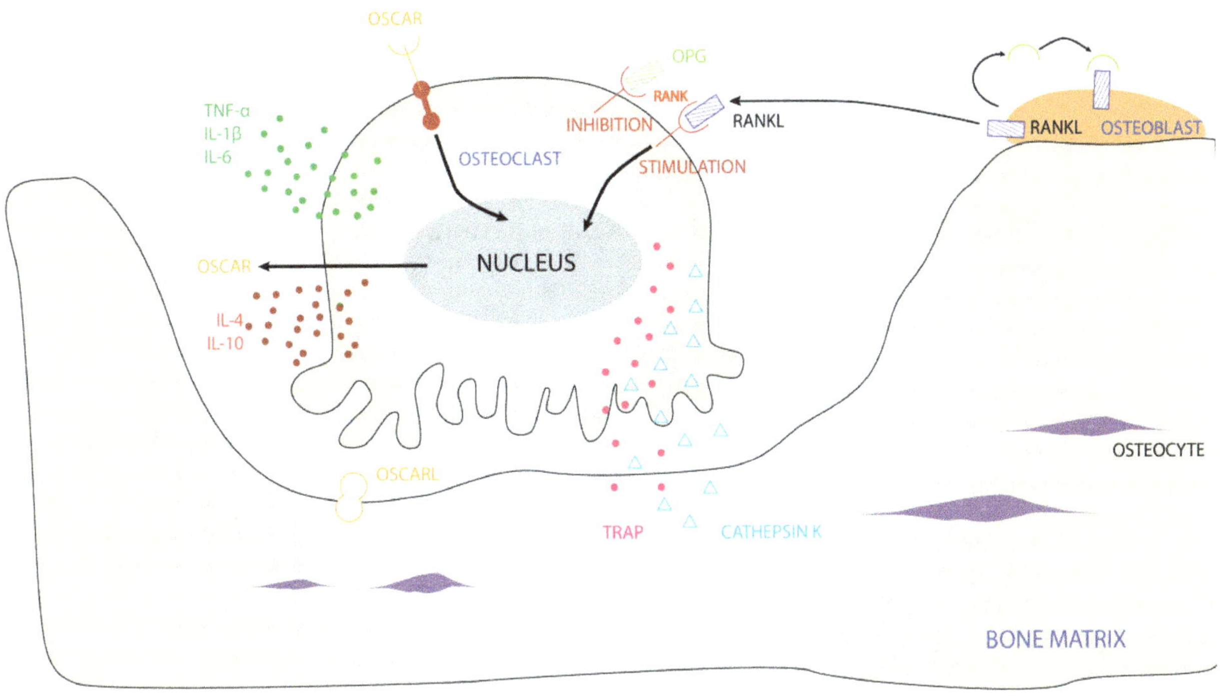

Figure 2. Key factors affecting osteoblast and osteoclast survival and functions.

4. Fatty acids in the bone marrow

FAs are carboxylic acids and often contain a long, unbranched aliphatic chain. FAs are categorized as saturated (SFAs), monounsaturated (MUFAs), and polyunsaturated (PUFAs) based on their structural and chemical properties. SFAs do not contain any double bonds or other functional groups along the chain, which is fully saturated with hydrogen atoms. The principal dietary SFAs are palmitic acid (16:0) and stearic acid (18:0), which are composed of 16 and 18 carbon atoms, respectively. MUFAs contain one pair of carbon atoms linked by a *cis* double bond. Oleic acid (18:1n−9), which contains 18 carbon atoms with a double bond at the 9th carbon from the methyl end of the FA molecule, is the major dietary MUFA and represents 55–83% of the total FAs in virgin olive oil [20]. Carbon chains containing 2 or more *cis* double bonds, with the first double bond located between either the 3rd and 4th or the 6th and 7th carbon atoms from the methyl end of the FA molecule, that belong to the n−3 or n−6, respectively, PUFA families. These families cannot be synthesized by the human body (double bonds can be introduced into all positions of the FA chain with the exception of the n−3 and n−6 positions); and therefore, must be obtained from the diet as α-linolenic acid (18:3n−3) and linoleic acid (18:2n−6) or their long-chain PUFA derivatives. Of these FAs, eicosapentaenoic acid (EPA, 20:5n−3), docosahexaenoic acid (DHA, 22:6n−3), dihomo-γ–linolenic acid (20:3n−6), and arachidonic acid (AA, 20:4n−6) are the most metabolically significant [21].

FA compositions of total lipids present in bone marrows change with the species studied. Thus, palmitic acid, stearic acid, and oleic acid are predominant in rats and cows [22], whereas palmitic acid, oleic acid, and linoleic acid are the main FAs in bone marrows of humans, dogs, guinea pigs, and rabbits [23]. Bone FA profile usually reflects the FA composition of the diet. For example, when animals were fed diet supplemented with linoleic acid or α-linolenic acid, concentrations of these two FAs were higher in femoral cortical bone and marrow [24]. Recent animal and human intervention studies reported that dietary FAs affect bone health. In general, high intakes of long-chain omega-3 PUFAs rather than long-chain omega-6 PUFAs are beneficial for bone mass [25], whereas SFAs intake is harmful [26].

5. Direct action of exogenous fatty acids on bone cells

At the level of bone cell biology, *de novo* biosynthesized FAs or FAs taken up by cells are mostly incorporated into both phospholipids located in cell membranes and triglycerides in cytoplasmic lipid droplets. On the other hand, membrane FA composition has demonstrated to modulate intracellular signaling pathways and many cell functions such as membrane fluidity and permeability [20]. Thus, FAs may influence the bone formation/resorption balance by affecting the functionality of OBs and OCs.

In vitro studies have demonstrated that exogenous FAs supplemented to the OBs or OCs culture media can affect their survival and functions. Data indicate that SFAs, mainly palmitic and stearic acids, are pivotal for OBs by inducing both autophagy and apoptosis [27, 28]. PUFAs also alter OB proliferation and functions [29, 30], while oleic acid seems to be neutral in OBs [31].

Few studies, summarized in **Table 1**, have focused at exogenous FA effects on OCs and the data are partially contradictory, at least for SFAs. Indeed, SFAs, mostly myristic, palmitic, and stearic acids 16:0 were first reported to inhibit osteoclastogenesis [32], and recently, to enhance it by inhibiting apoptosis of mature OCs [33]. The actions of exogenous FAs on bone cells include their ability to modulate different signaling pathways that are involved in general cell growth, differentiation, inflammation, and apoptosis processes. FAs can also alter expression/activation of different nuclear transcription factors which play an important role in bone metabolism, such as nuclear factor κB (NF-κB, crucial for many bone cell processes and for OC activity), and peroxisome proliferator–activated receptor γ (PPARγ, role in bone-fat relationship) [35]. To start cell signaling, FAs play via protein sensors located either in cytosol (i.e., FA-binding proteins (FABPs) and PPARs) or at cell surface (i.e., specific receptors that belong to the family of G-protein–coupled receptors (GPCs)). These extracellular receptors are likely to play an important role in bone physiology since they are expressed at the surface of OBs and OCs [32]. As outlined in **Table 2**, there are currently six receptors known to be linked by FAs of different carbon chain length and degree of saturation. GPR120 has been reported to be expressed in OBs; however, these cells do not express GPR40, 41, or 43 [32]. In a review of the effects of exogenous FAs on osteoclast OC development at concentrations of 0.1–10 μg/ml, the most potent effects were observed in response to palmitic and stearic acids, implying that signaling through GPR120 mediates, at least in part, the direct osteoclastogenic actions of medium and long-chain SFAs [32].

On the other hand, limited evidence exists as to the actions of PUFAs on OC development. Two studies have reported inhibitory actions of linoleic acid on osteoclastogenesis in bone

FAs and concentration used [Reference]	Cell model	Effects on the studied markers	Main outcomes
4:0, 8:0, 12:0, 14:0, 16:0, 18:0, 18:1, 18:2n–6, 18:3n–3, 20:4n–6, 20:5n–3, and 22:6n–3 (0.3–115 μM) [32]	1,25-Dihydroxyvitamin D3-stimulated murine bone marrow–derived macrophages and RANKL-stimulated murine macrophage cell line RAW264.7	↓TRAP positive cells by SFAs and no clear-cut differences between n-3 and n-6 PUFAs	SFAs inhibit osteoclastogenesis, probably via receptors expressed at the surface of OCs
12:0 and 16:0 (20–100 μM) [33]	RANKL/M-CSF-stimulated murine bone marrow–derived macrophages	↓ Annexin V staining ↑ TRAP positive cells (SFAs) ↑ MIP-1α production (SFAs) ↑ NF-κB, TLR4, and MyD88 activation (SFAs) ↓TRAP positive cells (PUFAs)	SFAs enhance cell survival in mature OCs
18:2n–6 (1–100 μM) [34]	RANKL-stimulated murine macrophage cell line RAW264.7	↓TRAP positive cells ↓Bone resorption area	Linoleic acid inhibit OC differentiation, possibly by modulating the downstream molecules of RANKL signaling

Table 1. Effect of exogenous free fatty acids (FAs) on osteoclast functions and survival (Adapted from Ref. [4]).

Receptor	Ligand(s)	Sites of expression	Function
GPR40	C6–C22 FAs, saturated and unsaturated	Pancreatic islets Gut Brain Monocytes **Osteoclasts**	Glucose-stimulated insulin secretion
GPR41	C1–C6 FAs	Adipocytes Bone marrow Spleen Lymph node PBMCs **Osteoclasts**	Leptin production
GPR43	C1–C6 FAs	Adipocytes Colon PBMCs **Osteoclasts**	Adipogenesis lipolysis inhibition
GPR84	C9–C14 FAs	PBMCs Lung	Regulation of inflammatory response
GPR119	Lysophosphatidylcholine and oleoylethanolamide	L cells Pancreas	GLP-1 and insulin secretion
GPR120	C14–C18 Saturated and C16–C22 unsaturated	L cells **Osteoclasts** **Osteoblasts**	GLP-1 secretion

Boldface, the main bone cells (Osteoclasts and osteoblasts)

Table 2. Fatty acid (FA) receptors (Adapted from Ref. [36]).

marrow cultures and RAW264.7 cells [32, 34]. A subsequent report found that DHA, but not EPA, substantially decreased OC development in RANKL-treated in bone marrow cultures and RAW264.7 cells [37, 38]. The mechanism(s) by which PUFAs modulate bone cell function are uncertain, but may include direct incorporation into cell membranes, with subsequent alteration of levels of intracellular prostanoids and eicosanoids [37].

6. Effect of postprandial triglyceride–rich lipoproteins on bone cells

The postprandial state, the period that comprises and follows a meal, plays an important, yet underappreciated role in the genesis of numerous pathological conditions. After fatty food consumption, dietary FAs are largely incorporated into nascent triglyceride-rich lipoproteins (TRLs), which are released from the small intestine into the blood. It has been previously shown that SFAs, MUFAs, and PUFAs have dissimilar postprandial effects on risk factors for chronic diseases [39], suggesting that short-term outcomes in response to dietary FA adjustment could be useful to finely tune fat consumption, even for preventing diet-related chronic diseases [40]. However, *in vivo* studies on markers of osteoclastogenesis during the postprandial state in humans or *in vitro* studies on interaction of human postprandial TRLs with monocyte-derived OCs were unknown. In fact, there are only a few labs studying the link between the postprandial state and osteoclastogenesis. One of them has demonstrated that serum obtained from healthy subjects following the consumption of a meal containing almonds may inhibit OC maturation and function in primary human OC precursor cells, providing direct evidence to support the association between regular almond consumption and a reduced risk of osteoporosis [41]. Inspired in these findings, our group demonstrated for the first time in 2016 that the RANKL/OPG ratio is postprandially modulated by the predominant FAs in dietary fats, being particularly increased after the ingestion of an SFA-enriched meal when compared to the ingestion of MUFA-enriched meals [42]. *In vitro,* we also observed an increase of OC marker gene expression and a decrease of OPG gene expression in human monocyte-derived OCs in response to postprandial TRL-SFAs, further supporting the notion that dietary saturated fats may promote osteoclastogenesis through pathways involving the metabolism of intestinal lipoproteins. Importantly, TRL-MUFAs and TRL-PUFAs did not alter these osteoclastogenic markers or OPG, suggesting that the substitution of dietary saturated fats by monounsaturated fats (in combination with omega-3 PUFAs) may be useful to prevent excessive osteoclastogenesis associated to postprandial events.

In spite of the increasing evidence of the pivotal role of FAs on bone physiology as biological modulators of osteoclastogenesis, nutritional interventions might be a reliable therapeutic target to induce positive effects on skeletal health. Further, careful preclinical and clinical studies are likely to shed additional light on this important area of bone biology.

Conflicts of interest

The authors state no conflict of interest.

Acknowledgements

This study was supported by the research Grant AGL2016-80852 (Spanish Ministry of Economy, Industry, and Competitiveness). BB and SL acknowledge financial support from "V Own Research Plan" (University of Seville) and the Spanish Research Council (CSIC)/Juan de la Cierva, respectively. A major portion of Sections 4 and 6 in this chapter borrows from authors' previous publications [20, 42].

Author details

Sergio Montserrat-de la Paz[1*†], Rocio Abia[1], Beatriz Bermudez[2], Sergio Lopez[1] and Francisco JG Muriana[1†]

*Address all correspondence to: delapaz@us.es

1 Laboratory of Cellular and Molecular Nutrition, Spanish National Research Council (CSIC), Seville, Spain

2 Department of Cell Biology, University of Seville, Seville, Spain

†These authors contributed equally to this work.

References

[1] Bermeo S, Gunaratnam K, Duque G. Fat and bone interactions. Curr Osteoporos Rep. 2014;12:235-242.

[2] Devlin MJ, Rosen CJ. The bone-fat interface. Basic and clinical implications of marrow adiposity. Lancet Diabetes Endocrinol. 2015;3:141-147.

[3] Yeung DK, Griffith JF, Antonio GE, et al. Osteoporosis is associated with increased marrow fat content and decreased marrow fat unsaturation: a proton MR spectroscopy study. J Magn Reson Imaging. 2005;22:279-285.

[4] During A, Penel G, Hardouin P. Understanding the local actions of lipids in bone physiology. Prog Lipid Res. 2015;59:126-146.

[5] Segeletz S, Hoflack B. Proteomic approaches to study osteoclast biology. Proteomics. 2016;16:2545-2556.

[6] Tolar J, Teitelbaum SL, Orchard PJ. Osteopetrosis. N Engl J Med. 2004;351:2839-2849.

[7] Prideaux M, Findlay DM, Atkins GJ. Osteocytes: the master cells in bone remodeling. Curr Opin Pharmacol. 2016;28:24-30.

[8] Vaananen HK, Zhao H, Mulari M. The cell biology of osteoclast function. J Cell Sci. 2000;113:377-381.

[9] Salo J, Lehenkari P, Mulari M, et al. Removal of osteoclast bone resorption products by transcytosis. Science. 1997;276:270-273.

[10] Saftig P, Hunziker E, Wehmeyer O, et al. Impaired osteoclastic bone resorption leads to osteopetrosis in cathepsin-K-deficient mice. Proc Natl Acad Sci USA. 1998;95:13453-13458.

[11] Dougall WC, Glaccum M, Charrier K, et al. RANK is essential for osteoclast and lymph node development. Genes Dev. 1999;13:2412-2424.

[12] Arai F, Miyamoto T, Ohneda O, et al. Commitment and differentiation of osteoclast precursors cells by the sequential expression of c-fms and receptor activator of nuclear factor kappa B (RANK) receptors. J Exp Med. 1999;190:1741-1754.

[13] Yoshida H, Hayashi S, Kunisada T, Ogawa M, et al. The murine mutation osteopetrosis is in the coding region of the macrophage colony stimulating factor gene. Nature. 1990;345:442-444.

[14] Boyle WJ, Simonet WS, Lacey DL. Osteoclast differentiation and activation. Nature. 2003;423:337-342.

[15] Li J, Sarosi I, Yan XQ, et al. RANK is the intrinsic hematopoietic cell surface receptor that controls osteoclastogenesis and regulation of bone mass and calcium metabolism. Proc Natl Acad Sci USA. 2000;97:1566-1571.

[16] Rivollier A, Mazzorana M, Tebib J, et al. Immature dendritic cell transdifferentiation into osteoclasts: a novel pathway sustained by the rheumatoid arthritis microenvironment. Blood. 2004;104:4029-4037.

[17] Thomas GP, Baker SU, Eisman JA, et al. Changing RANKL/OPG mRNA expression in differentiating murine primary osteoblasts. J Endocrinol. 2001;170:451-460.

[18] Hofbauer LC, Heufelder AE. Role of receptor activator of nuclear-kappaB ligand and osteoprotegerin in bone cell biology. J Mol Med. 2001;79:243-253.

[19] Lomaga MA, Yeh WC, Sarosi I, et al. TRAF6 deficiency results in osteopetrosis and defective interleukin-1, CD40, and LPS signaling. Gene Dev. 1999;13:1015-1024.

[20] Lopez S, Bermudez B, Montserrat-de la Paz S, et al. Membrane composition and dynamics: a target of bioactive virgin olive oil constituents. Biochimica et Biophysica Acta - Biomembranes (BBA-Biomembranes). Biochim Biophys Acta. 2014;1838:1638-1656.

[21] Montserrat-de la Paz S, Bermudez B, Cardelo MP, et al. Olive oil and postprandial hyperlipidemia: Implications for atherosclerosis and metabolic syndrome. Food Funct. 2016;7:4734-4744.

[22] Fazzalari NL. Bone remodeling: a review of the bone microenvironment perspective for fragility fracture (osteoporosis) of the hip. Semin Cell Dev Biol. 2008;19:467-472.

[23] Yeung DK, Lam SL, Griffith JF, et al. Analysis of bone marrow fatty acid composition using high-resolution proton NMR spectroscopy. Chem Phys Lipids. 2008;151:103-109.

[24] Li Y, Greiner RS, Salem Jr N, et al. Impact of dietary n-3 FA deficiency on rat bone tissue FA composition. Lipids. 2003;38:683-686.

[25] Poulsen RC, Moughan PJ, Kruger MC. Long-chain polyunsaturated fatty acids and the regulation of bone metabolism. Exp Biol Med. 2007;232:1275-1288.

[26] Corwin RL, Hartman TJ, Maczuga SA, et al. Dietary saturated fat intake is inversely associated with bone density in humans: analysis of NHANES III. J Nutr. 2006;136:159-165.

[27] Elbaz A, Wu X, Rivas D, et al. Inhibition of fatty acid biosynthesis prevents adipocyte lipotoxicity on human osteoblasts *in vitro*. J Cell Mol Med. 2009;14:982-991.

[28] Wang D, Haile A, Jones LC. Dexamethasone-induced lipolysis increases the adverse effect of adipocytes on osteoblasts using cells derived from human mesenchymal stem cells. Bone. 2013;53:520-530.

[29] Maurin AC, Chavassieux PM, Vericel E, et al. Role of polyunsaturated fatty acids in the inhibitory effect of human adipocytes on osteoblastic proliferation. Bone. 2002;31:260-266.

[30] Watkins BA, Li Y, Lippman HE, et al. Modulatory effect of omega-3 polyunsaturated fatty acids on osteoblast function and bone metabolism. Prostaglandins Leukot Essent Fatty Acids. 2003;68:387-398.

[31] Kim JE, Ahn MW, Baek SH, et al. AMPK activator, AICAR, inhibits palmitate-induced apoptosis in osteoblast. Bone. 2008;43:394-404.

[32] Cornish J, MacGibbon A, Lin JM, et al. Modulation of osteoclastogenesis by fatty acids. Endocrinology. 2008;149:5688-5695.

[33] Oh SR, Sul OJ, Kim YY, et al. Saturated fatty acids enhance osteoclast survival. J Lipid Res. 2010;51:892-899.

[34] Rahman MM, Bhattacharya A, Fernandes G. Conjutated linoleic acid inhibits osteoclast differentiation of RAW264.7 cells by modulating RANKL signalling. J Lipid Res. 2006;47:1739-1748.

[35] Boyce BF, Yao Z, Xing L. Functions of nuclear factor kappaB in bone. Ann NY Acad Sci. 2010;1192:367-375.

[36] Grey A. Fatty acids and bone. Clinic Rev Bone Miner Metab. 2009;7:210-215.

[37] Rahman MM, Bhattacharya A, Fernandes G. Docosahexaenoic acid is more potent inhibitor of osteoclast differentiation in RAW264.7 cells than eicosapentaenoic acid. J Cell Physiol. 2008;214:201-209.

[38] Sun D, Krishnan A, Zaman K, et al. Dietary n-3 fatty acids decreases osteoclastogenesis and loss of bone mass in ovariectomized mice. J Bone Miner Res. 2003;18:1206-1216.

[39] Lopez S, Bermudez B, Ortega A, et al. Effects of meals rich in either monounsaturated or saturated fat on lipid concentrations and on insulin secretion and action in subjects with high fasting triglyceride concentrations. Am J Clin Nutr. 2011;93:494-499.

[40] Ortega A, Varela LM, Bermudez B, et al. Dietary fatty acids linking postprandial metabolic response and chronic diseases. Food Funct. 2012;3:22-27.

[41] Platt ID, Josse AR, Kendall CWC, et al. Postprandial effects of almond consumption on human osteoclast precursors – an *ex vivo* study. Metabolism. 2011;60:923-929.

[42] Naranjo MC, Indara I, Bermudez B, et al. Acute effects of dietary fatty acids on osteclastogenesis via RANKL/RANK/OPG system. Mol Nutr Food Res. 2016;60:2505-2513.

7

Fatty Acids and their Analogues as Anticancer Agents

Jubie Selvaraj

Abstract

Recent research supports the beneficial effects of dietary polyunsaturated fatty acids (PUFAs) on inhibiting tumour development. Long-chain fatty acids modulate the tumour cell response to chemotherapeutic drugs. Investigators recently claimed high dietary intake of omega-6 polyunsaturated fatty acids such as linoleic acid especially in association with a low intake of omega-3 polyunsaturated fatty acids such as docosahexaenoic acid to increase risks for cancers of the breast, colon and possibly prostate. In addition to these facts, a number of investigations have demonstrated that a modified fatty acid analogues are promising molecules in cancer prevention and have potential in the treatment of cancer. Although billions of dollars have been spent on research and development on anticancer drugs, the disease remains uncontrolled. It is expected that anticancer agents preferentially kill tumour cells without causing adverse effects on normal cells. But this is rarely achieved with the existing cancer therapy. Hence, polyunsaturated fatty acids have come under the category of nutraceuticals/functional foods; their exploration in the treatment of cancer may be considered as safe. This chapter describes the effects of long-chain fatty acids and their analogues in cancer chemotherapy.

Keywords: fatty acids, cancer, PUFA, fatty acid synthase, omega-3

1. Introduction to fatty acids

Plants, animals and microbes generally contain even number of carbon atoms in straight chains, with a carboxylic group at one end and double bonds with *cis* configuration on the another end. The chain length of the common fatty acids varies between 14 and 22, but on occasions can span between 2 and 36 or even more in animal tissues. Fatty acids found in animal tissues have one to six double bonds, whereas those in algae have up to five bonds. Higher plants rarely have more than three, whereas microbial fatty acids occasionally have more than one. The fatty acids, which are derived from triglycerides or phospholipids, have a chain of

4–28 carbons. Fatty acids, which are not attached to other molecules, are known as free fatty acids which on breakdown yield large quantities of ATP. Many cell types use either glucose or fatty acids for this purpose. In particular, heart and skeletal muscle prefer fatty acids [1].

Fatty acids may be monounsaturated, polyunsaturated or saturated (**Figure 1**). They help in moving oxygen through the blood stream to all parts of the body, aid cell membrane development and strengthen the organs and tissue. They also help in healthy skin and prevent early ageing and more importantly help rid the arteries of cholesterol build-up.

2. Types of fatty acids

2.1. Saturated fatty acids

Saturated fatty acids are straight-chain compounds with 14, 16 and 18 carbon atoms. The most abundant saturated fatty acids found in animal and plant tissues are esterified with odd- and even-numbered homologues with 2–36 carbon atoms. A list of common saturated fatty acids together with their trivial names and shorthand designations is given in **Table 1**.

2.2. Monoenoic fatty acids

Monoenoic fatty acids are straight-chain fatty acids containing 10–30 carbon atoms with one *cis*-double bond. The double bond can be in different positions and this is specified in the systematic nomenclature in relation to the carboxyl group (**Table 2**).

2.3. Polyunsaturated fatty acids

Polyunsaturated fatty acids (PUFAs) are fatty acids which contain multiple double bonds and are subdivided into families according to their derivation from specific biosynthetic precursors. In each instance, the families contain between two and six *cis*-double bonds separated

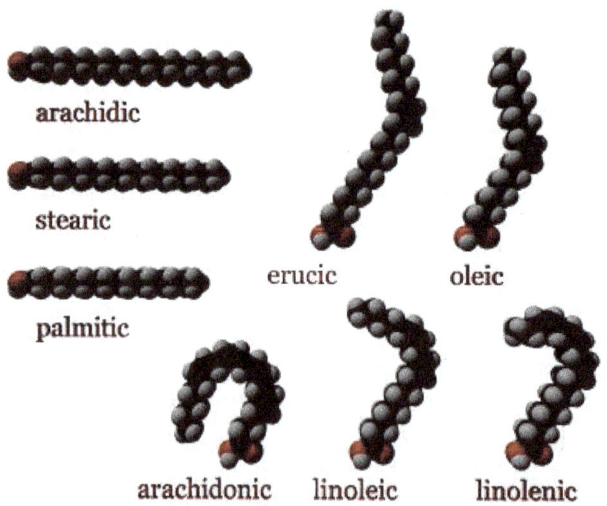

Figure 1. Naturally occurring fatty acids.

S. no.	Systematic name	Shorthand designation	Trivial name
1.	Ethanoic	2:0	Acetic
2.	Butanoic	4:0	Butyric
3.	Hexanoic	6:0	Caproic
4.	Octanoic	8:0	Caprylic
5.	Nonanoic	9:0	Pelargonic
6.	Decanoic	10:0	Capric
7.	Undecanoic	11:0	–
8.	Dodecanoic	12:0	Lauric
9.	Tridecanoic	13:0	–
10.	Tetradecanoic	14:0	Myristic
11.	Pentadecanoic	15:0	Myristic
12.	Hexadecanoic	16:0	Palmitic
13.	Heptadecanoic	17:0	Margaric
14.	Octadecanoic	18:0	Stearic
15.	Nonadecanoic	19:0	Margaric
16.	Arachidic	20:0	Eicosanoic
17.	Heneicosanoic	21:0	–
18.	Docosanoic	22:0	Behenic
19.	Tetracosanoic	24:0	Lignoceric

Table 1. Saturated fatty acids of general formula $CH_3(CH_2)_n$ COOH.

by single methylene\ groups, and have the same terminal structure [2]. A list of some of the important PUFAs is presented in **Table 3**.

2.4. Branched-chain and cyclopropane fatty acids

Branched-chain fatty acids, which occur widely in nature, are present as minor components except in bacteria, where they appear to replace unsaturated fatty acids functionally. The branch consists of a single methyl group, either on the penultimate (*iso*) or on the antepenultimate (*anteiso*) carbon atoms [3, 4].

2.5. Oxygenated and cyclic fatty acids

A large number of hydroperoxy, hydroxyl and epoxy fatty acids (eicosanoids) are formed enzymatically as intermediates in the biosynthesis of prostanoids. A large number of hydroxy fatty acids occur in seed oils, and the best known of these is ricinoleic acid which is the principle constituent of castor oil. Polyhydroxy fatty acids are present in plant cutins, shellacs and many seed oils.

S. no.	Systematic name	Shorthand designation	Trivial name
1.	*cis*-9-Tetradecenoic	14:1(n-5)	Myristoleic
2.	*cis*-9-Hexadecenoic	16:1(n-7)	Palmitoleic
3.	*trans*-3-Hexadecenoic	–	–
	cis-6-Octadecenoic	18:1(n-12)	Petraselenic
4.	*cis*-9-Octadecenoic	18:1(n-9)	Oleic
5.	*cis*-11-Octadecenoic	18:1(n-7)	*cis*-Vaccenic
6.	*trans*-11-Octadecenoic	–	Elaidic
7.	*cis*-9-Eicosenoic	20:1(n-11)	Gadoleic
8.	*cis*-11-Octadecenoic	18:1(n-9)	Gondic
9.	*cis*-13-Docosenoic	22:1(n-9)	Erucic
10.	*cis*-15-Tetracosenoic	24:1(n-9)	Nervonic

Table 2. Monoenoic fatty acids of general formula $CH_3(CH_2)mCH{=}CH(CH_2)nCOOH$.

2.6. Omega-3 and omega-6 fatty acids

The biological fatty acids are of different lengths, the last position is labelled as *omega* (ω). *Omega-3* fatty acids are long-chain polyunsaturated fatty acids (18–22 carbon atoms) with the first of many double bonds beginning with the third carbon atom. However, *omega-6* fatty acids have the first of many double bonds beginning with the sixth carbon atom. Alpha-linolenic acid (ALA) and linoleic acid (LA) are the parent compounds of the omega-3 family and omega-6 family of fatty acids, respectively.

S. no.	Systematic name	Shorthand designation	Trivial name
1.	9,12-Octadecadienoic*	18:2(n-6)	Linoleic
2.	6,9,12-Octadecatrienoic	18:3(n-6)	γ-Linolenic
3.	8,11,14-Eicosatrienoic	18:3(n-6)	Homo-γ-linolenic
4.	5,8,11,14-Eicosatetraenoic	20:4(n-6)	Arachidonic
5.	4,7,10,13,16-Eicosapentaenoic	20:5(n-6)	–
6.	9,12,15-Octadecatrienoic	18:3(n-6)	α-Linolenic
7.	5,8,11,14,17-Eicosapentaenoic	20:5(n-3)	EPA
8.	7,10,13,16,19-Docosapentaenoic	22:5(n-3)	–
9.	4,7,10,13,16,19-Docosahexaenoic	22:5(n-3)	DHA
10.	5,8,11-Eicosatrienoic	20:3(n-9)	Mead's acid

*The double bond configuration in each instance is *cis*.

Table 3. Polyunsaturated fatty acids of general formula $CH_3(CH_2)m(CH{=}CHCH_2)x(CH_2)$ n COOH.

Although the International panel of lipid experts says the ideal ratio of *omega-3* to *omega-6* essential fatty acids is approximately 1:1, still we follow the ratio 20:1 in our diet [5]. Long-chain polyunsaturated fatty acids cannot be formed *de novo* but can be synthesized from the essential fatty acids like linoleic acid and alpha-linolenic acid. These two essential fatty acids are desaturated and lengthened progressively by microsomal enzyme systems to form highly unsaturated, long-chained fatty acids such as arachidonic acid and docosahexaenoic acid (DHA). The *omega-3 and omega-6* fatty acids are not interconvertible. Dietary fish and fish oil supplements are a direct source of *omega-3* fatty acids and dietary oils have large quantity of *omega-6* fatty acids [6].

3. Polyunsaturated fatty acids as anticancer agents

Yonesawa and co-workers carried out the inhibitory effect of conjugated eicosapentaenoic acid (cEPA) on mammalian DNA polymerase and topoisomerase activities and human cell proliferation. They found that the inhibitory effect of cEPA was stronger than that of the non-conjugated EPA and suggested the therapeutic potential of cEPA as a leading anticancer compound that poisons mammalian DNA polymerase (POLS) [7]. The work carried by *Unduri* revealed the tumouricidal and antiangiogenic actions of gamma-linolenic acid (GLA) and its derivatives. It was found that GLA being an endogenous naturally occurring molecule had no significant side effects [8]. Paul *et al.* reported that the long-chain eicosapentaenoic acid (EPA) and docosahexaenoic acids (DHA) have been consistently shown to inhibit the proliferation of breast and prostate cancer cell lines *in vitro* and to reduce the risk and progression of these tumours in animal experiments. Many investigations revealed that the above-said fatty acids inhibit cyclooxygenase-2 and the oxidative metabolism of arachidonic acid (AA) to PGE_2. EPA and DHA also have been shown to inhibit lipoxygenase which metabolizes AA to hydroxyl eicosatetraenoic acids and leucotrienes which suppress apoptosis, stimulate angiogenesis and stimulate tumour cell division (**Figure 2**). Further, they explained that the n-3 PUFAs potentially affect carcinogenesis by specific mechanisms [9]. These mechanisms are as follows: (1) alteration of the response of immune system to cancer cells through the suppression of arachidonic acid (AA, 20:4n-6)-derived eicosanoid biosynthesis; (2) alteration of metabolism, cell growth and differentiation; (3) alteration of oestrogen metabolism, which leads to reduced oestrogen-stimulated cell growth; (4) alteration of free radicals and productivity; and (5) alteration of the mechanisms involving insulin sensitivity and membrane fluidity. Interest in the use of supplementary omega-3-fatty acids to reduce the risk of cancer and other chronic-debilitating conditions, including cardiovascular disease and cognitive impairment, stems from several long-standing avenues of registration [9, 10]. Furthermore, the anticancer activity of fatty acids is well evidenced by Helmut *et al.* in experimental and human studies, which summarize that a high intake of omega-3 PUFAs and monounsaturated fatty acids is protective in breast, colon and prostate cancers [11].

The author and her research group isolated methyl gamma linolenate (**GLA-ME**) (**1**) from *Spirulina platensis* and the compound showed strong cytotoxicity against A-549 cells [13] when compared with the standard drug Rutin. Rutin is a bioflavanol which is a well-established

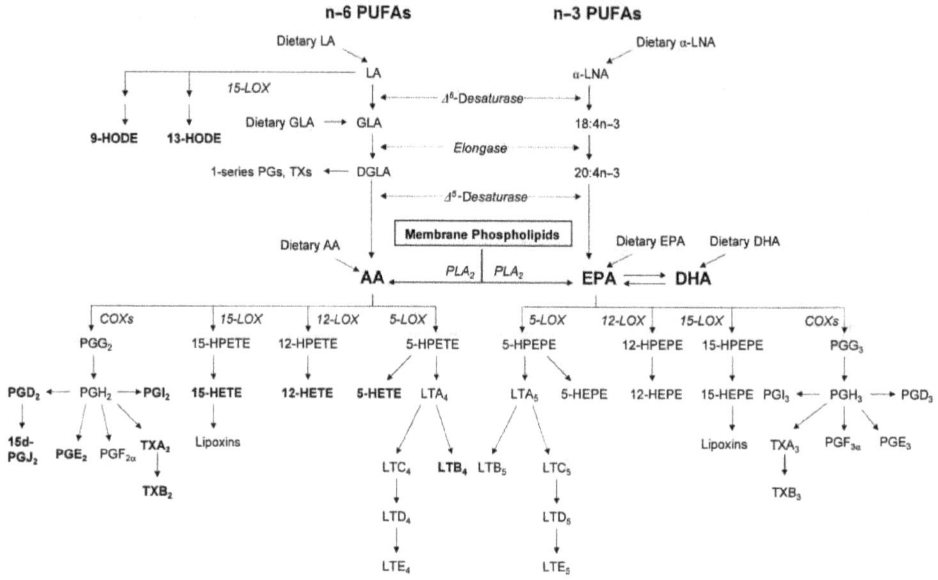

Figure 2. Overview of the metabolism of n-6 and n-3 polyunsaturated fatty acids (PUFAs) into eicosanoids involved in inflammation and carcinogenesis [12].

promising anticancer agent, and its mechanism may be due to the induction of apoptosis [14]. The comparative results are given in **Figure 3** and **Table 4**, respectively. The probable mechanism may be due to the induction of apoptosis of tumour cells by augmenting free radical generation. It is evidenced by the research work carried out by Unduri *et al.* [8]. They also reported that the induction of apoptosis of tumour cells by GLA is due to its action at the gene/oncogene level and by altering BCl-2 expression. Hence, it may be concluded that the cytotoxicity shown by GLA-ME may be due to the induction of apoptosis effect. However, a detailed study of this mechanism is in progress.

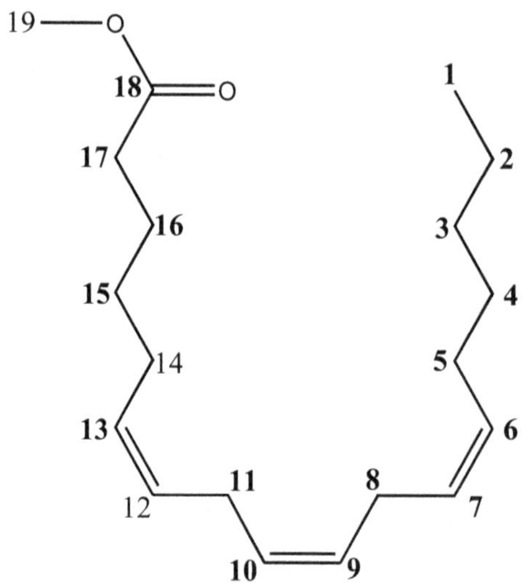

1

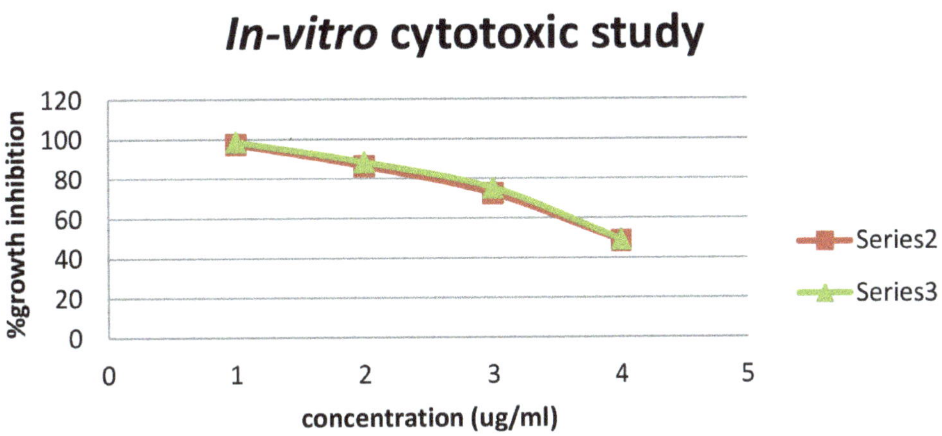

Figure 3. *In vitro* cytotoxic studies □: GLA-ME, Δ: standard rutin.

S. no.	Compound	Concentration (μM)	% growth inhibition	CTC_{50}
1.	GLA-;ME	3.333	97.45	0.468
2.		1.666	86.39	
3.		0.833	72.38	
4.		0.416	48.45	
5.	Rutin	3.333	98.65	0.442
6.		1.666	88.41	
7.		0.833	75.25	
8.		0.416	49.05	

Table 4. Determination of cytotoxicity by SRB method.

4. Polyunsaturated fatty acids as adjunct to chemotherapeutic agents

Kong and co-workers found out that gamma linolenic acid modulates the response of multi-drug-resistant K562 leukaemic cells to anticancer drugs. The study also revealed that GLA could modulate the response to anticancer drugs in P-gp overexpressing multidrug-resistant cells, which could be due to decrease P-gp expression [15]. In another study, Julie and co-workers reported that alpha linolenic acid and docosahexaenoic acid alone combined with trastuzumab reduced HER2 overexpressing breast cancer cell growth but differentially regulated HER2-signalling pathways. Their finding is different in classic mechanisms whereby n-3 PUFAs exert their effect in breast cancer. The results strongly suggest that DHA reduces growth factor receptor signalling as indicated by reductions in the phosphorylation of AKT and MAPK while the opposite effect is seen for the plant-based n-3 PUFA ALA [16]. Effenberger and co-workers synthesized novel N-acylhydrazones of doxorubicin which were derived from saturated, unsaturated and methyl or bornyl terminated fatty acids. The mode of cytotoxic action of the hydrazones was largely apoptotic. They led to a distinct long-term decrease of bcl-2 MRNA expression, the precise apoptotic mechanism and the involvement of caspases varied for the individual cell lines and test compounds. The apoptosis of 518A2 melanoma cells treated with some compounds was characterized by an early onset of initiator caspase-9 activity. By contrast, apoptosis elicited in 518A2 or in HL-60 cells by remaining compounds was accompanied by high-initiator caspase-8 activity. The genuine slump of the bcl-2 mRNA expression may be the reason for the observed quick and steep hike of the ratio of bax mRNA to bcl-2 mRNA in 518A2 cells. Apoptosis induced by doxorubicin **(2)** and its derivatives **(3)** and **(4)** in HL-60 and 518A2 cells also proceeds with a swift and distinct loss of mitochondrial membrane potential regardless of the divergent caspase kinetics. This was a proof that fatty acid analogues are more than just lipophilic shuttle groups [17].

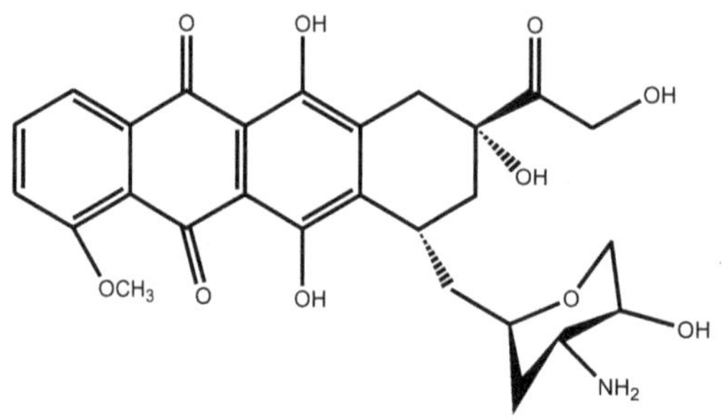

2

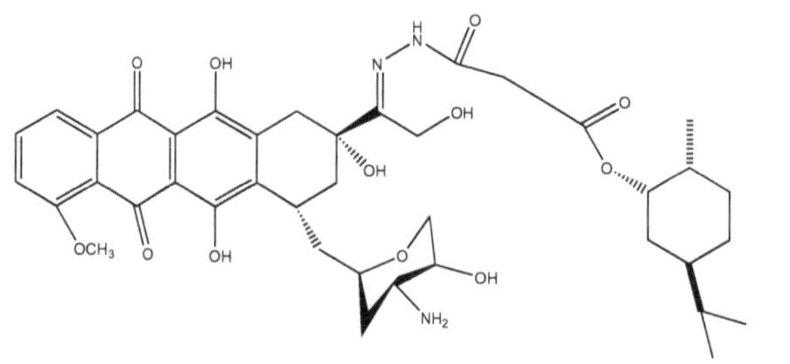

3

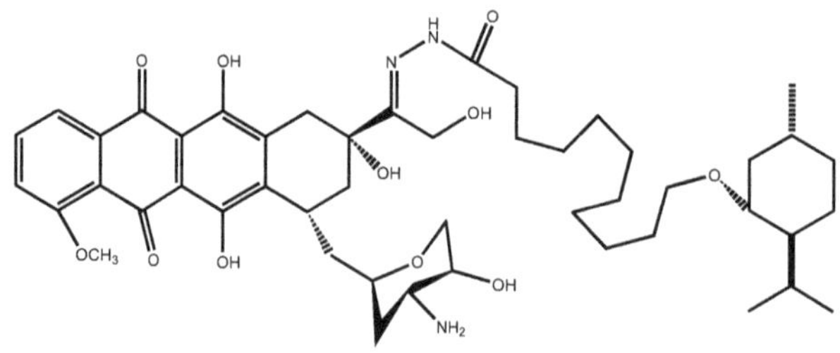

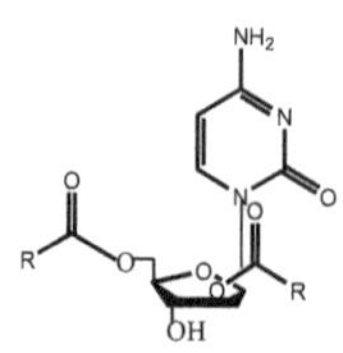

4

Piyali *et al.* studied the antiproliferative activity of somatostatin analogue with N-terminal acylation with long-chain fatty acids in human breast adenocarcinoma cell lines. The antiproliferative activity of the somatostatin analogue RC-160 (D-Phe-Cys-Tyr-D-Trp-Lys-Val-Cys-Trp-NH_2) is limited by its short serum half-life. To circumvent this limitation, fatty acids of chain lengths ranging from 4 to 18 were individually conjugated to the N-terminal residue of RC-160. Although the affinity of palmitoyl –RC-160 towards somatostatin receptors remains unaltered when compared to the –RC-160, it exhibited significantly higher antiproliferative activity on MCF-7 cells. On further increase in the lipopeptide chain, the bioactivity of lipophilized –RC-160 was reduced. Increasing the peptide hydrophobicity beyond this range reduced the bioactivity of lipophilized –RC-160. Accordingly, stearoyl –RC-160 manifested lower antineoplastic activity and receptor-binding affinity relative to palmitoyl –RC-160 and RC-160 itself. It was observed that an increase in bioactivity was manifested within an optimum range of the lipopeptide. The probable mechanisms may be alterations of the signalling pathways. Lipophilization of RC-160 with long-chain fatty acids like palmitic acid improves its stability and antiproliferative activity, thereby improving the scope of enhancing its therapeutic index [18].

5. Fatty acid analogues as anticancer agents

A number of investigations have demonstrated that a variety of modified fatty acid analogues are promising molecules in cancer prevention and have potential in the treatment of cancer. Bhupender *et al.*synthesized fatty acyl amide derivatives of doxorubicin (**5**) and evaluated their *in vitro* anticancer activities. The results indicated that the designed molecule with comparable antileukaemia activity to cytarabine with sustained release effect is possible by structure modification [19].

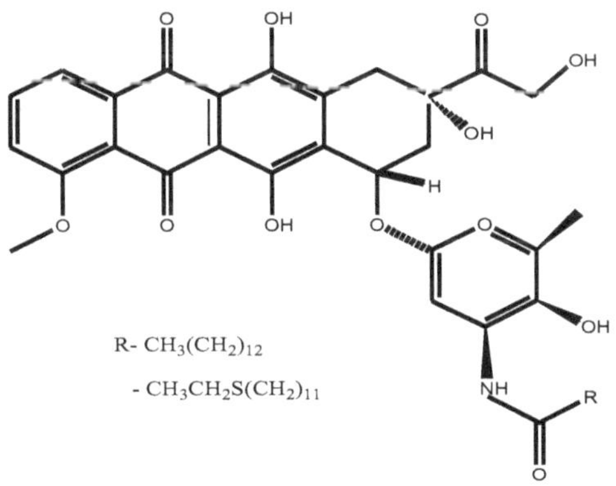

5

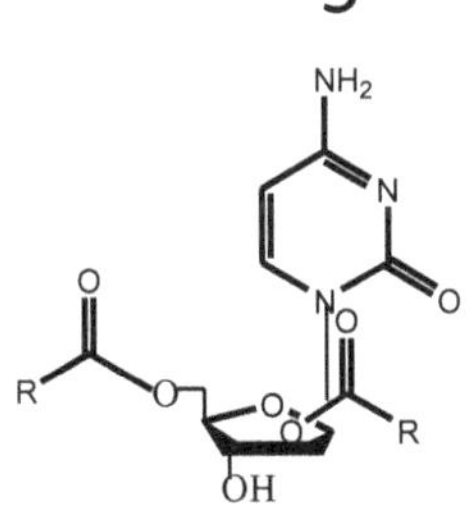

6

They also synthesized fatty acyl ester derivatives (**6**) of cytarabine and evaluated them for anti-leukaemia activity. Some of 2′,5′-dimyristoyl derivatives of cytarabine were found to inhibit the growth of CCRF-CEM cells [20]. Liu *et al.* reported the synthesis and antitumour evaluation of N^4 fatty acyl amino derivatives of cytarabine. The bioavailability of cytarabine is low due to its low lipophilicity. In order to improve the lipophilicity and bioavailability of cytarabine, a series of fatty acyl amino acid cytarabine analogues (**7**) were synthesized. It was found that the derivatives synthesized were more lipophilic than cytarabine. The antitumour activity determined in HL-600 and HeLa cells showed that the derivatives were more active in HeLa cells than cytarabine while most of them demonstrated similar activity to cytarbine in HL-60 cells. The length of fatty acids in the derivatives seemed to have an impact on the activity observed [21].

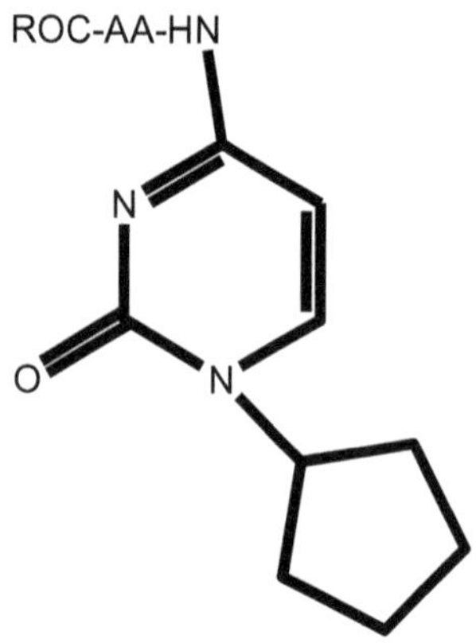

(7)

AA=Amino acids R-sugar

Zhang Chun-hong and co-workers synthesized new panaxadiol fatty acid esters (**8**) and evaluated them for their antitumour activity. Tumour cell used was Vero cell line. Positive control was 5-FU, blank was an RPMI1640 culture medium, negative control was an RPMI1640 culture medium and the solvent for drugs to be tested. The compounds show the strongest antitumour activity [22].

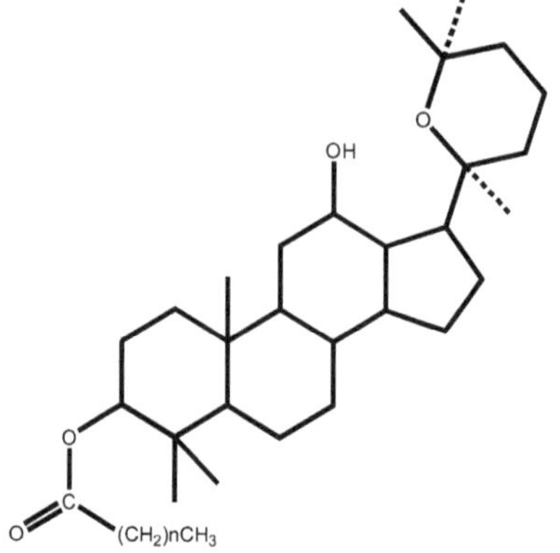

8

Earlier, the author of the present chapter has reported some novel fatty acid heterocyclic conjugates and their anticancer evaluation on human lung carcinoma cell lines [23, 24]. The compounds have shown comparable cytotoxicity towards human lung carcinoma cell lines. The compound (**9**), fatty acid chain substituted 1,3,4-oxadiazole showed maximum cytotoxic activity. It was observed that the presence of toxophoric –N=C-O- linkage in 1,3,4 oxadiazole nucleus may be responsible for the antitumour activity. Further, 1,3,4 oxadiazole is a good bioisostere of amide and ester functionalities with substantial improvement in biological activity in hydrogen-bonding interactions with different targets responsible for the tumour development. The 1,2,4-triazole substituted fatty acid analogues (**10**) displayed promising cytotoxicity towards human lung carcinoma cell lines. It was also observed that the length of the fatty acids plays a vital role in antitumour activity.

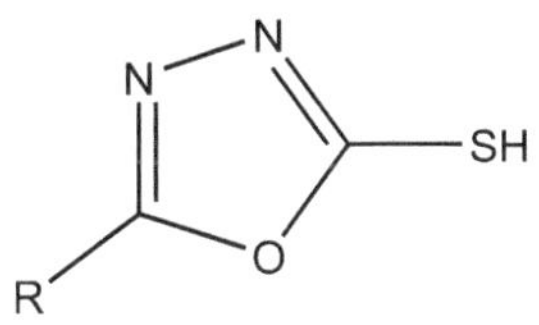

9

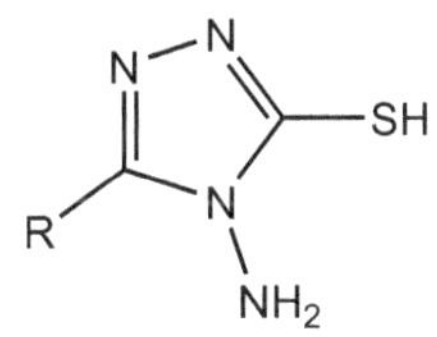

10

R-long chain fatty acids

6. Fatty acid synthase as a potential target in cancer

Human fatty acid synthase (HFAS) is a multifunctional enzyme that is essential for the endogenous synthesis of long-chain fatty acid from its precursor acetyl Co-A and malonyl Co-A (**Figure 4**). Blocking HFAS activity causes cytotoxicity [25]. The unique carboxyl terminal thioesterase (TE) domain of fatty acid chain plays a critical role in regulating the chain length of fatty acid releases. Also, the up-regulation of HFAS in a variety of cancer makes the thio esterase domain a candidate target for therapeutic treatment [26]. It was evident from the literature that the long alkyl/alkenes tail of the fatty acids can bind into the long groove tunnel site of thio-esterase domain of FAS which may be one of the factors of anticancer activities of fatty acids [27].

Employing these strategies, the author and her research group carried out the *in silico* studies on fatty acid analogues. The group designed new derivatives of stearic acid and palmitic acid and studied their *in silico*-binding affinities towards key enzyme human fatty acid synthase-thio-esterase domain (PDB code 2PX6). The literature clearly says that an identification of oncogenic antigen-519 (OA-519) from human breast carcinoma cells as FAS has made it an important diagnostic and prognostic marker for breast cancer patients [28, 29]. By superposing the scaffold structure of all our designed analogues, it is seen that these analogues bind in the same orientation and similar position in terms of the common structure, that is, long aliphatic chain (**Figure 5**). It complies with the fact that the substrate-binding site of HFAS is made up of hydrophobic groove. The docking studies revealed that there are two hydrogen-bonding interactions between the OH group of triazolo thiadiazole of

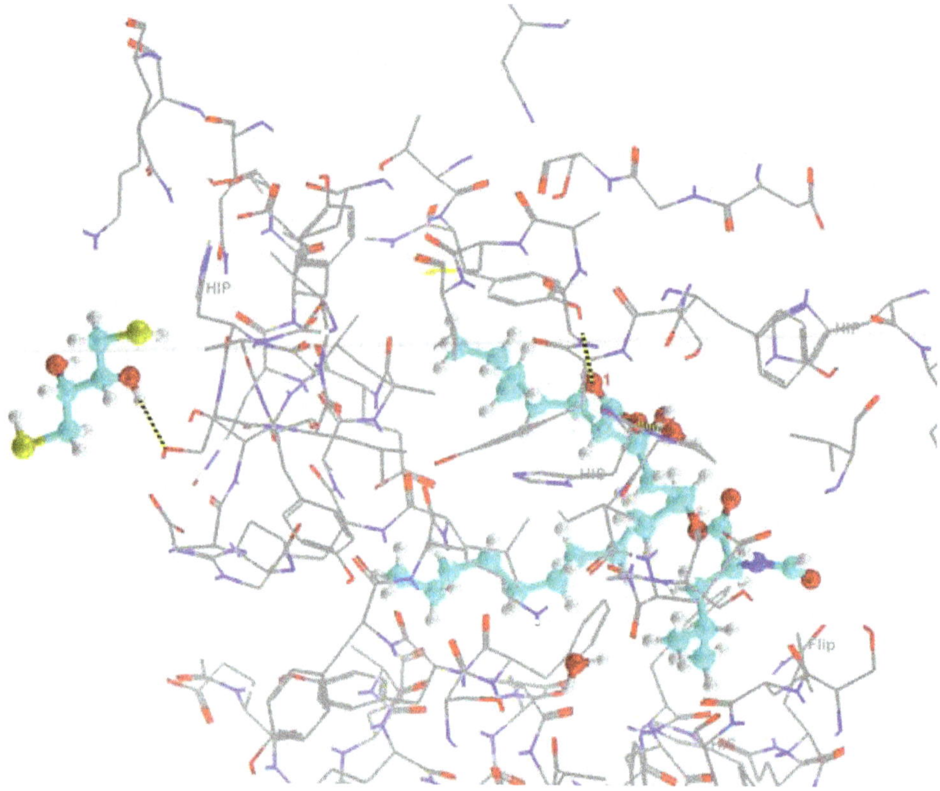

Figure 4. Human fatty acid synthase (PDB id: 2PX6).

synthesized analogues and HIS-2481 and SER-2308 residues (**Figure 6**). These interactions revealed the important binding mode, since these two residues are present in the *"catalytic triad"* of FAS-TE domain [30]. Further, the long alkyl/alkenyl chain of our synthesized analogues fits into the hydrophobic groove of the substrate-binding site. The docking pose and hydrogen-bonding interactions of one of the representative compounds are shown in **Figures 5** and **6**, respectively.

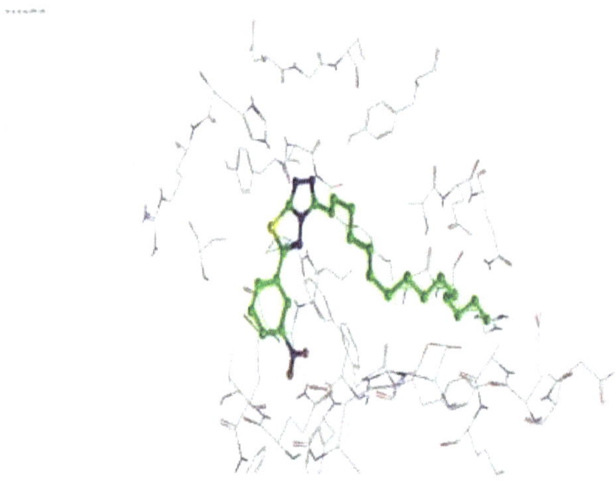

Figure 5. Docking pose.

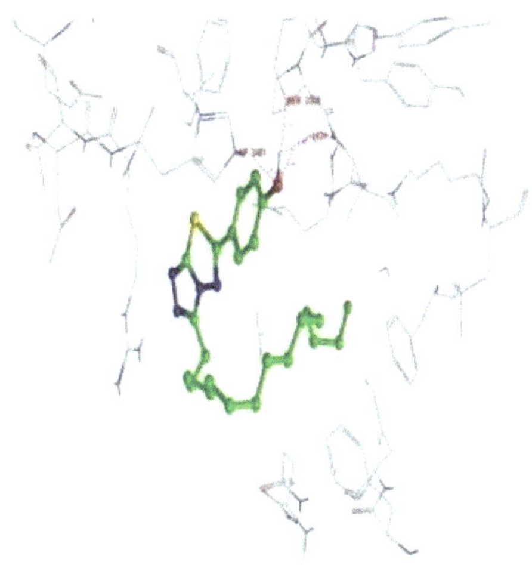

Figure 6. Hydrogen-bonding interactions.

Babak Oskouian and co-workers reported the overexpression of fatty acid synthase in $SKBR_3$ breast cancer cell line and. The objective of this study was to use a breast cancer-derived cell line, $SKBR_3$, as a model to define the underlying mechanism for overexpression of FAS in cancer cells [31]. Silva *et al.* reported a clinic pathological study of $ErbB_2$ and Ki-67 in head and neck squamous cell carcinoma (SCC) and the overexpression of fatty acid synthase enzyme. They showed FAS expression in HNSCC and pointed out ki-67 as a useful prognostic marker for these tumours [32]. Michelle Agostini *et al.* reported the proliferation of human oral squamous carcinoma cells and fatty acid synthase. FAS is overexpressed in several human cancers, such as prostate, breast, bladder, liver, lung, melanoma and oral squamous cell carcinoma [33].

7. Concluding remarks

As part of a conclusion to our discussion, the various studies have shown that fatty acids not only augment the tumouricidal action of anticancer drugs but also enhance the uptake of anticancer drugs leading to an increase in the intracellular concentration of the anticancer drugs. The omega-3 fatty acids have become adjutants to chemotherapeutic agents. Although the production of the above-said fatty acids is a big challenge, a possibility would be gradually implementing the production of these fatty acids in clinical use. Such novel uses of fatty acids in cancer therapy would provide the lipid field with a new avenue to impact public health.

Author details

Jubie Selvaraj

Address all correspondence to: jubiejawahar@gmail.com

Department of Pharmaceutical Chemistry, JSS College of Pharmacy (A Constituent Institution of JSS University-Mysuru), Rock Lands, Udhagamandalam, Tamil Nadu, India

References

[1] Gunstone FD. The Lipid Handbook. London: Chapman and Hall; 1986

[2] Robinson PG. Common names and abbreviated formulae for fatty acids. Journal of Lipid Research. 1982;**23**:1251-1253

[3] Garton GA. Aspects of the chemistry and biochemistry of branched-chain fatty acids. Chemistry and Industry (London). 1985:295-300

[4] Lough AK. The chemistry and biochemistry of phytanic, pristanic and related acids. Progress in the Chemistry of Fats and Other Lipids. 1973;**14**:1-48

[5] Simopoulos AP. The importance of the ratio of omega-6/omega-3 essential fatty acids. Biomedicine & Pharmacotherapy. 2002;**56(8)**:365-379

[6] William WC. Gas Chromatography and Lipids. Bridgwater, Somerset; The Oily Press; 1989. pp. 9-10

[7] Yonezawa Y, Hada T, Uryu K, Tsuzuki T, Eitsuka T, Miyazawa T. Inhibitory effect of conjugated eicosapentaenoic acid on mammalian DNA polymerase and topoisomerase activities and human cancer cell proliferation. Biochemical Pharmacology. 2005;**70**:453-460

[8] Unduri ND. Tumoricidal and anti angiogenic actions of gamma linolenic acid and its derivatives. Current Pharmaceutical Biotechnology. 2006;**7**:457-466

[9] Paul DT, Thomas ER, Alicja W. Intakes of fish and marine fatty acids and the risks of cancers of the breast and prostate and of other hormone-related cancers: A review of the epidemiologic evidence. The American Journal of Clinical Nutrition. 2003;**77**:532-543

[10] Carol JF, Bruce FK, Stephen DH. Omega-3-fatty acids for breast cancer prevention and survivorship. Breast Cancer. 2015;**17**:62

[11] Helmut B, Jagadeesan N, Robert WO. Dietary polyunsaturated fatty acids and cancers of the breast and colorectum: Emerging evidence for their role as risk modifiers. Carcinogenesis. 1999;**20(12)**:2209-2218

[12] Susanna CL, Maria K, Sundberg M, Alicja W. Dietary long chain n-3 fatty acids for the prevention of cancer; a review of potential mechanisms. The American Journal of Clinical Nutrition. 2004;**79**:935-945

[13] Jubie S, Dhanabal SP, Chaitanya MVNL. Isolation of methyl gamma linolenate from *Spirulina platensis* using flash chromatography and its apoptosis inducing effect. BMC Complementary and Alternative Medicine. 2015;**15**:263

[14] Ren W, Qiao Z, Wang H, Zhu L, Zhang L. Flavanoids: Promising anticancer agents. Medical Research Reviews. 2003;**23**:519-534

[15] Kong X, Ge H, Chen L, Liu Z, Yin Z, Li P. Gamma linolenic acid modulates the response of multidrug resistant K562 leukemic cells to anticancer drugs. Toxicology *In Vitro*. 2009;**23**:634-639

[16] Julie KM, Suskpreet K, Shikhil K, Ashleigh KAW, Lilan UT. α-Linolenic acid and docosahexaenoic acid, alone and combined with trastuzumab reduce HER2-overexpressing breast cancer cell growth but differentially regulate HER2 signalling pathways. Lipids in Health and Diseases. 2015;**14**:91

[17] Effengberger K, Breyer S, Schobert R. Modulation of doxorubicin activity in cancer cells by conjugation with fatty acyl and terpenyl hydrazones. European Journal of Medicinal Chemistry. 2010;**489**:1947-1954

[18] Piyali D, Anu S, Rama M. N-Terminal acylation of somatostatin analog with long chain fatty acids enhances its stability and antiproliferative activity in human breast carcinoma cells. Biological and Pharmaceutical Bulletin. 2002;**28(1)**:29-36

[19] Bhupender SC, Nicole SJ, Deendayal M, Anil Kumar, Keykavous P. Fatty acyl amide derivatives of doxorubicin: Synthesis and *in vitro* anticancer activities. European Journal of Medicinal Chemistry. 2011;**46**:2037-2042

[20] Bhupender SC, Deendayal M, Keykavous P. Synthesis and evaluation of fatty acyl ester derivatives of cytarabine as anti-leukemia agents. European Journal of Medicinal Chemistry. 2010;**45**:4601-4608

[21] Liu B, Cui C, Duan W, Zhao M, Feng S, Wang L, Liu H, Cui C. Synthesis and evaluation of antitumour activities of N^4 fatty acyl amino acid derivatives of 1-β-arabinofuranosylcytosine. European Journal of Medicinal Chemistry. 2009;**44**:3596-3600

[22] Zhang CH, Li XG, GaoYG, Zhang LX, FU XQ. Synthesis and primary research on antitumor activity of three new panaxadiol fatty acid esters. Chemical Research in Chinese Universities. 2007;**23(2)**:176-182

[23] Jubie S, Pawan Kumar Y, Chandrasekar MJN. Novel fatty acid analogues as fatty acid synthase-thio esterase domain inhibitors; Synthesis and their cytotoxicity screening. Letters in Drug Design and Discovery. 2015;**12**:495-499

[24] Jubie S, Dhanabal SP, Afzal Azam M, Muruganandham N, Kalirajan R, Elango K. Synthesis and characterization of some novel fatty acid analogues: A preliminary investigation against human lung carcinoma cell line. Lipids in Health and Diseases. 2013;**12**:45

[25] Teresa P, Helena A, Sílvia C, Glòria O, Carlos T, Sílvia OG, et al. A novel inhibitor of fatty acid synthase shows activity against HER2+ breast cancer xenografts and is active in anti-HER2 drug-resistant cell lines. Breast Cancer Research. 2011;**13**:R131

[26] Kuhajda, FP, Pizer ES, Li JN, Mani, NS, Frehywot GL, Townsend CA. Synthesis and antitumour activity of an inhibitor of fatty acid synthase. Proceedings of National Academic Sciences of the United States of America. 2000;**97(7)**:3450-3454

[27] Chakravarty B, Gu ZW, Chairala SS, Wakil SJ, Quiocho FA. Human fatty acid synthase: Structure and substrate selectivity of the thioesterase domain. Proceedings of National Academic Sciences of the United States of America. 2004;**101**:15567-15572

[28] Kuhajda FP, Piantadosi S, Pasternack GR. Haptoglobin-related protein (Hpr) epitopes in breast cancer as a predictor of recurrence of the disease. New England Journal of Medicine. 1989;**321**:636-641

[29] Kuhajda FP, Katumuluwa AI, Pasternack GI. Expression of haptoglobin related protein and its potential role as a tumor antigen. Proceedings of National Academic Science USA. 1989;**86**:1188-1192

[30] Nardini M, Dijkstra BW. α/β hydrolase fold enzymes: The family keeps growing. Current Opinion in Structural Biology. 1999;**9(6)**:732-737

[31] Babak O. Overexpression of fatty acid synthase in SKBR3 breast cancer cell line is mediated via a transcriptional mechanism. Cancer Letters. 2000;**149**:43-51

[32] Sabrina DS, Michelle A, Ines NN, Ricardo DC, Fabio AA, Marcio AL, Luiz PK, Edgard G. Expression of fatty acid synthase, ErbB2 and Ki-67 in head and neck squamous cell carcinoma. A clinicopathological study. Oral Oncology. 2004;**40**:688-696

[33] Michelle A, Sabrina DS Karina GZ, Ricardo DC, Jacks J, Massimo L, Edgard G. Fatty acid synthase is required for the proliferation of human oral squamous carcinoma cells. Oral Oncology. 2004;**40**:728-735

8

Fatty Acids in Veterinary Medicine and Research

Siobhan Simpson, Alison Mostyn and

Catrin S. Rutland

Abstract

Fatty acid regulation is an essential process for all animals. A number of studies have shown that diet affects the levels/availability of fatty acids in the body but increasingly an evidence shows that disease states can alter the amounts within the body too. Fatty acid levels and availability have been altered by a number of diseases, disorders and reactions including inflammatory responses, heart disease and heart failure and wound repair. They are also essential during the growth and development stages of animals. The amount of research into the consequences of different fatty acid intake and levels in various disease states and during development has increased in both humans and animals. This review presents an overview of the research undertaken to date and highlights the importance, uses and benefits of understanding the roles of fatty acids in both the healthy animals and animals under differing disorders and diseases.

Keywords: heart disease, Inflammation, development, nutrition, cancer, pregnancy

1. Introduction to fatty acids

Fatty acids consist of a carboxylic acid with a hydrocarbon chain tail, the length of which varies between fatty acids, as does the presence or absence of double bonds between the carbon atoms and their location [1]. Some fatty acids are ingested in the diet whereas others are synthesized into other fatty acids by elongation and desaturation enzymes [2–4], see **Figures 1** and **2**. In mammals, fatty acids are obtained from the diet prior to metabolism or incorporation as components of cells [5–8]. *n*-6 polyunsaturated fatty acids (PUFAs) and *n*-3 PUFAs are the two major groups of fatty acids; the first is obtained from fats and oils, and the latter from

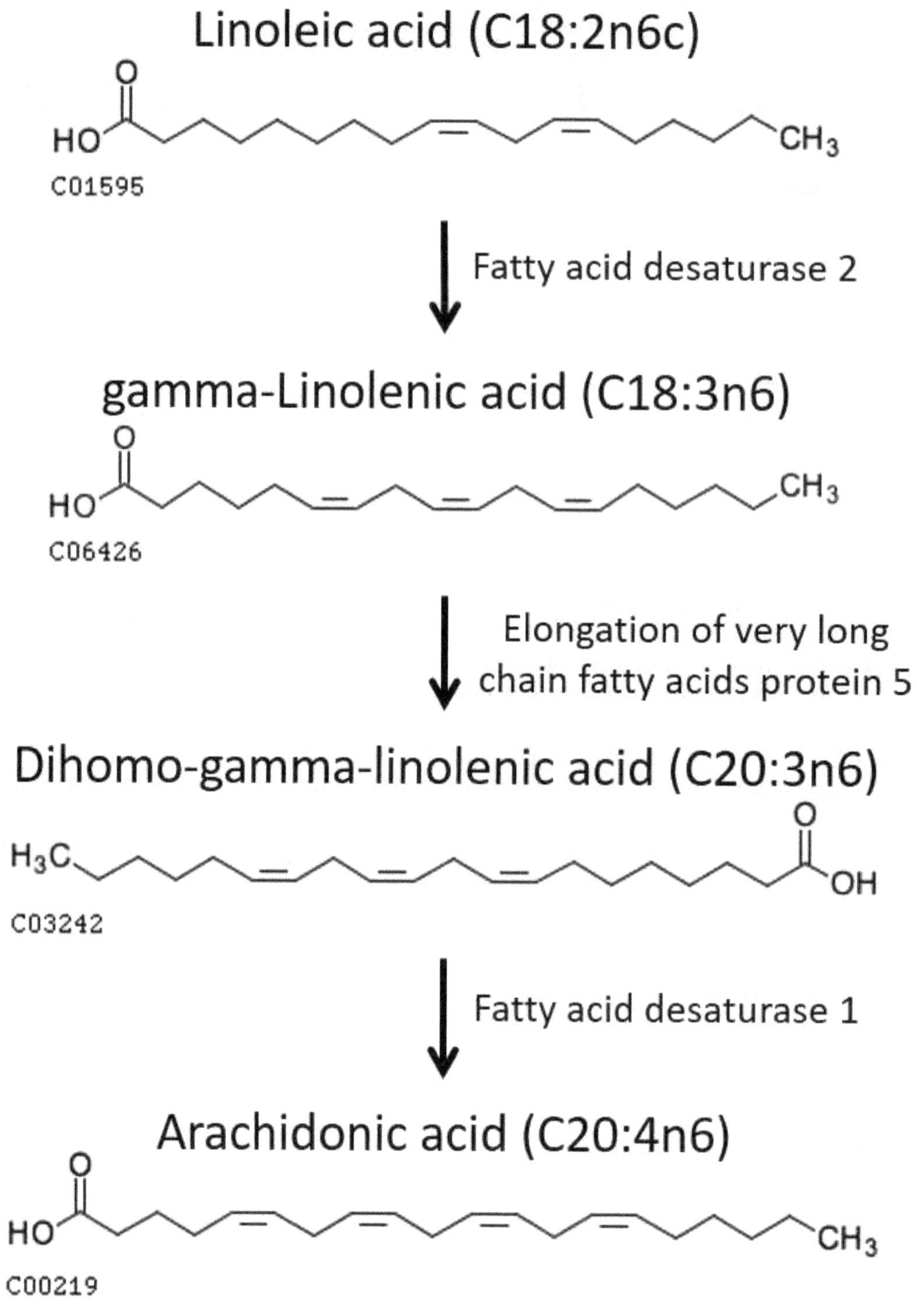

Figure 1. Schematic of linoleic and arachidonic acid biosynthetic pathway derived from KEGG pathway maps [2].

fish and seafood products [6]. It is essential that the precursors of both *n*-6 and *n*-3 PUFAs are extracted by mammals from their diet as they are not able to convert these fatty acids (FAs) between the two major pathways [9].

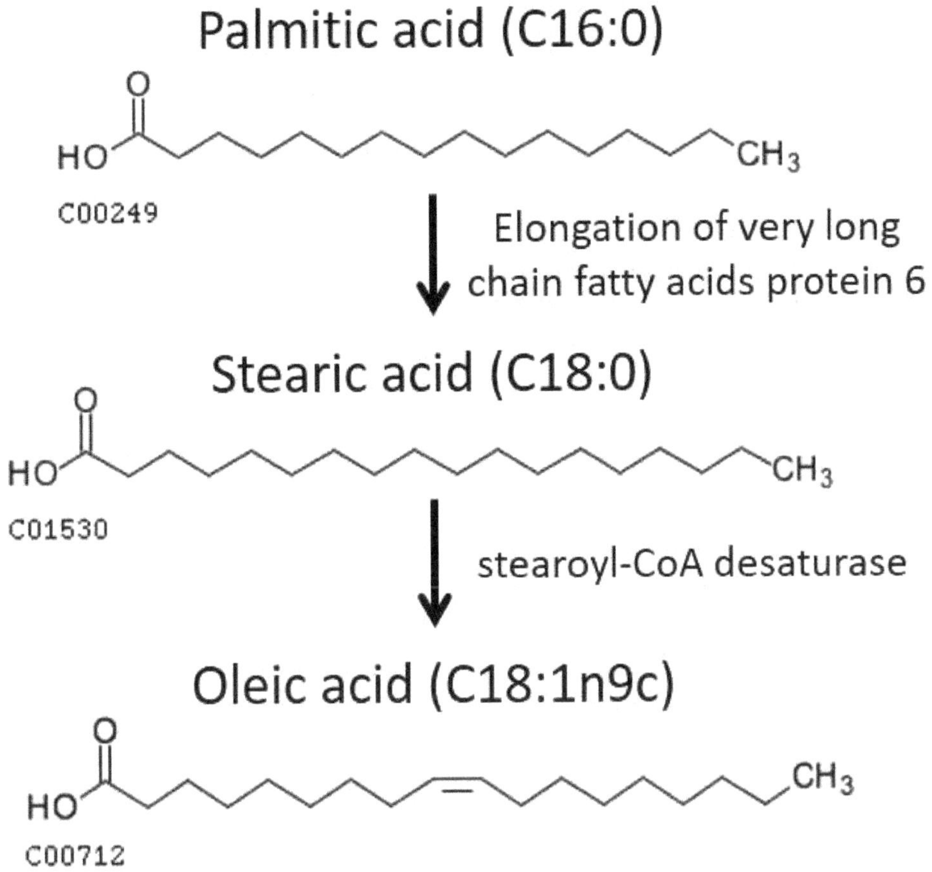

Figure 2. Schematic of palmitic and oleic acid biosynthesis pathway derived from KEGG pathway maps [2].

2. Inflammation, disease and the immune system

Fatty acids are crucial components of the immune system, providing the structural basis of all cell membranes, acting as signaling molecules, and providing a major substrate for energy production [1, 8, 10]. Many diseases involve inflammatory responses either as a reaction to disease or in the initiation of the disease process; although inflammation itself is not always detrimental, for instance, it is an important aspect of wound repair [11–14]. Elevated markers of inflammation are frequently detected in heart failure and cancers although this could be due to the response to disease, or the underlying cause of disease [15–19].

Fatty acid-derived eicosanoids are important contributors to the inflammatory response [13, 20, 21]. The *n*-6 PUFA arachidonic acid is a precursor of the most important pro-inflammatory eicosanoids, while the *n*-3 PUFA derivatives, eicosapentaenoic acid and docosahexaenoic acid metabolites are considered less inflammatory [20]. Arachidonic acid is released from cell membranes by phospholipase A_2 enzymes in response to pro-inflammatory stimuli [22–25]. Cyclooxygenase, lipoxygenase and cytochrome P450 enzymes then convert free arachidonic acid into eicosanoids [26–29]; however, these enzymes are rate limiting as they similarly convert other fatty acids to their metabolites [20]. It has been suggested that if cyclooxygenase, lipoxygenase and cytochrome P450 enzymes are exposed to increased levels of *n*-3 fatty acids, the result is fewer arachidonic acid-derived eicosanoids [20, 30].

Due to the difference in the inflammatory response between fatty acid metabolites, it is hypothesized that the fatty acid profiles could differ between diseased and healthy individuals. Indeed, fatty acid profiles have been shown to be altered in blood and tissues in individuals with a range of conditions compared to unaffected individuals in both humans and dogs. These conditions include Crohn's disease, heart disease, skin disease and cancer [31–34], and are discussed in greater detail below.

2.1. The role of fatty acids in Crohns' disease

An interesting inflammatory response disorder is inflammatory bowel disease, including Crohn's disease. A number of animal studies, including guinea pigs and rats, have shown novel results in the adipocytes, lipid rafts and fatty acid-derived messenger molecules which indicated that aberrant fatty acid composition could play a role in Crohn's disease [35–38]. This research led directly into looking at the role of FAs in human cases of Crohns' disease, a disorder which is linked to both inflammation and the immune system. Perinodal adipose tissue (PAT) is a specialized adipose tissue depot which surrounds lymph nodes and acts in a paracrine manner—delivering specific FAs and adipokines directly to the node. Research has demonstrated that PAT associated with the lymph node is present in most animals and humans [39]. Crohn's disease is associated with altered mesenteric PAT FA content, suggesting impaired delivery of FAs to lymphocytes [40]. For many years, patients with Crohn's disease have been advised to take dietary fish oils that are rich in *n*-3 PUFAs, but interestingly patients have naturally (prior to taking supplements) presented with higher levels of *n*-3 PUFAs than observed in controls with concurrent deficiencies in arachidonic acid (20:4*n*-6) [41–43]. More recent evidence suggests that higher levels of *n*-6 PUFAs, including linoleic acid (18:2*n*-6) were most effective at relieving inflammatory symptoms [43]. The biosynthetic links between arachidonic acid (20:4*n*-6) and linoleic acid (18:2*n*-6) are shown in **Figure 1** and help to understand why an increased linoleic acid intake could reverse the decrease in arachidonic acids observed in patients. A number of animal species develop differing forms of inflammatory bowel disease, therefore understanding whether FAs are affected for the differing types of animals and differing breeds could help to indicate differing dietary or treatment requirements.

2.2. The role of fatty acids in cardiovascular function and disease

A number of links have been made between fatty acid levels and heart disease and heart failure. Human patients with significant left ventricular dilation have a larger percentage of oleic acid and a smaller percentage of arachidonic acid in their blood serum compared to patients with moderate left ventricular dilation [33]. It is also important to highlight that none of the patients involved in the study had a confirmed diagnosis, and although valve disease and coronary artery disease were excluded as the underlying cause of left ventricular dilation, infarction was not. Infarction may have skewed the fatty acid results due to the strong inflammatory nature of myocardial infarction [33, 44].

In cats with hypertrophic cardiomyopathy, differing levels of FAs were observed when compared to cats with no hypertrophic symptoms [45]. Hypertrophic cardiomyopathy cats had higher levels of docosahexaenoic acid, palmitic acid and total *n*-3 PUFAs and lower levels of linoleic acid. Differential levels of docosatetraenoic acid have been observed in canine myocardial tissue in dogs affected by dilated cardiomyopathy [46]. Mobile lipid content within the myocardium was significantly increased in a 24-hour coronary occlusion canine heart, not only throughout the body but also 'local' increases were observed around the heart with cardiac levels up to 10 times higher than the rest of the body [47–50]. It has been suggested that increased fatty acid levels alongside a decrease in creatine can lead to diastolic dysfunction, as observed in humans with diabetic cardiomyopathy [51, 52]. Despite the observations in dogs and humans, a study in rats showed increased fatty acids and decreased creatine but no associated diastolic dysfunction was observed [53]. With differing observations between species, more research is needed in order to understand the mechanisms and circumstances under which diastolic alteration occurs. Increased levels of palmitoleic acid have been associated with heart failure, higher levels of behenic acid and stearic acid have been associated with lower risk of developing atrial fibrillation, women with higher circulating pentadecanoic acid are less likely to have a myocardial infarction, hypertensive rats have higher circulating eicosedienoic acid and in renal patients higher circulating C20:5n3 is associated with good cardiac functional measures [54–60].

Although the fatty acids themselves play a key role in cardiovascular health and disease, other molecules within the fatty acid utilization cascades play important roles too. Heart-type fatty acid-binding protein (H-FABP) is expressed in cardiomyocytes and despite the name, it is also expressed in renal and skeletal muscle cells [61]. Heart-type fatty acid-binding protein (H-FABP) is used as a prognosis tool biomarker in human cardiac disease as it indicates myocardial stretch and injury in chronic heart failure even in children. Higher levels of H-FABP are associated with a poorer long-term outcome in both adults and children [61–65]. Although little work has been carried out in other species, this is an area of research which has potential, in addition to investigating whether H-FABP levels are raised prior to infarction and/or heart disease. A rat model has shown that H-FABP is increased following cardiac injury [66]. It also enables detection via a number of differing methods including EIA, ELISA, fully automated latex-agglutination assay and qualitative lateral-flow assay microparticle enhanced immunoassay [61].

External factors such as diet and surgery can play large roles in fatty acid composition and cardiovascular health. A study looking at differing feeding regimes in obese rats in comparison with lean rats showed that *n*-3 acyl chains, unsaturated and polyunsaturated fatty acids, were all significantly higher in *obese* rats than in the *lean* ones [53]. What was also interesting was the fact that mild, short-term diet changes (food intake was restricted by 20% for two weeks) did not alter the cardiac fatty acid profiles. The obese mice also showed symptoms of early stage obese cardiomyopathy; although interestingly the symptoms of this started to improve upon calorie restriction, an important finding as it showed that mild calorie restriction can be of benefit under these circumstances. Fatty acids are not only an important indicator of heart disease in animals, but also important in situations such as surgery. Increased free fatty acid levels also have been noted in response to heart surgery in pigs especially when heparin is co-administered [67]. In the surgery cases, it was found that the young patients were more affected than older patients and the levels were more likely to rise if cyanosis and prolonged ischemia were present.

Although most of the work into cardiovascular health has concentrated on disease and disorders, a number of suggestions for healthy levels have been put forward as ways of preventing disease. There is some evidence that higher levels of circulating arachidic acid are associated with lower risk of atrial fibrillation and diabetes [57, 68]. Another example is docosahexaenoic acid (*n*-3 PUFA) which has been implicated as having beneficial effects in a wide range of diseases including heart disease and neurological dysfunction [55, 69].

2.3. Fatty acids and skin disease

There are two main ways in which differing fatty acid profiles contribute to skin disease—as part of inflammation and affecting membrane fluidity. These are not mutually exclusive and it is possible that fatty acids are affecting the development of skin disease via both. People with atopic eczema have been shown to have a different fatty acid profile in their skin than people without atopic eczema. In particular, they have shorter fatty acids within their skin than unaffected individuals. This difference is suggested to lead to impaired skin barrier [70]. Atopic eczema is an inflammatory disease and thus processes of inflammation as discussed earlier will be active in the disease process [71]. As with other cases where a difference in fatty acid profiles has been established between individuals with disease and healthy individuals, it is not clear whether the fatty acid change causes the disease or is a response to disease, or possibly both, but it is a potential novel treatment route. Similar to people with atopic eczema, pruritic dogs have been shown to have a different fatty acid profile compared to dogs with healthy skin [72]. More recently, dogs with atopic dermatitis whose diets were supplemented with *n*-3 PUFA improved significantly more than those given the placebo [73]. As with human skin disease, it is not clear as to how this works, but it is an additional treatment option and area for further research.

2.4. Cancer associations with fatty acids

Cancer is the result of aberrant cellular processes. Many genes and proteins are differentially expressed in tumor tissue compared to nontumor tissue [74–77]. Thus, it is intuitive that fatty acid profiles are likely to be altered in tumors compared to nontumor tissue and this has indeed been demonstrated in breast and prostate cancer [78, 79].

There have been studies showing that differential dietary intake of fatty acids can either reduce or increase risk of disease, including cancer. A meta-analysis of studies relating breast cancer risk with *n*-3 PUFA intake showed that overall increasing *n*-3 PUFA intake reduced the risk of developing breast cancer [78]. In transgenic mice in which males develop prostate cancer, *n*-3 PUFA intake from marine sources suppressed tumorigenesis [80]. This is also the case in people where there is reduced risk of developing prostate cancer with increased intake of marine *n*-3 PUFAs [81–83]. Longer chain *n*-3 PUFAs from non-marine sources, however, are associated with an increased risk of prostate cancer [79, 82, 83].

While ultimately work is required in whole organisms, cell lines are a valuable starting point for research. Of particular note in relation to veterinary medicine and fatty acids are two studies on canine tumor cell lines. The first is that of canine lymphoma cell lines; in this study, stearidonic acid was shown to sensitize cells to anticancer drugs, even when the cells were previously resistant to drugs [84]. The second study utilized fatty acids themselves as antitumor agents. In this study, a specific fatty acid, *trans*-10, *cis*-12 conjugated linoleic acid, was shown to inhibit cell growth and induce apoptosis in canine osteosarcoma cell lines and canine lipomas [85, 86].

3. The effects of fatty acids on fertility and during pregnancy and development

Many animal and human studies have established that restriction of a range of nutrients within the maternal diet throughout pregnancy results in offspring that are programmed to be at increased risk of later hypertension and metabolic disease including diabetes and obesity [87–90]. This theory has become known as the "developmental origins of health and disease" (DOHaD) hypothesis. Fatty acid intake has been shown to have effects even before pregnancy as severe undernutrition of specific fatty acids has resulted in low reproductive rates in males and females. For example, in male cats, a linoleic deficient diet results in tubular degeneration of the testes and low fertility rates, and in females, the litters were not viable [91, 92].

Other studies have shown birth defects in offspring from females fed on low fatty acid diets but it also showed that arachidonate was a key contributor to viable offspring [93, 94]. In contrast, excess macronutrient intake has been implicated in the incidence of the metabolic syndrome is emerging in a number of rodent [95–97] and sheep studies [98]. Studies linking maternal over-nutrition to adverse offspring health in later life are conspicuously lacking, despite a huge effort in understanding the influence of maternal nutrition and its link to obesity. A number of rodent studies have established that a high-fat maternal diet leads to impaired offspring glucose and lipid metabolism [95–97, 99], but the influence of increasing other dietary components has not been investigated, perhaps due to the assumption that a high-fat or "junk food" diet is more prevalent in the western world. Rodent studies of increased fat intake during pregnancy are often associated with an overall decrease in food intake which limits their usefulness [97]. The timing of a nutritional insult is also important in determining the outcome for offspring, differential results have been observed in studies investigating early or late gestational nutritional insults in both animal [100, 101] and human studies [102]. As well as a high-fat diet increasing adipocyte and ectopic lipid accumulation, it may also decrease glycogen deposition in skeletal muscle. Increased plasma free fatty acids impair insulin-stimulated glucose disposal, including glycogenesis and glucose uptake—resulting in reduced skeletal muscle glycogen content [103]. Type-2 diabetes in humans is associated with a reduction in glycogen synthase and tissue glycogen [104], it is unknown whether a sub-optimal maternal diet will result in similar changes in offspring. Recent work has demonstrated that there are physiological [105–107] and emerging molecular differences between pigs with low, normal or high birth weights [108–111]. Extensive physiological examinations of low and high birth weight pigs, at 12 months of age showed that low birth weight pigs had increased fat depth and glucose intolerance and insulin resistance [105]. Also of interest is that, peroxisome proliferator-activated receptor (PPAR)α expression in skeletal muscle is positively correlated to birth weight in these pigs [110]. In younger pigs (7 or 14 days of postnatal age) designated low, normal or high to birth weight, molecular differences have been observed in adipose tissue and skeletal muscle genes known to regulate lipid metabolism including uncoupling proteins (UCPs), PPARα and γ, fatty acid-binding protein (FABP) 3 and 4 and the glucocorticoid receptor (GR) [108, 109, 111].

The role of PPARs is not just restricted to animals subjected to over-nutrition. Studies of maternal low protein diets in rats have demonstrated that post-weaning, offspring had significantly increased hepatic PPARα expression due to decreased methylation as a result of differences in overall dietary fat intake [112]. PPARs are a nuclear hormone receptor family that have attracted much interest due to their involvement in adipogenesis, lipid metabolism, insulin sensitivity, inflammation and blood pressure [113]. PPARγ regulates transcription of genes involved in lipid metabolism by binding to responsive elements in the promoters of respective genes. This transcription regulation stimulates fatty acid storage in adipose tissue by increasing the storage capacity and the quantity of fatty acids that enter adipocytes and also plays a key role in adipocyte differentiation, promoting the formation of mature lipid-laden adipocytes [114]. The activities of PPARγ are regulated by fatty acids (which are thought to be the endogenous ligands) [115]. PPARγ is often referred to as the "genetic sensor" for fat and a number of dietary studies have demonstrated an increase following high-fat feeding [116, 117], which may provide benefits to the animal by protecting against lipotoxic species [117]. PPARα also acts as a ligand-activated transcription factor and is expressed in tissues which have a high rate of fatty acid catabolism such as skeletal muscle and liver. The fibrate group of drugs has long been utilized as a synthetic ligand for PPARα, but endogenous ligands are still under investigation. Long-chain fatty acyl-CoAs and saturated fatty acids however are known to activate PPARα at micromolar ranges [118]. PPARα has a key role in stimulating lipid oxidation pathways to prevent storage of fats as well as increasing insulin sensitivity and glucose tolerance. The expression of PPARs may represent one of the molecular factors driving excess tissue lipid uptake, storage and production in animals that experienced a sub-optimal environment in utero, in particular low birth weight offspring; ectopic lipid storage, especially intramyocellular, is associated with glucose intolerance and type-2 diabetes [104, 119].

The regulation of fatty acids is also an important factor during the lactation period. A number of studies have shown that the relative fatty acid content of milk differs depending on the species. Donkeys have milk more similar to humans than cows, with lower levels of saturated fats and higher essential fatty acids than cows, more akin to humans [120, 121]. Milk, from humans, dog, and guinea pig are mostly comprised from long-chain fatty acids (48–54 acyl carbon atoms), cow, sheep, and goat, have more short-chain acids (28–54 acyl carbon atoms) and horses tended to have medium-chain fatty acids (26–54 carbon atoms range) [122]. Maternal diet can also have an impact on the fatty acid contents of her milk. This has been shown in many species from mice and sheep to humans [123–125]; the pregnancy status of the mother also vastly changes milk fatty acid composition [126]. These are important factors when assessing whether the mother is receiving an appropriate diet, assessing whether she is pregnant or not and whether milk replacement formulae contain the appropriate levels of fatty acids.

4. Fatty acid-binding proteins and lipid modulation

Fatty acids are now recognized as crucial components of cellular signaling cascades, in particular, those regulating lipid metabolism, as described above with PPARs. Research into fatty acids as signaling molecules is in its infancy, but it is well known that fatty acids are ligands for transcription factors. Fatty acids are carried through tissue membranes and in the cytosol by chaperones known as fatty acid-binding proteins (FABPs), of which there are a number of tissue-specific isoforms [127]. Knock-out mice not expressing the adipocyte-specific FABP4 exhibited protection from the metabolic effects (e.g. insulin resistance and hypercholesterolaemia) of a high-fat diet, suggesting FABP4 modulates a number of components of the metabolic syndrome [127]. In skeletal muscle, a fat-rich diet increases the expression of the cytosolic and plasma membrane specific FABP [128].

Insulin resistance is characterized by a decrease in the enzymes and proteins involved in lipid oxidation [129]. Lipogenesis and adipogenesis are modulated by the enzymes acetyl-CoA carboxylase 1 and 2 (ACC1 and ACC2, respectively) and AMP-activated protein kinase (AMPK); both enzymes are potential drug targets to treat obesity and the metabolic syndrome and AMPK has been suggested as a target for metformin [130, 131]. Briefly, ACC1 controls fatty acid biosynthesis and ACC2 controls fatty acid oxidation. ACC1 catalyses the conversion of acetyl-COA to malonyl-CoA, therefore modulating the rate limiting step of long-chain fatty acid biosynthesis in adipose tissue. ACC2 is expressed in skeletal muscle, where the product malonyl-CoA inhibits fatty acid oxidation. The AMPKα subunit is activated during periods of metabolic stress (e.g. increased AMP/ATP ratio) by phosphorylation and inhibits the activity of ACC1 and 2, thus promoting fatty acid oxidation, glucose uptake and inhibits lipid synthesis [132] and thereby reducing ectopic lipid storage. An isocaloric high-fat diet has been shown to inhibit AMPK in rats [133]. Despite great potential for modulation by maternal diet, there are few DOHAD studies of ACC and AMPK expression; however, early studies of an obese pregnant ewe model have shown decreased AMPK signaling in fetal offspring muscle [98].

5. Future fatty acid research and medicine

Although artificially induced disease often only replicates a small aspect of disease and does not reflect the typically longer time scales involved in natural disease progression in both humans and animals [134, 135], these studies can be valuable when compared to naturally occurring diseases in order to understand mechanisms and development. All of the 'natural population' studies discussed in this chapter may have their own caveats too. Differences in diet, age, sex and even pre-clinical symptoms and diagnosis can all affect the results observed in both disease and fatty acid states. This chapter has concentrated on development, cardiovascular disease, cancer and immunity but differing fatty acids have been implicated or associated with in a number of diseases and disorders ranging from human, rodent and canine epilepsy through to canine ADHD and reproductive ability [92, 136, 137].

Fatty acid profiling has important potential applications as a diagnosis tool across the species, especially in cases where pre-clinical symptoms are difficult to observe. Although it is not always necessarily known if differences in fatty acid profiles are contributing to the initiation of disease or are a response to disease processes, these differences could be drug targets [26, 138–140]. In addition, there are genes that contribute to fatty acid profile composition and if a particular part of the pathway is shown to be different in individuals with disease compared to healthy individuals, these could be likely genes for candidate gene studies in the future [141, 142]. The scientific methodologies available for looking at lipid levels have also progressed over the years; just one example is the use of proton magnetic resonance spectroscopy of protons (H-MRS) to assess cardiac lipids in a non-invasive manner [52]. This is a valuable tool for animal health and welfare, and there are additional uses in looking at metabolism and fatty acids. Much of the present work involves looking at genes and lipid levels of animals intended for the meat industry. An example is the evidence that differing polymorphisms in genes can result in differing meat quality traits. This includes fatty acid synthase (FASN) which was found to correlate with meat weight loss during the first salting of dry-cured ham production [143], meat quality including marbling in cattle [144] and playing a role in the mammary gland and milk in goats and cattle [145, 146], in addition to many other roles. Differing H-FABP polymorphisms/expression levels have also been related to growth rate and size of beef cattle and chickens and could therefore provide useful markers for breeding [147, 148].

Research into the links between fatty acids and differing developmental stages and disease states is increasing in both humans and animals and provides the potential for innovative diagnostic and treatments tools.

Acknowledgements

This work was supported by the Biotechnology and Biological Sciences Research Council [grant number BB/J014508/1], by generous funding to Catrin S. Rutland from the BBSRC University of Nottingham Doctoral Training Programme.

Author details

Siobhan Simpson[1†], Alison Mostyn[1,2†] and Catrin S. Rutland[1*]

*Address all correspondence to: catrin.rutland@nottingham.ac.uk

1 Faculty of Medicine, School of Veterinary Medicine and Science, University of Nottingham, UK

2 Faculty of Medicine, School of Health Sciences, University of Nottingham, UK

† Joint first authors

References

[1] Fahy E, et al. A comprehensive classification system for lipids. Journal of Lipid Research. 2005;**46**(5):839-862

[2] Kanehisa M, et al. KEGG as a reference resource for gene and protein annotation. Nucleic Acids Research. 2016;**44**(D1):D457-D462

[3] Vessby B, et al. Indices of fatty acid desaturase activity in healthy human subjects: Effects of different types of dietary fat. British Journal of Nutrition. 2013;**110**(5):871-879

[4] Wang Y, et al. Tissue-specific, nutritional, and developmental regulation of rat fatty acid elongases. Journal of Lipid Research. 2005:**46**(4):706-715

[5] Farvid MS, et al. Dietary linoleic acid and risk of coronary heart disease: A systematic review and meta-analysis of prospective cohort studies. Circulation. 2014 Oct 28;**130**(18):1568-78. doi: 10.1161/CIRCULATIONAHA.114.010236. Epub 2014 Aug 26

[6] Meyer BJ, et al. Dietary intakes and food sources of omega-6 and omega-3 polyunsaturated fatty acids. Lipids. 2003;**38**(4):391-398.

[7] Burns CP, et al. Effect of modification of plasma membrane fatty acid composition on fluidity and methotrexate transport in L1210 murine leukemia cells. Cancer Research. 1979;**39**(5):1726-1732

[8] Kröger J, et al. Erythrocyte membrane fatty acid fluidity and risk of type 2 diabetes in the EPIC-Potsdam study. Diabetologia. 2014;**58**(2):282-289

[9] Kang JX, et al. Transgenic mice: Fat-1 mice convert n-6 to n-3 fatty acids. Nature. 2004;**427**(6974):504

[10] Wolters M, et al. Reference values of whole-blood fatty acids by age and sex from European children aged 3-8 years. International Journal of Obesity. 2014;**38**(S2):S86-S98

[11] Danesh J, et al. Low grade inflammation and coronary heart disease: Prospective study and updated meta-analyses. British Medical Journal. 2000;**321**(7255):199-204

[12] Gurtner GC, et al. Wound repair and regeneration. Nature. 2008;**453**(7193):314-321

[13] Harris RE. Cyclooxygenase-2 (cox-2) blockade in the chemoprevention of cancers of the colon, breast, prostate, and lung. Inflammopharmacology. 2009;**17**(2):55-67

[14] Mohler, ER, et al. Bone formation and inflammation in cardiac valves. Circulation. 2001;**103**(11):1522-1528

[15] Aukrust P, et al. Elevated circulating levels of C-C chemokines in patients with congestive heart failure. Circulation. 1998;**97**(12):1136-1143

[16] Il'yasova D, et al. Circulating levels of inflammatory markers and cancer risk in the health aging and body composition cohort. Cancer Epidemiology Biomarkers & Prevention. 2005;**14**(10):2413-2418

[17] Leyva F, et al. Uric acid in chronic heart failure: A marker of chronic inflammation. European Heart Journal. 1998;**19**(12):1814-1822

[18] Pankuweit S, et al. The HLA class II allele DQB1 0309 is associated with dilated cardiomyopathy. Gene. 2013;**531**(2):180-183

[19] Shiels MS, et al. Circulating inflammation markers and prospective risk of lung cancer. Journal of the National Cancer Institute. 2013 Dec 18;**105**(24):1871-80. doi: 10.1093/jnci/djt309. Epub 2013 Nov 18.

[20] Bagga D, et al. Differential effects of prostaglandin derived from ω-6 and ω-3 polyunsaturated fatty acids on COX-2 expression and IL-6 secretion. Proceedings of the National Academy of Sciences. 2003;**100**(4):1751-1756

[21] Vane JR. Inhibition of prostaglandin synthesis as a mechanism of action for aspirin-like drugs. Nature New Biology. 1971;**231**(25):232-235

[22] Bonventre JV, et al. Reduced fertility and postischaemic brain injury in mice deficient in cytosolic phospholipase A2. Nature. 1997;**390**(6660):622-625

[23] Burch RM, Luini A, Axelrod J. Phospholipase A2 and phospholipase C are activated by distinct GTP-binding proteins in response to alpha 1-adrenergic stimulation in FRTL5 thyroid cells. Proceedings of the National Academy of Sciences. 1986;**83**(19):7201-7205

[24] Lee IT, et al. Cooperation of TLR2 with MyD88, PI3K, and Rac1 in lipoteichoic acid–induced cPLA(2)/COX-2–dependent airway inflammatory responses. The American Journal of Pathology. 2010;**176**(4):1671-1684

[25] Xu J, et al. Prostaglandin E2 production in astrocytes: Regulation by cytokines, extracellular ATP, and oxidative agents. Prostaglandins, Leukotrienes, and Essential Fatty Acids. 2003. **69**(6):437-448

[26] McAdam BF, et al. Systemic biosynthesis of prostacyclin by cyclooxygenase (COX)-2: The human pharmacology of a selective inhibitor of COX-2. Proceedings of the National Academy of Sciences. 1999;**96**(1):272-277

[27] Morrison AR, Pascoe N. Metabolism of arachidonate through NADPH-dependent oxygenase of renal cortex. Proceedings of the National Academy of Sciences. 1981;**78**(12):7375-7378

[28] Oliw EH, et al. Arachidonic acid metabolism in rabbit renal cortex. Formation of two novel dihydroxyeicosatrienoic acids. Journal of Biological Chemistry. 1981;**256**(19):9924-9931

[29] Palmer RMJ, et al. Chemokinetic activity of arachidonic acid lipoxygenase products on leuocyctes of different species. Prostaglandins. 1980;**20**(2):411-418

[30] Todoric J, et al. Adipose tissue inflammation induced by high-fat diet in obese diabetic mice is prevented by n−3 polyunsaturated fatty acids. Diabetologia. 2006;**49**(9):2109-2119

[31] Azordegan N, et al. Carcinogenesis alters fatty acid profile in breast tissue. Molecular and Cellular Biochemistry. 2012;**374**(1):223-232

[32] Baylin A, Campos H. Arachidonic acid in adipose tissue is associated with nonfatal acute myocardial infarction in the central valley of Costa Rica. The Journal of Nutrition. 2004;**134**(11):3095-3099

[33] Rupp H, et al. Inverse shift in serum polyunsaturated and monounsaturated fatty acids is associated with adverse dilatation of the heart. Heart. 2010;**96**(8):595-598

[34] Taugbøl O, Baddaky-Taugbøl B, Saarem K. The fatty acid profile of subcutaneous fat and blood plasma in pruritic dogs and dogs without skin problems. Canadian Journal of Veterinary Research. 1998;**62**(4):275-278

[35] Pond CM, Mattacks CA. The source of fatty acids incorporated into proliferating lymphoid cells in immune-stimulated lymph nodes. British Journal of Nutrition. 2003;**89**(3): 375-382

[36] Mattacks CA, Sadler D, Pond CM. The effects of dietary lipids on dendritic cells in perinodal adipose tissue during chronic mild inflammation. British Journal of Nutrition. 2004;**91**(6):883-892

[37] Riol-Blanco L, et al. The neuronal protein Kidins220 localizes in a raft compartment at the leading edge of motile immature dendritic cells. European Journal of Immunology. 2004;**34**(1):108-118

[38] Morelli AE, Thomson AW. Dendritic cells under the spell of prostaglandins. Trends in Immunology. 2003;**24**(3):108-111

[39] Pond CM. Paracrine interactions of mammalian adipose tissue. Journal of Experimental Zoology. Part A, Comparative Experimental Biology. 2003;**295**(1):99-110

[40] Westcott E, et al. Fatty acid compositions of lipids in mesenteric adipose tissue and lymphoid cells in patients with and without Crohn's disease and their therapeutic implications. Inflammatory Bowel Diseases. 2005;**11**(9):820-827

[41] Miura S, et al. Modulation of intestinal immune system by dietary fat intake: Relevance to Crohn's disease. Journal of Gastroenterology and Hepatology. 1998;**13**(12):1183-1190

[42] Trebble TM, et al. Peripheral blood mononuclear cell fatty acid composition and inflammatory mediator production in adult Crohn's disease. Clinical Nutrition. 2004;**23**(4): 647-655

[43] Gassull MA, et al. Fat composition may be a clue to explain the primary therapeutic effect of enteral nutrition in Crohn's disease: Results of a double blind randomised multicentre European trial. Gut. 2002;**51**(2):164-168

[44] Frangogiannis NG. The inflammatory response in myocardial injury, repair and remodeling. Nature Reviews Cardiology. 2014;**11**(5):255-265

[45] Hall DJ, et al. Comparison of serum fatty acid concentrations in cats with hypertrophic cardiomyopathy and healthy controls. Journal of Feline Medicine and Surgery. 2014;**16**(8):631-636

[46] Smith CE, et al. Myocardial concentrations of fatty acids in dogs with dilated cardiomyopathy. American Journal of Veterinary Research. 2005;**66**(9):1483-1486

[47] Evanochko WT, et al. Proton NMR spectroscopy in myocardial ischemic insult. Magnetic Resonance in Medicine. 1987;**5**(1):23-31.

[48] Bouchard A, et al. Visualization of altered myocardial lipids by 1H NMR chemical-shift imaging following ischemic insult. Magnetic Resonance in Medicine. 1991;**17**(2):379-389

[49] Straeter-Knowlen IM, et al. 1H NMR spectroscopic imaging of myocardial triglycerides in excised dog hearts subjected to 24 hours of coronary occlusion. Circulation. 1996;**93**(7):1464-1470

[50] Reeves RC, et al. Demonstration of increased myocardial lipid with postischemic dysfunction (myocardial stunning) by proton nuclear magnetic-resonance spectroscopy. Journal of the American College of Cardiology. 1989;**13**(3):739-744

[51] Diamant M, et al. Diastolic dysfunction is associated with altered myocardial metabolism in asymptomatic normotensive patients with well-controlled type 2 diabetes mellitus. Journal of the American College of Cardiology. 2003;**42**(2):328-335

[52] Faller KM, et al. (1)H-MR spectroscopy for analysis of cardiac lipid and creatine metabolism. Heart Failure Reviews. 2013;**18**(5):657-668

[53] Ruiz-Hurtado G, et al. Mild and short-term caloric restriction prevents obesity-induced cardiomyopathy in young Zucker rats without changing in metabolites and fatty acids cardiac profile. Frontiers in Physiology. 2017;**8**:42

[54] Djoussé L, et al. Plasma phospholipid concentration of cis palmitoleic acid and risk of heart failure. Circulation: Heart Failure. 2012;**5**(6):703-709

[55] Eide IA, et al. Plasma levels of marine n-3 fatty acids and cardiovascular risk markers in renal transplant recipients. European Journal of Clinical Nutrition. 2016;**70**(7):824-830

[56] Fosshaug LE, et al. Altered levels of fatty acids and inflammatory and metabolic mediators in epicardial adipose tissue in patients with systolic heart failure. Journal of Cardiac Failure. 2015;**21**(11):916-923

[57] Fretts AM, et al. Plasma phospholipid saturated fatty acids and incident atrial fibrillation: The cardiovascular health study. Journal of the American Heart Association: Cardiovascular and Cerebrovascular Disease. 2014;**3**(3):e000889

[58] Kim OY, et al. Altered heart and kidney phospholipid fatty acid composition are associated with cardiac hypertrophy in hypertensive rats. Clinical Biochemistry. 2013;**46**(12): 1111-1117

[59] Warensjö E, et al. Biomarkers of milk fat and the risk of myocardial infarction in men and women: A prospective, matched case-control study. The American Journal of Clinical Nutrition. 2010. **92**(1):194-202

[60] Wu JHY, et al. Fatty acids in the de novo lipogenesis pathway and risk of coronary heart disease: The Cardiovascular Health Study. The American Journal of Clinical Nutrition. 2011;**94**(2):431-438

[61] Pelsers MMAL, Hermens WT, Glatz JFC. Fatty acid-binding proteins as plasma markers of tissue injury. Clinica Chimica Acta. 2005;**352**(1-2):15-35

[62] Zoair A, et al. Serum level of heart-type fatty acid binding protein (h-FABP) before and after treatment of congestive heart failure in children. Pediatric Cardiology. 2015;**36**(8):1722-1727

[63] Niizeki T, et al. Combination of heart-type fatty acid binding protein and brain natriuretic peptide can reliably risk stratify patients hospitalized for chronic heart failure. Circulation Journal. 2005;**69**(8):922-927

[64] Niizeki T, et al. Persistently increased serum concentration of heart-type fatty acid-binding protein predicts adverse clinical outcomes in patients with chronic heart failure. Circulation Journal. 2008;**72**(1):109-114

[65] Cabiati M, et al. High peripheral levels of h-FABP are associated with poor prognosis in end-stage heart failure patients with mechanical circulatory support. Biomarkers in Medicine. 2013;**7**(3):481-492

[66] Ozdemir C, et al. Heart-type fatty acid binding protein and cardiac troponin I may have a diagnostic value in electrocution: A rat model. Journal of Forensic and Legal Medicine. 2016;**39**:76-79

[67] Wittnich C, Belanger MP. What factors contribute to the elevation of serum free fatty acids (FFA) in newborns in the cardiac surgical setting? Canadian Journal of Physiology and Pharmacology. 2017

[68] Forouhi NG, et al. Differences in the prospective association between individual plasma phospholipid saturated fatty acids and incident type 2 diabetes: The EPIC-InterAct case-cohort study. The Lancet, Diabetes & Endocrinology. 2014;**2**(10):810-818

[69] D'Ascoli TA, et al. Association between serum long-chain omega-3 polyunsaturated fatty acids and cognitive performance in elderly men and women: The Kuopio Ischaemic Heart Disease Risk Factor Study. European Journal of Clinical Nutrition. 2016;**70**(8):970-975

[70] van Smeden J, et al. The importance of free fatty acid chain length for the skin barrier function in atopic eczema patients. Experimental Dermatology. 2014;**23**(1):45-52

[71] Ring J, et al. Guidelines for treatment of atopic eczema (atopic dermatitis) Part II. Journal of the European Academy of Dermatology and Venereology. 2012;**26**(9):1176-1193

[72] Taugbol O, Baddaky-Taugbol B., Saarem K. The fatty acid profile of subcutaneous fat and blood plasma in pruritic dogs and dogs without skin problems. Canadian Journal of Veterinary Research-Revue Canadienne De Recherche Veterinaire. 1998;**62**(4):275-278

[73] Muller MR, et al. Evaluation of cyclosporine-sparing effects of polyunsaturated fatty acids in the treatment of canine atopic dermatitis. Veterinary Journal. 2016;**210**:77-81

[74] Calon A, et al. Stromal gene expression defines poor-prognosis subtypes in colorectal cancer. Nature Genetics. 2015;**47**(4):320-329

[75] Fehrmann RSN, et al. Gene expression analysis identifies global gene dosage sensitivity in cancer. Nature Genetics. 2015;**47**(2):115-125

[76] Gillet JP, et al. A gene expression signature associated with overall survival in patients with hepatocellular carcinoma suggests a new treatment strategy. Molecular Pharmacology. 2016;**89**(2):263-272

[77] van't Veer LJ, et al. Gene expression profiling predicts clinical outcome of breast cancer. Nature. 2002;**415**(6871):530-536

[78] Zheng JS, et al. Intake of fish and marine n-3 polyunsaturated fatty acids and risk of breast cancer: Meta-analysis of data from 21 independent prospective cohort studies. BMJ-British Medical Journal. 2013;**346**

[79] Gann PH, et al. Prospective study of plasma fatty acids and risk of prostate cancer. Journal of the National Cancer Institute. 1994;**86**(4):281-286

[80] Akinsete JA, et al. Consumption of high omega-3 fatty acid diet suppressed prostate tumorigenesis in C3(1) Tag mice. Carcinogenesis. 2012;**33**(1):140-148

[81] Augustsson K, et al. A prospective study of intake of fish and marine fatty acids and prostate cancer. Cancer Epidemiology, Biomarkers & Prevention. 2003;**12**(1):64-67

[82] Leitzmann MF, et al. Dietary intake of n-3 and n-6 fatty acids and the risk of prostate cancer. The American Journal of Clinical Nutrition. 2004;**80**(1):204-216

[83] Pelser C, et al. Dietary fat, fatty acids, and risk of prostate cancer in the NIH-AARP diet and health study. Cancer Epidemiology, Biomarkers & Prevention. 2013;**22**(4):697-707

[84] Pondugula SR, et al. Stearidonic acid, a plant-based dietary fatty acid, enhances the chemosensitivity of canine lymphoid tumor cells. Biochemical and Biophysical Research Communications. 2015;**460**(4):1002-1007

[85] Wong J, et al. Isomer specificity of conjugated linoleic acid on suppression of osteosarcomas. Journal of Nature and Science. 2015;**1**(4):e67

[86] Long E, et al. Fatty acid profile and gene expression in canine lipoma – a promising treatment target? Advances in Animal Biosciences. 2013;P032

[87] Cettour-Rose P, et al. Redistribution of glucose from skeletal muscle to adipose tissue during catch-up fat. A link between catch-up growth and later metabolic syndrome. Diabetes. 2005;**54**:751-756

[88] Barker DJP, et al. Fetal nutrition and cardiovascular disease in adult life. The Lancet. 1993;**341**:938-941

[89] Phillips DIW, et al. Size at birth and plasma leptin concentrations in adult life. International Journal of Obesity. 1999;**23**:1-5

[90] Whorwood CB, et al. Maternal undernutrition during early to midgestation programs tissue-specific alterations in the expression of the glucocorticoid receptor, 11beta-hydroxysteroid dehydrogenase isoforms, and type 1 angiotensin II receptor in neonatal sheep. Endocrinology. 2001;**142**(7):2854-2864

[91] MacDonald ML, et al. Effects of linoleate and arachidonate deficiencies on reproduction and spermatogenesis in the cat. The Journal of Nutrition. 1984;**114**(4):719-726

[92] Risso A, et al. Effect of long-term fish oil supplementation on semen quality and serum testosterone concentrations in male dogs. International Journal of Fertility & Sterility. 2016;**10**(2):223-231

[93] Morris JG. Do cats need arachidonic acid in the diet for reproduction? Journal of Animal Physiology and Animal Nutrition. 2004;**88**(3-4):131-137

[94] Morris JG, Rogers QR, O'Donnell JA. Lysine requirement of kittens given purified diets for maximal growth. Journal of Animal Physiology and Animal Nutrition. 2004;**88**(3-4):113-116

[95] Siemelink M, et al. Dietary fatty acid composition during pregnancy and lactation in the rat programs growth and glucose metabolism in the offspring. Diabetologia. 2002; **45**(10):1397-1403

[96] Taylor PD, et al. Impaired glucose homeostasis and mitochondrial abnormalities in offspring of rats fed a fat-rich diet in pregnancy. American Journal of Physiology. 2005;**288**(1):R134-R139

[97] Armitage JA, Taylor PD, Poston L. Experimental models of developmental programming: Consequences of exposure to an energy rich diet during development. Journal of Physiology. 2005;**565**(1):3-8

[98] Zhu MJ, et al. AMP-activated protein kinase signalling pathways are down regulated and skeletal muscle development impaired in fetuses of obese, over-nourished sheep. Journal of Physiology. 2008;**586**(10):2651-2664

[99] Cerfa ME, et al. Hyperglycaemia and reduced glucokinase expression in weanling offspring from dams maintained on a high-fat diet. British Journal of Nutrition. 2006;**95**:391-396

[100] Symonds ME, et al. Maternal nutrient restriction during placental growth, programming fetal adiposity and juvenile blood pressure control. Archives of Physiology and Biochemistry. 2003;**111**(1):45-52

[101] Chadio SE, et al. Impact of maternal undernutrition on the hypothalamic-pituitary-adrenal axis responsiveness in sheep at different ages postnatal. Journal of Endocrinology. 2007;**192**(3):495-503

[102] Roseboom T, de Rooij S, Painter R. The Dutch famine and its long-term consequences for adult health. Early Human Development. 2006;**82**(8):485-491

[103] Kelley DE, et al. Interaction between glucose and free fatty acid metabolism in human skeletal muscle. Journal of Clinical Investigation. 1993;**92**(1):91-98

[104] Savage DB, Petersen KF, Shulman GI. Disordered lipid metabolism and the pathogenesis of insulin resistance. Physiological Reviews. 2007;**87**(2):507-520

[105] Poore KR, Fowden AL. The effect of birth weight on glucose tolerance in pigs at 3 and 12 months of age. Diabetologia. 2002;**45**:1247-1254

[106] Poore KR, et al. The effects of birth weight in basal cardiovascular function in pigs at 3 months of age. Journal of Physiology. 2002;**539**:969-978

[107] Poore KR, Fowden AL. Insulin sensitivity in juvenile and adult large white pigs of low and high birthweight. Diabetologia. 2004;**47**:340-348

[108] Mostyn A, et al. Influence of size at birth on the endocrine profiles and expression of uncoupling proteins in subcutaneous adipose tissue, lung and muscle of neonatal pigs. American Journal of Physiology. 2005;**288**(6):R1536-R1542

[109] Mostyn A, et al. The influence of size at birth on lipid storage and peroxisome proliferator activated receptor g (PPARg) expression in adipose tissue (AT). Early Human Development. 2006;**82**:634

[110] Marten N, et al. The influence of piglet birth weight on peroxisome proliferator activated receptor α (PPARα) expression in skeletal muscle. Proceedings of the Neonatal Society. 2006, http://www.neonatalsociety.ac.uk/abstracts/martenn_2006_pigletperoxisomeproliferator.shtml

[111] Mostyn A, et al. Differences in FABP3 and FABP4 messenger RNA expression in skeletal muscle and subcutaneous adipose tissue between normal birth weight and low and high birth weight porcine offspring at days 7 and 14 of postnatal life. Proceedings of the Nutrition Society. 2007;**66**:57A

[112] Lillycrop KA, et al. Dietary protein restriction of pregnant rats induces and folic acid supplementation prevents epigenetic modification of hepatic gene expression in the offspring. Journal of Nutrition. 2005;**135**(6):1382-1386

[113] Evans RM, Barish GD, Wang Y. PPARs and the complex journey to obesity. Nature Medicine. 2004;**10**(4):1-7

[114] Rosen ED, Spiegelman BM. PPARg: A nuclear regulator of metabolism, differentiation, and cell growth. The Journal of Biological Chemistry. 2001;**276**(41):37731-37734

[115] Bays H, Mandarino L, DeFronzo RA. Role of the adipocyte, free fatty acids and ectopic fat in the pathogenesis of type 2 diabetes mellitus: Peroxisomeal proliferator-activated receptor agonists provide a rational therapeutic approach. Journal of Clinical Endocrinology and Metabolism. 2004;**89**(2):463-478

[116] Vidal-Puig A, et al. Regulation of PPAR gamma gene expression by nutrition and obesity in rodents. Journal of Clinical Investigation. 1996;**97**(11):2553-2561

[117] Medina-Gomez G, et al. PPAR gamma 2 prevents lipotoxicity by controlling adipose tissue expandability and peripheral lipid metabolism. PLoS Genetics. 2007;**3**(4):e64

[118] Xu HE, et al. Molecular recognition of fatty acids by peroxisome proliferator-activated receptors. Molecular Cell. 1999;**3**(3):397-403

[119] Unger RH. Weapons of lean body mass destruction: The role of ectopic lipids in the metabolic syndrome. Endocrinology. 2003;**144**(12):5159-5165

[120] Chilliard Y, et al. Ruminant milk fat plasticity: Nutritional control of saturated, polyunsaturated, trans and conjugated fatty acids. Annales De Zootechnie. 2000;**49**(3):181-205

[121] Martemucci G, D'Alessandro AG. Fat content, energy value and fatty acid profile of donkey milk during lactation and implications for human nutrition. Lipids in Health and Disease. 2012;**11**

[122] Breckenr WC, Kuksis A. Molecular weight distributions of milk fat triglycerides from seven species. Journal of Lipid Research. 1967;**8**(5):473-478

[123] Innis SM. Impact of maternal diet on human milk composition and neurological development of infants. American Journal of Clinical Nutrition. 2014;**99**(3):734s-741s

[124] Oosting A, et al. Rapid and selective manipulation of milk fatty acid composition in mice through the maternal diet during lactation. Journal of Nutritional Science. 2015;**4**

[125] Capper JL, et al. Polyunsaturated fatty acid supplementation during pregnancy alters neonatal behavior in sheep. Journal of Nutrition. 2006;**136**(2):397-403

[126] Barreiro R, et al. Effects of bovine pregnancy on the fatty acid composition of milk: The significance for humans needs. Food Additional & Contaminants. Part A, Chemistry, Analysis, Control, Exposure Risk & Assessment. 2017:1-9

[127] Makowski L, Hotamisligil GS. Fatty acid binding proteins—the evolutionary crossroads if inflammatory and metabolic responses. J Nutr. 2004 Sep;**134**(9):2464S-2468S

[128] Roepstorff C, et al. Studies of plasma membrane fatty acid-binding protein and other lipid-binding proteins in human skeletal muscle. Proceedings of the Nutrition Society. 2004;**63**:239-244

[129] Panarotto D, et al. Insulin resistance affects the regulation of lipoprotein lipase in the postprandial period and in an adipose tissue-specific manner. European Journal of Clinical Investigation. 2002;**32**:84-92

[130] Zhou G, et al. Role of AMP-activated protein kinase in mechanism of metformin action. Journal of Clinical Investigation. 2001;**108**(8):1167-1174

[131] Tong L. Acetyl-coenzyme A carboxylase: Crucial metabolic enzyme and attractive target for drug discovery. Cellular and Molecular Life Sciences. 2005;**62**:1784-1803

[132] Hardie DG, Hawley SA. AMP-activated protein kinase: The energy charge hypothesis revisited. BioEssays. 2001;**23**(12):1112-1119

[133] Liu Y, et al. High-fat diet feeding impairs both the expression and activity of AMPKa in rats' skeletal muscle. Biochemical and Biophysical Research Communications. 2006;**339**(2):701-707

[134] Denayer T, Stöhr T, Van Roy M. Animal models in translational medicine: Validation and prediction. New Horizons in Translational Medicine. 2014;**2**(1):5-11

[135] van der Worp HB, et al. Can animal models of disease reliably inform human studies? PLoS Medicine. 2010;**7**(3):e1000245

[136] DeGiorgio CM, Taha AY. Omega-3 fatty acids (-3 fatty acids) in epilepsy: Animal models and human clinical trials. Expert Review of Neurotherapeutics. 2016;**16**(10):1141-1145

[137] Puurunen J, et al. A non-targeted metabolite profiling pilot study suggests that tryptophan and lipid metabolisms are linked with ADHD-like behaviours in dogs. Behavioral and Brain Functions. 2016;**12**:27 doi: 10.1186/s12993-016-0112-1

[138] Buzzai M, et al. The glucose dependence of Akt-transformed cells can be reversed by pharmacologic activation of fatty acid [beta]-oxidation. Oncogene. 2005;**24**(26):4165-4173

[139] Kantor PF, et al. The antianginal drug trimetazidine shifts cardiac energy metabolism from fatty acid oxidation to glucose oxidation by inhibiting mitochondrial long-chain 3-ketoacyl coenzyme A thiolase. Circulation Research. 2000;**86**(5):580-588

[140] Serisier S, et al. Fenofibrate lowers lipid parameters in obese dogs. The Journal of Nutrition. 2006;**136**(7):2037S-2040S

[141] Purushotham A, et al. Hepatocyte-specific deletion of SIRT1 alters fatty acid metabolism and results in hepatic steatosis and inflammation. Cell Metabolism. 2009;**9**(4):327-338

[142] Stavinoha MA, et al. Diurnal variations in the responsiveness of cardiac and skeletal muscle to fatty acids. American Journal of Physiology—Endocrinology and Metabolism. 2004;**287**(5):E878-E887

[143] Wang DF, et al. Effects of nutritional level of concentrate-based diets on meat quality and expression levels of genes related to meat quality in Hainan black goats. Animal Science Journal. 2015;**86**(2):166-173

[144] Jurie C, et al. Adipocyte fatty acid-binding protein and mitochondrial enzyme activities in muscles as relevant indicators of marbling in cattle. Journal of Animal Science. 2007;**85**(10):2660-2669

[145] Zhu, J.J., et al., Inhibition of FASN reduces the synthesis of medium-chain fatty acids in goat mammary gland. Animal, 2014. **8**(9):1469-1478

[146] Chamberlain MB, DePeters EJ. Impacts of feeding lipid supplements high in palmitic acid or stearic acid on performance of lactating dairy cows. Journal of Applied Animal Research. 2017;**45**(1):126-135

[147] Liang W, et al. Investigation of the association of two candidate genes (H-FABP and PSMC1) with growth and carcass traits in Qinchuan beef cattle from China. Genetics and Molecular Research. 2014;**13**(1):1876-1884

[148] Li DL, et al. Growth, carcase and meat traits and gene expression in chickens divergently selected for intramuscular fat content. British Poultry Science. 2013;**54**(2):183-189

9

Short-Chain Fatty Acids are Antineoplastic Agents

Mohammad Salah Abaza, Aneela Afzal and

Mohammad Afzal

Abstract

Human diet contains a mixture of saturated and unsaturated fatty acids. These are either long, medium or short chain fatty acids. As commonly believed, all fatty acids are not detrimental to human health. In addition to energy reserves, long chain fatty acids are known as acylating agents for many biomolecules such as cholesterol, terpenoids as well as steroid hormones. They are also involved in acylation of polyphenols such as flavonoids making them palatable for better absorption and biological activities. Polyunsaturated fatty acids (PUFAs) are known for their numerous beneficial health effects including cancer and inflammation. PUFA, particularly ω3 fatty acids, have attracted attention as anticancer agents and particularly for colorectal cancer. PUFAs exhibit immunomodulatory activities controlling inflammosome and are used as adjuvants together with standard anticancer drugs. A reciprocal interaction of short chain fatty acids with PUFAs has been suggested for their anticancer activities. Thus, in colon cancer cells, sodium butyrate (NaB) interacts with docosahexaenoic acid inducing cell differentiation or catalyze apoptosis. These results encouraged us to investigate NaB, a C4 acid, as an adjuvant to standard proteasome inhibitors. Our results show that NaB sensitizes colon cancer cell lines for treatment with proteasome inhibitors.

Keywords:histone deacylating agents, proteasome inhibitors, short-chain fatty acids, sodium butyrate, polyunsaturated fatty acids

1. Background

Cancer appears when the cellular growth network is disturbed and tumor cells resist apoptosis, resulting in uncontrolled growth and progression of tumor cells. Nucleosome acylation and deacylation of histones play a critical role in the tumorigenesis progression by regulating chromatin structure and function. The histone acetyltransferases (HATs) and deacylases (HDACs) create a fine equilibrium between acylation and deacylation of histones (**Figure 1**). Once this equilibrium is disturbed, it leads to cancer promotion and progression. A number of synthetic compounds, such as cyclic tetrapeptides, benzamides, suberoylanilide hydroxamic acid, and associated branched hydroxamic acid derivatives, are expended as inhibitors of HDAC. Short-chain fatty acids, cogitated as novel drugs, have also been used as HDAC inhibitors. These compounds lead to an accumulation of acylated histones in healthy and tumor cells, arresting the cell cycle in the G1 and/or G2 phases, and provoking apoptosis in cancer cells. Therefore, HDAC inhibitors in controlled doses are recognized as innovative antitumor and anti-inflammatory drugs, and sodium butyrate and sodium valproate, a C8 FA, have been used as HDAC inhibitors.

Fatty acids, contingent to the number of carbons, can be classified into three groups:

(a) C2:0–C6:0, short-chain fatty acids (SCFAs) are dietetic and colonic fermentation products. SCFAs have promise as antitumor agents in numerous types of cancer cells.

(b) C8:0–C14:0, intermediate chain fatty acids (ICFAs) have antimicrobial physiognomies, but some reports have also recognized them as antineoplastic mediators.

(c) C16:0–C24:0, classified as long-chain fatty acids (LCFA), provoke oxidative stress commanding apoptosis in tumor cells.

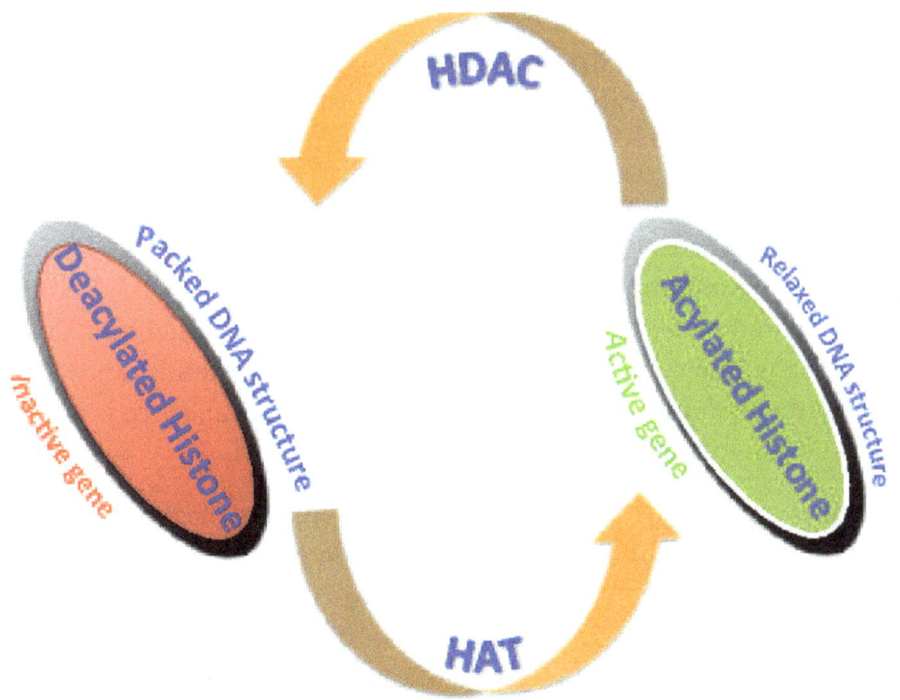

Figure 1. Acylation and deacylation of histones.

Several studies have shown that a combined treatment with MCFA and SCFA is more effective for inducing cell death through apoptosis. Additionally, numerous studies have reported the use of sodium butyrate and propionate as antineoplastic agents. In this chapter, the anticancer effect of water soluble NaB and its conceivable potential to augment the anticancer effect of proteasome inhibitors, as well as the principal mechanism of action of butyrate and/or proteasome inhibitors on human colorectal tumor cells, will be discussed.

The human diet is commonly deficient in ω3-fatty acids that are common components of fish and fish products. Fish oil, high in ω3-fatty acids, is known to have anticancer activities through apoptosis of cancer cells, and numerous reports have appeared to support this claim [1–8]. Thus, docosahexaenoic (22:6, *n*-3) and eicosapentaenoic (20:5, *n*-3) acids are effective antitumor adjuvants that provoke apoptosis in several types of tumor cells without an injury to natural cells [9–11]. However, most unsaturated fatty acids can undergo oxidative stress (OS), which has been implicated in many pathological conditions. OS modifies many biological molecules/pathways, often resulting in serious consequences. Lipid peroxidation, in addition to DNA and protein oxidation, is one such modification of lipids that involves unsaturated fatty acids, resulting in the formation of fatal peroxyl radicals and activating many transcription factors. These include NF-κB, AP-1, p53, HIF-1α, PPAR-γ, β-catenin/Wnt, and Nrf2, which lead to cancer progression. In turn, a stimulation of these transcription factors activates over 500 genes, including cytokines and growth factor. This generates a strong relationship between OS, inflammation, and tumorigenesis. The fatty acid peroxyl radicals are a source of reactive hydroxyl-aldehydes, such as 4-hydroxynonenal, 4-oxononenal, and malondialdehyde. These radicals elaborate Millard reactions with proteins and other N-biomolecules, triggering mutations with an outbreak of cancer. In many ways, the metabolism of tumor cells differs from normal cells. One of the main differences between tumor cells and normal cells is the lower level of natural antioxidants desirable for defense from OS. Certain saturated fatty acids, such as odd carbon and branched carbon fatty acids, are also known to have anticancer activities [12].

Tumor cells adjust to an array of nutrient stocks for their subsistence. During an impaired glucose metabolism, tumor cells, for nutrients, switch to lipid metabolism. Cancer cells with their multiple metabolic compartments, depletion of sugars, amino acids, and lipids are considerably higher compared with their counterpart normal cells. Channeling their biosynthetic nutrients between and within cells, tumor cells contribute to their endurance and evolution. Therefore, one of the therapeutic targets to control the growth of cancer cells is to focus on the metabolic modifications between tumor and normal cells [13].

The involvement of phospholipase D1 (PLD1) in controlling the plasticity of cancer cell has been reported [14]. Cai et al. have reported that oxidation of fatty acids is the main basis of energy metabolism and inhibition of PLD1 results in a downregulation of lipid energy metabolism in tumor cells [14]. Numerous other inhibitors of lipid synthesis have also yielded encouraging results in limiting the proliferation of cancer cells. In this context, 5-(tetradecyloxy)-2-furoic acid, an inhibitor of acetyl-CoA carboxylase, giving malonyl-CoA, which is an intermediary in the synthesis of fatty acids, has offered promising results for inhibiting cancer cell upturn and proliferation. Other lipid synthesis inhibitors, such as fatty acid synthase (FASN), cerulenin, and irgasan, suppress the proliferation of MiaPaCa-2 and AsPC-1 cells through depletion of fatty acids and apoptosis of cancer cells [15]. Leucine deficit also inhibits FASN in breast cancer cells [16]. An overexpression of FASN in neoplasms is widely reported in the literature, as its inhibition by certain synthetic imides, such as N-phenylmaleimides [17].

Histone deacylase (HDAC) promotes deacylation by hydrolyzing histone lysine residues and plays a significant role in the regulation of gene expression. However, HDAC is overexpressed in several forms of cancer and is a target for several anticancer drugs. HDAC inhibitors prohibit the deacylation of not only histone but also nonhistone proteins and promote cell survival and anticancer activity. The inhibitors of HADC, presently approved by the FDA, include vorinostat, romidepsin, belinostat, and panobinostat, and several other HDAC inhibitors are already in clinical trial. Manal et al. have reported that SCFAs are novel HDAC inhibitors with advanced anticancer characteristics [18]. However, carnitine palmitoyl transferase 1 (CPT1), which is involved in the transport of long-chain fatty acids for β-oxidation, has been reported to be a specific target for anticancer therapies that is more selective than HDAC [19].

Gonadotropin-releasing hormone-III, when acylated with butyric acid at lysine position four, forms the bioconjugate, GnRH-III(4)Lys(Bu), and is reported to have significant benefits over free daunorubicin as an antitumor agent [20].

2. Butyrate

Butyrate is a short 4-carbon fatty acid and is one of the three observed in the mammalian colonic lumen [21]. It is known that anaerobic fermentation of carbohydrates and proteins in the lumen produces butyric acid [22]. Biological reaction modification, resulting in gene activation and growth control, by butyrate and its water-soluble sodium salt has been reported [23]. An inhibition of DNA synthesis may be responsible for arresting the proliferating cells and an induction of cell differentiation [24]. These results have led to contemplate that a short-chain fatty acid, such as butyrate, may be a useful agent with antiproliferative and antineoplastic significance for typical mucosal epithelial cells [25]. Since butyrate is a dietary short-chain fatty acid with low toxicity and growth inhibitory consequences, we decided to investigate its potential to augment the anticancer effect of a collection of proteasome inhibitors (MG115, MG132, PSI-1, PSI-2, and epoxomicin) on colorectal cancer cells [26].

For ubiquitin-dependence of cellular proteins, a proteasome of multicatalytic nature and a degradation of over 80% intracellular proteins have been proposed [27]. For the regulation of protein synthesis during cellular stress, including apoptosis, impaired DNA, hypoxia, signal transduction, and so on, ubiquitin is recognized to play a dynamic role [28]. In oncology, validation of proteasome as a clinical agent has been provided by the use of bortezomib, which is a boronic acid dipeptide [28] and is an effective agent for treating multiple myeloma and certain types of non-Hodgkin's lymphoma [29, 30]. Nevertheless, many patients do not respond to bortezomib. This is despite the fact that a regular use of bortezomib has many serious consequences including cardiac problems, excruciating neuropathy as well as thrombocytopenia [31–34]. For proteasome recovery, the treatment with bortezomib has been restricted biweekly [35]. Furthermore, in tumorigenesis, drug resistance to proteasome inhibitors [36] is a challenge.

Numerous types of tumors have been treated by an induction of apoptosis that can be triggered by several drugs and proteasome inhibitors. Again, drug toxicity and cell resistance are the cost that the patients have to bear [37]. The dietary sodium butyrate offers a valuable treatment with minor toxicity but a high degree of apoptic strength [38] making it a substance of choice for treating various types of tumors. We hypothesized that the anticancer characteristics of the proteasome inhibitors MG115, MG132, PSI-1, PSI-2 and epoxomicin in human colorectal carcinoma could be potentiated by NaB [26].

Human colorectal cancer cell line SW837 treated with NaB, MG115, and a combination of the two, for 24 h, showed a minor growth inhibition of the tumor cells (mean, 6 ± 0.4%) (**Figure 2A**). A distinct inhibition of SW837 cells (mean, 85 +_ 2%), with an increase in the treatment time to 72 h, was observed. While a modest inhibition (mean 31 ± 4%) was detected after a single treatment with MG115, in 72 hrs. A combination of two of the therapies NaB and MG115 had a vivid inhibitory effect on the growth of SW837 tumor cells (mean, 85 ± 2%) and it was comparable with NaB when applied alone. However, the combination treatment for 72 h produced a statistically significant ($P \leq 0.004$) inhibition of SW837 tumor cells compared with a single treatment with MG115. Increasing the treatment time to 120 h with the combination therapy, SW837, exhibited an inhibition of growth (mean, 90 ± 2%) compared with a sole action of NaB (mean, 87 ± 2%). Contrarily, after 120 h of treatment, MG115 alone showed only a humble inhibition (mean, 31 ± 5%). The growth inhibition of SW837 was statistically significant ($P \leq 0.002$) in a combination therapy of NaB and MG115 for 120 h compared with MG115 alone (**Figure 2A**).

Next, we tested the efficacy of MG132 and NaB on the growth of SW837 for 24 h. It was found that the growth of SW837 was nonsignificantly affected by the individual two therapies (**Figure 2B**). However, after 72 h of treatment, a combination of the two therapies significantly constrained the growth of SW837 (mean, 86 ± 2%). A growth inhibition (mean, 84 ± 2%) comparable to the combination treatment was observed with NaB. A solitary treatment with MG132 for 72 h resulted in nonsignificant inhibition of SW837 (mean, 40 ± 4%). The inhibition change in SW837, after a combination treatment of NaB/MG132 and a sole treatment with MG132, was statistically significant ($P \leq 0.022$). With a prolonged treatment of SW837 for 120 h, the combination of NaB/MG132 resulted in a distinctive reticence in the cell growth (mean, 89 ± 2%). However, a treatment with MG132 alone for 120 h resulted in a minuscule growth inhibition (mean, 47 ± 5%). The change in inhibition of SW837 after a combination treatment of NaB/MG132 and a sole treatment with MG132 was statistically significant ($P \leq 0.037$) (**Figure 2B**).

We also investigated the effect of another proteasome PSI-1 alone and in combination with NaB. A 24 h treatment of SW837 tumor cells resulted in (mean, 12 ± 1.0%; 1.3 ± 0.4% and 3.0 ± 0.4%) for NaB, PSI-1, and combination of the two agents, respectively (**Figure 2C**). A striking difference (mean, 96.0 ± 4.0%) was observed when SW837 was treated with a combination of NaB and PSI-1, compared with PSI-1 alone (mean, 54.0 ± 2.0%) for 72 h. The change in inhibition of SW837 after a combination treatment of NaB/PSI-1 and a sole treatment with PSI-1 was statistically significant ($P \leq 0.001$). The change in inhibition of SW837 cells after a combination treatment of NaB/PSI-1 and a sole treatment with PSI-1 was statistically nonsignificant ($P \leq 0.32$). A prolonged treatment of SW837 cells for 120 h produced comparable results **Figure 2C**. The combination action showed a higher inhibition for SW837 cells, compared

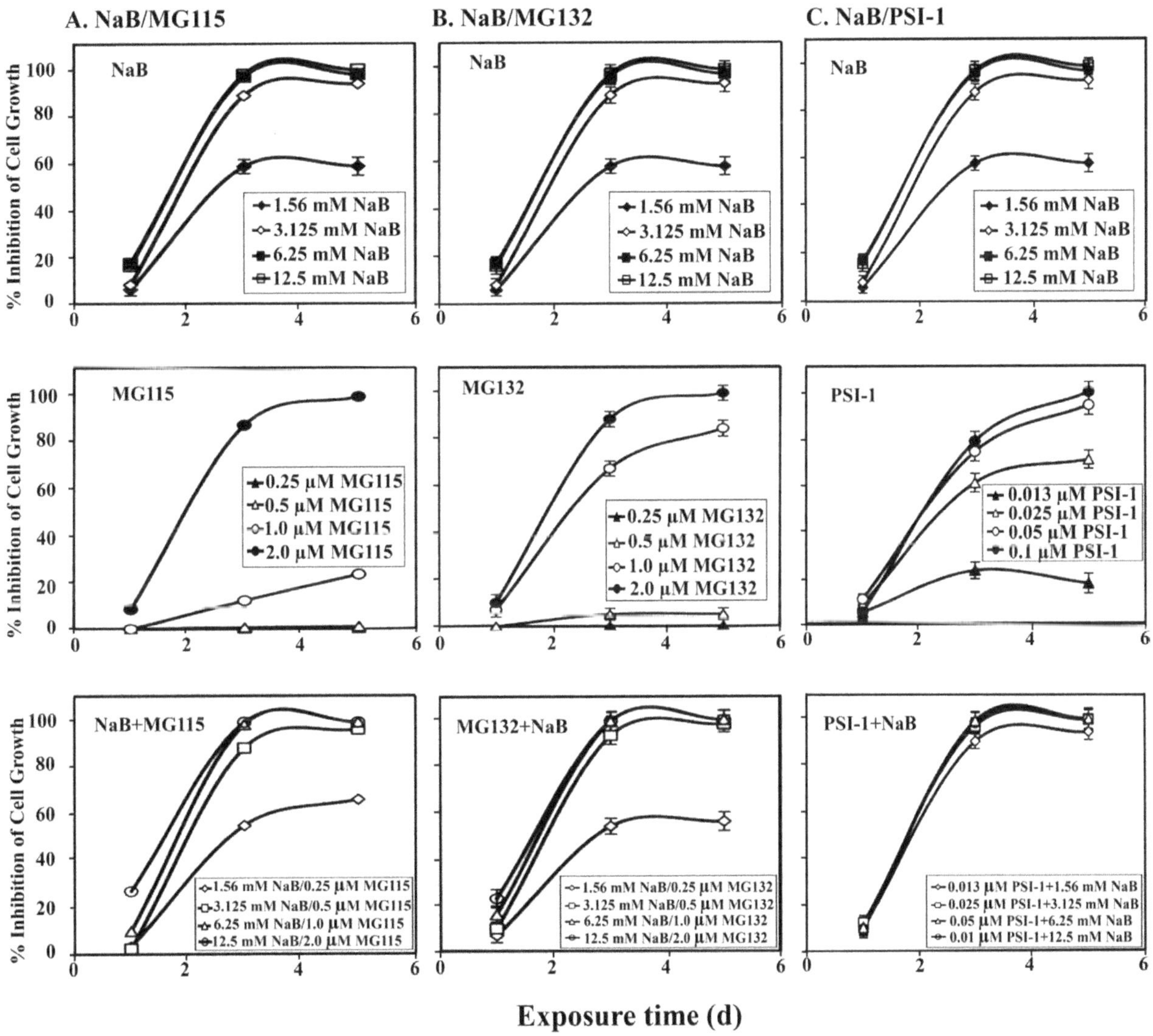

Figure 2. Enhancement of the anticancer effect of proteasome inhibitors MG115, MG132, and PSI-1 with NaB on human colorectal cancer SW837 cells. SW837 cells were plated (27 × 10^3 cells/well) into 96-well plates and incubated at 37°C in a non-CO_2 incubator. After 18 h, the cells were treated with NaB (1.56–12.5 mM), MG115 (0.25–2.0 µM), MG132 (0.25–2 µM), PSI-1 (0.013–0.1 µM), and the combinations of NaB and MG115 (A), MG132 (B), or PSI-1 (C) starting 18 h after seeding the cells in culture. Control cells were left untreated or treated with vehicle (DMSO) at a final concentration (0.1%). Cell growth was monitored by MTT assay.

with a usage of NaB alone (mean, 87.0 ± 2.0%). The other results of SW837 cells inhibition with NaB, PSI-2, epoxomicin, and their combinations are shown in **Figure 3**.

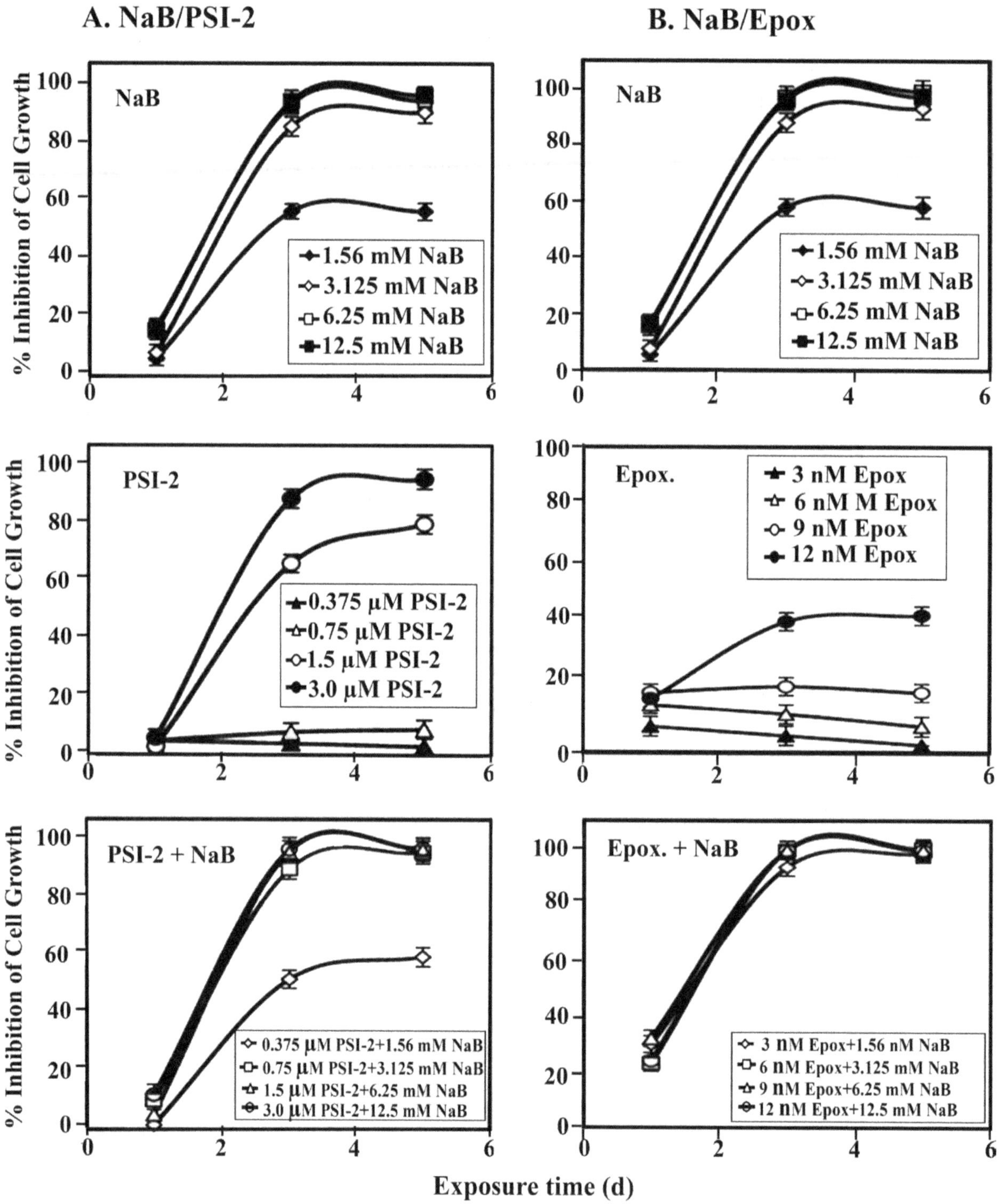

Figure 3. Enhancement of the anticancer effect of proteasome inhibitors PSI-2 and epoxomicin with NaB on human colorectal cancer SW837 cells. SW837 cells were plated (27×10^3 cells/well) into 96-well plates and incubated at 37°C in a non-CO_2 incubator. After 18 h, the cells were treated with NaB (1.56–12.5 mM), PSI-2 (0.375–3.0 µM), epoxomicin (3.0–12 nM), and the combinations of NaB and PSI-2 (A) or epoxomicin (B) starting 18 h after seeding the cells in culture. Control cells were left untreated or treated with vehicle (DMSO) at a final concentration (0.1%). Cell growth was monitored by MTT assay.

An analysis of our investigations has established that human colorectal cancer cell line, SW837, when treated with 3 mM NaB caused amassing of cells in the G1-phase (82.7%), and a corresponding reduction in the number of cells in G2/M- (2.61%) and S- (14.7%) phases (**Figure 4**). In addition, treatment with 1.0 μM MG115, 0.1 μM MG132, 0.1 μM PSI-1, 1.5 μM PSI-2, or 12 nM epoxomicin followed a buildup of cells in the S-phase (55.5, 33.7, 41.5, 42.7, and 32.2%, respectively) and G2-phase (12.2, 36.1, 44.3, 29.9, and 45.7%, respectively) with a consequent reduction in the total cells in the G1-phase (29.0, 29.7, 14.1, 27.4 and 22.3%, respectively).

A combination of NaB at 3 mM and MG115 or MG132 at 1.0 μM concentration caused the colorectal cancer cells arrest in the G1-phase (79.8 or 75.5%, respectively) and the G2-phase (6.73 or 14.4%, respectively). The upturn in the G1-phase complemented by an equivalent reduction in the S-phase of the cells (13.5 or 10%, respectively) (**Figure 4**). A combination treatment with NaB (3 mM) and PSI-1 (1.0 μM), PSI-2 (1.5 μM), or epoxomicin (12 nM) resulted in an increase in the number of cells in the G2-phase (45.2, 92.7, and 88.3%, respectively). This was accompanied by a parallel reduction in the quantity of cells in the G1-phase (55.4, 7.29, or 12.3%, respectively) and the S-phase (0.0%) (**Figure 4**).

Next, we turned to analyze the effect of antineoplastic agents on the DNA of treated cells by agarose gel electrophoresis. The DNA was extracted from the untreated and treated human colorectal cancer cells that displayed a discrete ladder pattern, displaying apoptosis. These consequences obviously displayed that the action of NaB, proteasome inhibitors, or their combination triggered the apoptotic trail. The magnitude of apoptosis of cancer cells, treated with a combination of NaB and proteasome inhibitors was prominent compared with the NaB or proteasome inhibitors alone (**Figure 5**).

The regulation of gene expression and inhibition of histone deacylases are regulated by NaB [39]. The hyperacetylation of histones and an amelioration of the availability of the transcription factors to nucleosomal DNA are due to inhibition of histone deacylases [40]. Hyperacetylation of nonhistone proteins, modification of DNA methylation, careful inhibition of histone phosphorylation, and alteration of intracellular kinase signaling may be the other cellular targets of NaB [39]. This multistage mechanism of butyrate explains the gene expression regulation and its impact on the crucial regulators of apoptosis and the cell cycle.

The synergistic apoptotic consequences of NaB and proteasome inhibitors may offer new opportunities in research to develop therapeutic strategies to contain human colorectal cancer. The proteasome inhibitors seem to act as apoptotic agents only in the rapidly dividing cells while shielding quiescent cells from apoptosis that may be activated by many diverse compounds [30]. For this specific action, proteasome inhibitors may be used as a substitute in the treatment of some proliferative disorders. Moreover, treatment with a combination of proteasome inhibitors, effective apoptotic agents, such as NaB, and other short-chain fatty acids may be a valuable therapeutic strategy for the treatment of proliferative diseases such as colorectal cancer. Thus, further research in this area may be very rewarding and offer hope to the suffering patients around the globe.

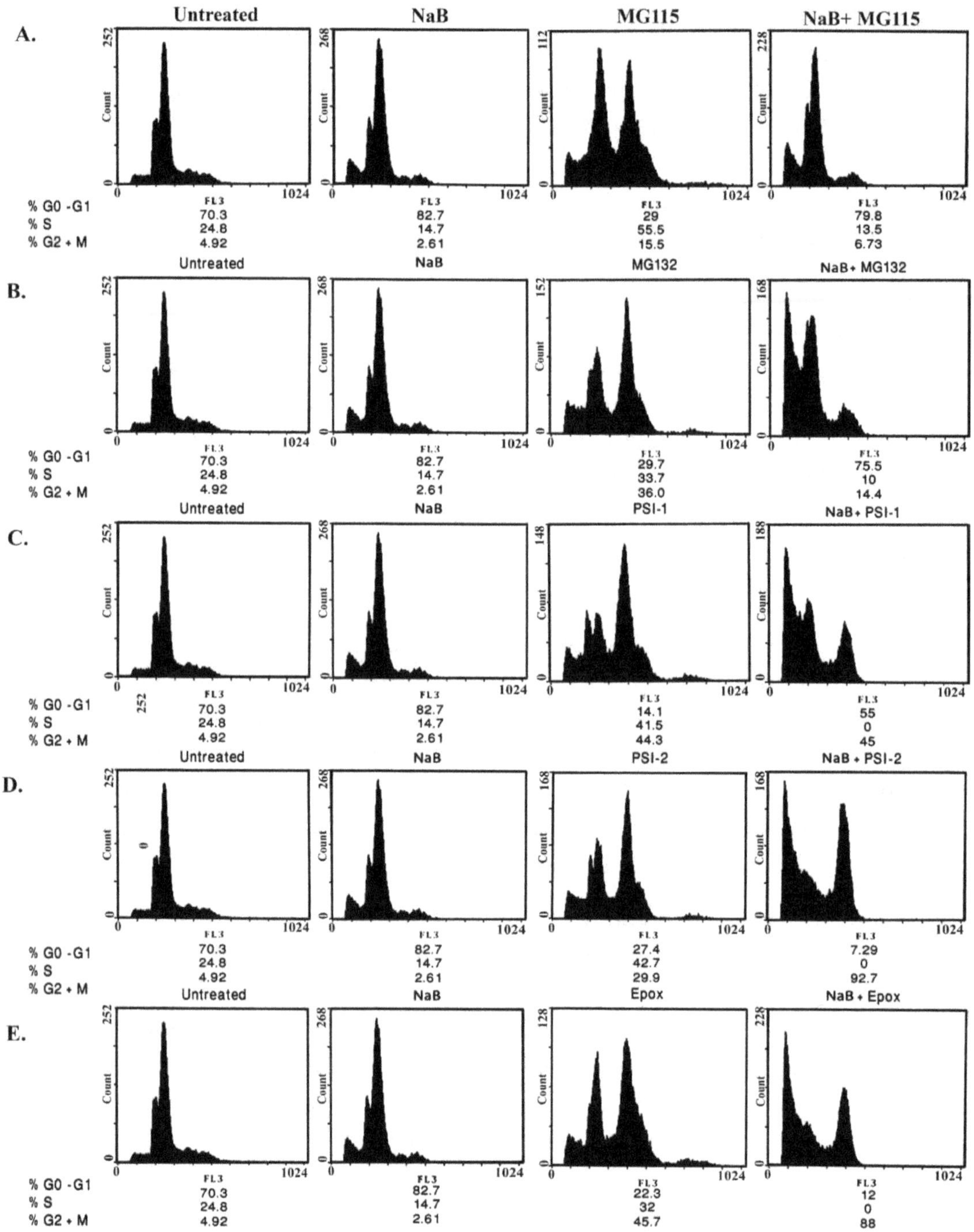

Figure 4. Cell cycle distribution of human colorectal cancer SW837 cell treated with NaB, proteasome inhibitors, and their combinations. SW837 cells were plated (5×10^5 cells/well) into 24-well plates and incubated at 37°C in a non-CO_2 incubator. After 18 h, the cells were treated individually with NaB (3.0 mM), MG115 (1.0 μM), MG132 (1.0 μM), PSI-1 (0.1 μM), PSI-2 (1.5 μM), and epoxomicin (12 nM) or treated with the combinations NaB/MG115 (3 mM/1.0 μM), NaB/MG132 (3.0 mM/1.0 μM), NaB/PSI-1 (3.0 mM/0.1 μM), NaB/PSI-2 (3.0 mM/1.5 μM), and Na/epoxomicin (3.0 mM/12 nM) for 72 h. At least duplicate samples were analyzed and 20,000 events were scored for each sample. The vertical axis represents the relative number of events and the horizontal axis represents the fluorescence intensity. The percentage of cells in different cell cycle phases was calculated using Phoenix statistical software package. A-E: Single and combined treatments with NaB and MG11S, MG132, PSI-1, PSI-2 or Epox, respectively.

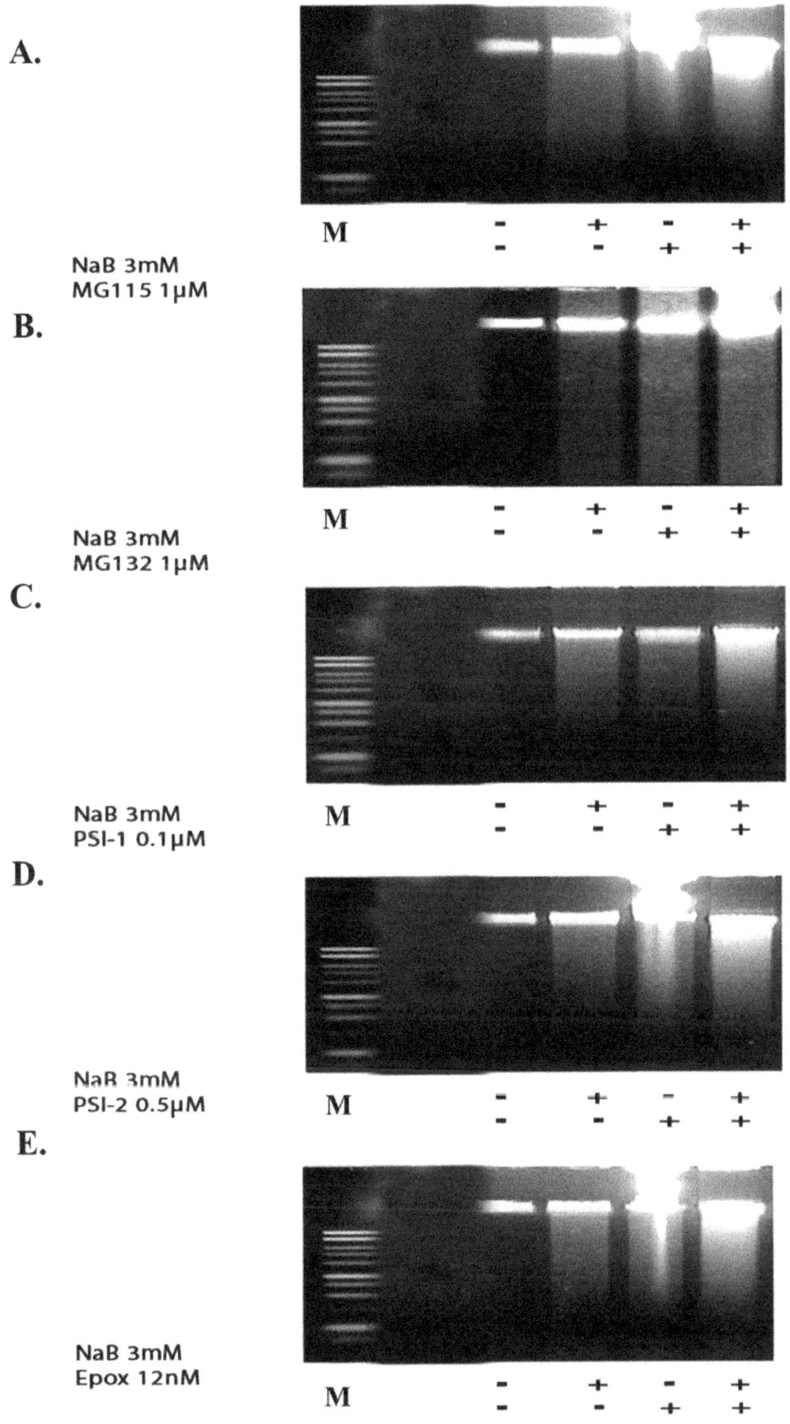

Figure 5. Assessment of apoptosis in human colorectal SW837 cells treated with NaB, proteasome inhibitors, and their combinations. SW837 cells were plated (5×10^5 cells/well) into 24-well plates and incubated at 37°C in a non-CO2 incubator. After 18 h, the cells were treated individually with NaB (3.0 mM), MG115 (1.0 μM), MG132 (1.0 μM), PSI-1 (0.1 μM), PSI-2 (1.5 μM), and epoxomicin (12 nM) or treated with the combinations NaB/MG115 (3 mM/1.0 μM), NaB/MG132 (3.0 mM/1.0 μM), NaB/PSI-1 (3.0 mM/0.1 μM), NaB/PSI-2 (3.0 mM/1.5 μM), and Na/epoxomicin (3.0 mM/ 12 nM) for 72 h. DNA fragments were extracted and analyzed on 1.0% agrose gel. A-E: Single (+) and combined (++) treatments with NaB and MG115, MG132, PSI-1, PSI-2 or Epox, respectively.

Author details

Mohammad Salah Abaza[1], Aneela Afzal[2] and Mohammad Afzal[1]*

*Address all correspondence to: afzalq8@gmail.com

1 Department of Biological Sciences, Faculty of Science, Kuwait University, Kuwait

2 Advanced Imaging Center, Oregon Health Sciences University, Portland, USA

References

[1] Yasueda A, Shinzaki S, Iijima H, Mizushima T, Nishimura J, Hiyama S, Ohno S, Ito T. Safety of emulsifying lipid formation containing omega-3 polyunsaturated fatty acdis for patients with Crohn's disease. Anticancer Research. 2016;**36**(7):3753-3759

[2] Neuwirthova J, Gal B, Smilek P, Urbankova P, Kostrica R. Anticancer effect of fish oil-a fable or the truth? Klinical Onkology. 2016;**29**(2):100-106. https://www.ncbi.nlm.nih.gov/pubmed/27081798.

[3] Sheng H, Chen X, Liu B, Li P, Cao W. Omega-3 polyunsaturated fatty acids enhance cisplatin efficacy in gastric cancer cells by inducing apoptosis via ADORA1. Anticancer Agents of Medical Chemistry. 2016;**16**(9):1085-1092.

[4] Rehman K, Mohd Amin, MC, Yuen NP, Zulfakar MH. Immunomodulatory effectiveness of fish oil and omega-3 fatty acids in human non-melanoma skin carcinoma cells. Journal of Oleological Science. 2016;**65**(3):217-224. https://www.ncbi.nlm.nih.gov/pubmed/26876681.

[5] Eltweri AM, Thomas AL, Metcalfe M, Calder PC, Dennison AR, Bowrey DJ. Potential applications of fish oils rich in omega-3 polyunsaturated fatty acids in the management of gastrointestinal cancer. Clinical Nutrition. 2017,**36**(1):64-78. doi:10.1016/j.clnu.2016.01.007.

[6] Chauvin L, Goupille C, Blanc C, Pinault M, Domingo I, Guimaraes C, Bougnoux P, Chevalier S, Maheo K. Long chain n-3 polyunsaturated fatty acids increase the efficacy of docetaxel in mammary cancer cells by downregulating Akt and PKCepsilon/delta-induced ERK pathways. Biochimica et Biophysics Acta. 2016;**1861**(4):380-390.

[7] Huang Q, Wen J, Chen G, Ge M, Gao Y, Ye X, Liu C, Cai C. Omega-3 polyunsaturated fatty acids inhibited tumor growth via preventing the decrease of genomic DNA methylation in colorectal cancer rats. Nutritional Cancer. 2016;**68**(1):113-119.

[8] Xue H, Ren W, Denkinger M, Schlotzer E, Wischmeyer PE. Nutrition modulation of cardiotoxicity and anticancer efficacy related to doxorubicin chemotherapy by glutamine and omega-3 polyunsaturated fatty acids. JPEN Journal of Parentereral and Enteral Nutrition. 2016;**40**(1):52-66.

[9] Song EA, Kim H. Docosahexaenoic acid induces oxidative DNA damage and apoptosis and enhances the chemosensitivity of cancer cells. International Molecular Science. 2016, **17**,1257. doi:10.3390/ijms17081257.

[10] Molfino A, Amabile MI, Monti M, Arcieri S, Rossi Fanelli F, Muscaritoli M. The role of docosahexaenoic acid (DHA) in the control of obesity and metabolic derangements in breast cancer. International Journal of Molecular Science. 2016;**17**(4):505. doi:10.3390/ijms17040505.

[11] Pettersen K, Monsen VT, Hakvag Pettersen CH, Overland HB, Pettersen, G, Samdal H, Tesfahun AN, Lundemo AG, Bjorkoy G, Schonberg SA. DHA-induced stress response in human colon cancer cells—Focus on oxidative stress and autophagy. Free Radical, Biological Medicine. 2016;**90**:158-172.

[12] Mika A, Stepnowski P, Kaska L, Proczko M, Wisniewski P, Sledzinski M, Sledzinski T. A comprehensive study of serum odd-and branched-chain fatty acids in patients with excess weight. Obesity. 2016;**24**(8):1669-1676. https://www.ncbi.nlm.nih.gov/pubmed/27355152.

[13] Martinez-Outschoorn UE, Peiris-Pages M, Pestell RG, Sotgia F, Lisanti M P. Cancer metabolism: A therapeutic perspective. National Review of Clinical Oncology. 2017,**14** (1):11-31. doi:10.1038/nrclinonc.2016.60.

[14] Cai M, He J, Xiong J, Tay LW, Wang Z, Rog C, Wang J, Xie Y, Wang G, Banno Y, Li F, Zhu M, Du G. Phospholipase D1-regulates autophagy supplies free fatty acids to counter nutrient stress in cancer cells. Cell Death Disease. 2016;**7**(11):e2448. https://www.ncbi.nlm.nih.gov/pubmed/27809301

[15] Nishi K, Suzuki K, Sawamoto J, Tokizawa Y, Iwase Y, Yumita N, Ikeda T. Inhibition of fatty acid synthesis induces apoptosis of human pancreatic cancer cells. Anticancer Research. 2016;**36**(9):4655-4660

[16] Xiao F, Wang C, Yin H, Yu J, Chen S, Fang J, Guo F. Leucine deprivation inhibits proliferation and induces apoptosis of human breast cancer cells via fatty acid synthase. Oncotarget. 2016; **7**(39):63679-63689. doi: 10.18632/oncotraget.11626

[17] Rosolen D, Kretzer IF, Winter E, Noldin VF, Rodrigues do Carmo IA, Filippin-Monteiro FB, Cechinel-Filho V, Creczynski-Pasa TB. N-Phenylmaleimides affect adipogenesis and present antitumor activity through reduction of FASN expression. Chemical Biology Interactions. 2016;**258**:10-20

[18] Manal M, Chandrasekar KJ, Gomathi Priya J, Nanjan MJ.. Inhibitors of histone deacylase as antitumor agents:A critical review. Bioorganic Chemistry. 2016;**67**:18-42

[19] Pucci S, Zonetti MJ, Fisco T, Polidoro C, Bocchinfuso G, Palleschi A, Novelli G, Spagnoli LG, Mazzarelli P. Carnitine palmitoyl transferase-1A (CPT1A): A new tumor specific target in human breast cancer. Oncotarget. 2016;**7**(15):19982-19996. https://www.ncbi.nlm.nih.gov/pubmed/26799588

[20] Kapuvari B, Hegedus R, Schulcz A, Manea M, Tovari J, Gacs A, Vincze, B, Mezo G. Improved in vivo antitumor effect of a daunorubicin—GnRH-III bioconjugate modified by apoptosis inducing agent butyric acid on colorectal carcinoma bearing mice. Investigational New Drugs. 2016;**34**(4):416-423

[21] Velazquez OC, Lederer HM, Rombeau JL. Butyrate and the colonocyte: Production, absorption, metabolism, and therapeutic implications. Advanced Experimental Medical Biology. 1997;**427**:123-134

[22] Macfarlane GT, Gibson GR. Microbiological aspects of the production of short-chain fatty acids in the large bowel. In: Physiological and Clinical Aspects of Short-Chain Fatty Acids. Cambridge: Cambridge University Press; 1995. pp. 87-105

[23] Pouillart PR. Role of butyric acid and its derivatives in the treatment of colorectal cancer and hemoglobinopathies. Life Science. 1998;**63**:1739-1760

[24] Kruth J, Defer N, Tichonicky L. Molecular and cellular action of butyrate. Comptes Rendus des Seances de la Societe de Biologie et de ses Filiales. 1992;**186**:12-25

[25] Velazquez OC, Rombeau JL. Butyrate: Potential role in colon cancer prevention and treatment. Advances in Experimental Medicine and Biology. 1997;**427**:169-181

[26] Abaza MSI. Augmentation of the anticancer effects of proteasome inhibitors by combination with sodium butyrate in human colorectal cancer cells. Experimental and Therapeutic Medicine. 2010;**1**:675- 693

[27] Ciechanover A. Proteolysis from the lysosome to ubiquitin and the proteasome. National Review of Molecular Cell Biology. 2005;**6**:79-87

[28] Adams J. The proteasome: A suitable antineoplastic target. National Review of Cancer, 2004;**4**:349-360

[29] Richardson PG, Sonneveld P, Schuster MW, Irwin D, Stadmauer EA, et al. Bortezomib or high-dose dexamethasone for relapsed multiple myeloma. New England Journal of Medicine. 2005;**352**:2487-2498

[30] Goy A, Younes A, McLaughlin P, Pro B, Romaguera JE, et al. Phase II study of proteasome inhibitor bortezomib inhibitor in relapsed or refractory B-cell non-Hodgkin's lymphoma. Journal of Clinical Oncology. 2005;**23**:667-675

[31] Hacihanefioglu A, Tarkun P, Gonullu E. Acute severe cardiac failure in a myeloma patient due to proteasome inhibitor bortezomib. International Journal of Hematology. 2008;**88**:219-222

[32] Enrico O, Gabriele B, Nadia C, Sara G, Daniele V, et al. Unexpected cardiotoxicity in hematological bortezomib-treated patients. British Journal of Haematology. 2007;**138**:396-397

[33] Richardson PG, Briemberg H, Jagannah S, Wen PY, Bariogie B, et al. Frequency, characteristics, and reversibility of peripheral neuropathy during treatment of advanced multiple myeloma with bortezomib. Journal of Clinical Oncology. 2006;**24**:3113-3120

[34] Lonial S, Waller EK, Richardson PG, PG, Jagannath S, Oriowski RZ, et al. Risk factors and kinetics of thrombocytopenia associated with bortezomib for relapsed, refractory multiple myeloma. Blood. 2005;**106**:3777-3784

[35] Papandreou CN, Daliani DD, Nix D, Yang H, Madden T, et al. Phase I trial of the proteasome inhibitor bortezomib in patients with advanced solid tumors with observations in androgen-independent prostate cancer. Journal of Clinical Oncology. 2004;**22**:2108-2121

[36] McConkey D, Zhu K. Mechanisms of proteasome inhibitor action and resistance in cancer. Drug Resist Updates. 2008;**11**:164-179

[37] Andrews PA, Howell SB. Cellular pharmacology of cisplatin: Perspectives on mechanisms of acquired resistance. Cancer Cell. 1990;**2**:35-43

[38] Miller AA, Kurschel E, Osieka P, Schmidt CG. Clinical pharmacology of sodium butyrate in patients with acute leukemia. European Journal of Cancer Clinical Oncology. 1987;**23**:1283-1287

[39] Daly K, Shirazi-Beechey SP. Microarray analysis of butyrate-regulated genes in colonic epithelial cells. DNA Cell Biology. 2006;**25**:49-62

[40] Davie JR. Inhibition of histone deacetylase activity by butyrate. Journal of Nutrition. 2003;**133** (Suppl 7):2493s

10

Correction of Fatty Acids Metabolism as Treatment Strategy of Autism

Afaf El-Ansary and Hanan Qasem

Abstract

Autism is a neurodevelopmental disorder clinically presented as abnormalities in social interaction and communication, repetitive behaviors, usually accompanied by various neurobehavioral disorders, such as learning disability, hyperactivity and anxiety.

It is well known that more than 50% of human brain weight is composed of lipids with a remarkably high content of long-chain polyunsaturated fatty acids (LCPUFA). Adequate supply of different fatty acids and lipids is critically needed by developing brain to achieve normal growth. Essential polyunsaturated fatty acids (PUFAs) are critical for normal prenatal brain development. There has been increasing evidence that impairment of PUFAs metabolic pathway could affect the normal function of nervous system which is related to pathogenesis of autism.

Studies have demonstrate that autistic patients may exhibit abnormal PUFAs metabolism, which manifests as varying impaired levels of lipid mediators such as prostaglandins, eicosanoids, and isoprostanes in serum and plasma of autistic patients.

Consequently, interventions related to metabolic correction of fatty acids, phospholipids, prostaglandins, eicosanoids, and isoprostanes as fatty acids–derived signaling molecules were discussed in details with special reference to Omega-3 Fatty Acids supplementation and its recognized role in the correction of oxidative stress, neuroinflammation, glutamate excitotoxicity as ascertained etiological mechanisms in autism.

Keywords: fatty acids, omega-3, omega-6, prostaglandins, eicosanoids, isoprostanes

1. Introduction

In the last decades, researchers have been focused on lipids to make clear idea about it in both physiological and disease sides. Until now, 600 molecular species have been discovered from human plasma described as lipidome [1, 2]. Lipidome provides a comprehensive classification of lipids with their structure and function. About 60% of dry human brain is composed approximately from lipids with over 20% polyunsaturated fatty acids [3, 4].

From biochemistry point, PUFAs are type of simple lipid that contain one or more double bonds in *Cis* configuration. PUFAs are two classes omega-3 and omega-6 and these classes do not convert to other forms and play important roles in biochemical changes in the body. Omega-3 and omega-6 are dietary essential fatty acids because they cannot be synthesized by human body beside they can prevent deficiency symptoms [5]. The main sources of Omega-3 fatty acids are vegetable oil and fishes. Docosahexaenoic acid (DHA, 22:6ω3) and eicosapentaenoic acid (EPA, 20:5ω3) are omega-3 fatty acid with 22 carbons and 6 double bonds (22:6n-3). While vegetables are the main sources of omega-6, arachidonic acid (AA; 20:4ω6) is an omega-6 fatty acid with 20 carbons and 4 double bonds. These two fatty acids are the predominant long-chain (20 and 22 carbons) PUFAs in human brain [6].

In the last years, the interest in the health consumption for Omega-3 has led to more researches and manufacturing of these fatty acid as supplement foods. The European committee has been suggested that the minimum requirement of omega-3 and omega-6 is approximately 0.5 and 1% of energy intake, respectively. PUFAs are now regarded as nutritionally essential fatty acids [7]. Deficiency in these fatty acids causes dermatitis, growth retardation and infertility. They play critical role as second messengers in the process of signal transductions, structural component of ceramide and specific role in membrane function. These essential fatty acids found in the diet in the form of α-linolenic acid LA (n-3) and linoleic acid LNA (n-6). These fatty acids contain 18 carbon atoms which can be metabolized to more highly unsaturated members of their family mainly arachidonic [8] and docosahexaenoic acid [1]. The pathway takes place mainly in the liver and may be occur in the other tissues as well. In endoplasmic reticulum, the conversion of LA to AA occurs, this step consists of sequential alternating elongation and desaturation reactions catalyzed by fatty acid elongase and desaturase. DHA metabolic process occurs via separate channeled carnitine-dependent mitochondrial pathway. The outer mitochondrial membrane could well be the sole site for DHA. PUFAs accumulate in brain during myelination process. The turnover of PUFAs is unknown, but studies suggest that it is high because the huge demand of them especially in developmental stage of brain. The most important PUFA for infants is DHA. Clinical studies have shown that infants who feed milk-containing DHA in it have higher neurodevelopmental scores compare to other who do not have DHA in their feed. AA and DHA do not accrete in adult brain and plasma AA and DHA only need to replace brain consumption. About 18 and 4 mg/day are the estimate AA and DHA that up taken by brain from plasma unesterified form, respectively. Phospholipases family is responsible for releasing of AA and DHA from brain phospholipid membrane.

2. The importance of the omega-6 and omega-3 ratio

The differences between omega-6 and omega-3 acids are very small and may be insignificant. They exert opposite effects, ω-3 PUFAs work as anti-inflammatory agent and ω-6 PUFAs as pro-inflammatory agent. These opposing effects are not easily explained. It was suggested that the variation between ω-6 and ω-3 PUFA is based on the molecular basis in particular, to recognize various PUFAs [9]. The dietary deficiency of ω-3 fatty acids, as well as the particular roles of ω-6 and ω-3, becomes an important subject, and their ratio takes a deeper look into the disease issues. The optimal recommended ratio between ω-6 and ω-3 fatty acids has many aspects. One aspect is the recommendation for total daily dietary intake in various phases of life (e.g., infancy, pregnancy, adulthood and old age). Another aspect is the optimal ratio of PUFAs as a food supplement or treatment [10]. PUFAs are used in the body in a variety of conditions, such as in dermatological diseases and in cardiovascular disorders. One particular area is the role of PUFAs in the brain and the utility of PUFA to protect and stabilize the neuronal membrane in health and in disease. PUFAs play a critical role in the central nervous system (CNS) and CNS conditions. Many researchers have demonstrated that various PUFAs mediate a lot of process in brain. Some studies examine the best ratio between ω-6 and ω-3 PUFAs to help body to do its role in good way. The required ratio of ω-6 and ω-3 may differ when used for different tissues or functions. 1:1 is the optimal ratio for preventing cardiovascular diseases. 4:1 is the optimal ratio for brain-mediated functions and has protective and stabilizing effects on the neuronal membrane. The ratio between those PUFAs should be stable to maintain human health [11] (**Figure 1**).

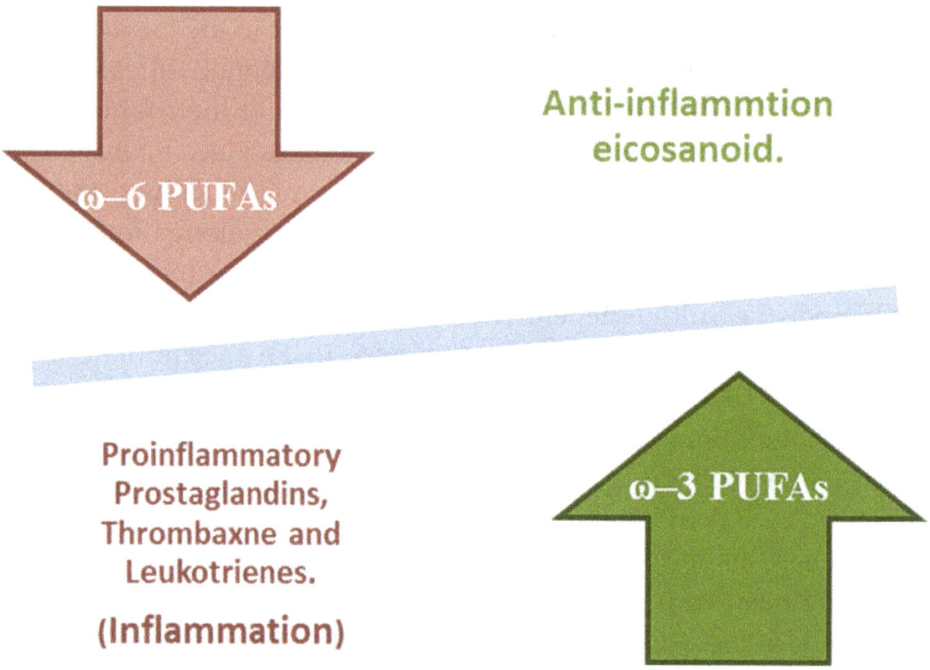

Figure 1. The optimal omega-6 to omega-3 PUFAs balance in the body.

Recently, it is well accepted that early alterations of the intestinal microbiota composition with environmental factors such as Cesarean delivery, bottle feeding, diet and abuse of antibiotic, can induce gut dysbiosis that might be linked to abnormal neurodevelopment and brain dysfunction [12]. The role of gut-brain axis in the etiology of autism is ascertained and related to intestinal dysbiosis as autistic feature [13]. Based on the fact that gut microbiota are greatly affected with diet, it was interested to discuss the role of ω-3/ω-6 PUFAs on microbial composition of the gut. While some studies demonstrate that ω-6-rich diet shows negative impact on gut microbiota through the induction of overgrowth of Bacteroidetes and Firmicutes as bacterial species related to gastrointestinal inflammation frequently occurs in autistic patients [14, 15], ω-3 was proved to induce the growth of bifidobacteria and Lactobacillus as bacterial species that dampening inflammatory responses [16].

3. Lipid mediators

The releasing of AA happens in response to inflammation, ischemia and excitotoxicity, while DHA release occurs in response to ATP, Bardykinin and cholinergic and serotonergic receptors. These 20 carbon atoms are precursor of lipid mediators that regulate inflammation and immune system. These mediators include eicosanoids and docosanoid and synthesized by many different enzymes and contribute or protect from the risk of inflammation [17, 18]. Cyclooxygenase [19], lipooxygenase (LOX) and cytochrome P450 are the main enzymes involved in lipid mediator's synthesis [17]. COX facilitates conversion of AA to prostaglandin E2 (PGE2). There are two types of COX: COX-1 and COX-2 and their expression differ according to tissues and body situation. Expression of COX-1 occurs in all tissues, while basal COX-2 expression in neurons or in response to inflammation [20, 21] (**Figure 2**).

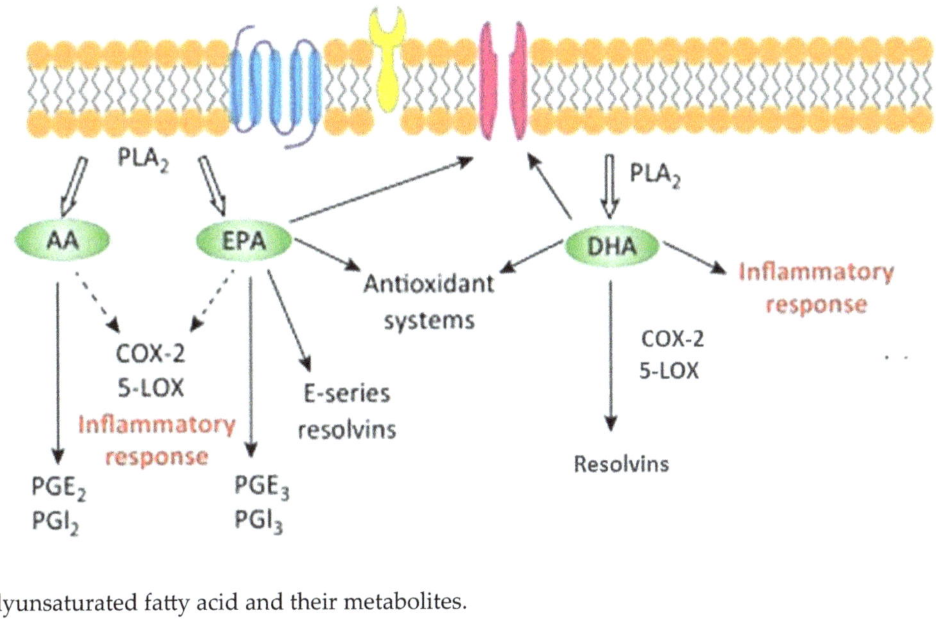

Figure 2. Polyunsaturated fatty acid and their metabolites.

Fatty acids and their mediators have numerous functions in the central nervous system (CNS), including a role in inflammation, glucose production, food intake and in analgesia, beside synaptic function; they activate or suppress neurotransmitter release including glutamate, GABA, monoamine neurotransmitters, opioids and acetylcholine. They also lead to microglia activation and the production of pro-inflammatory cytokines in the hippocampus.

Experimental studies have indicated that DHA is involved in learning and memory, but the real mechanisms underlying these effects are not well studied. It has protection effect by enhancing neuronal survival neurogenesis. DHA is the main PUFA in phosphatidylserine. It plays role in suppression of inflammatory cytokine expression and can invade macrophage and microglia. It also blocks macrophage and microglia-induced activation of NF-κB in the CNS of rodents with neuroinflammation [22].

AA and DHA are rapidly incorporated in the nervous tissue of retina and brain during the brain's growth spurt, which mainly takes place from the last trimester of pregnancy up to 2 years of age. Beyond development of the central nervous system, AA and DHA fatty acids may influence brain function throughout life by modifications of neuronal membrane fluidity, membrane activity-bound enzymes, number and affinity of receptors, function of neuronal membrane ionic channels, and production of neurotransmitters and brain peptides.

4. Abnormal fatty acid metabolism as etiological mechanism in autism

To understand the effect of DHA and AA on brain development and cognition, a lot of interests have been given to the role of PUFAs in infancy and early childhood life. Brain development in infants and children occurs in specific stages during early life. Unesterified ω-3 and ω-6 fatty acid content of the brain increase considerably during development. For proper CNS function high demand, sufficient supply of the essential PUFAs and proper ratio of AA to DHA are needed as critical process in the early life.

Many studies have observed a relationship between plasma or serum n-3 and n-6 PUFAs imbalances and neurodevelopmental disorders such as autism [23]. As mentioned above, DHA and AA play an important role in the nervous system, including retinal development and vision, neurogenesis and neuronal differentiation, neural plasticity, signal transduction, inflammation, learning and memory. These functions may be regulated by a number of gene products activated by PUFAs during development. Some clinical trials have been conducted on the beneficial effect of dietary ω-3 PUFA supplementation on behavior in various neurodevelopmental disorders, including autism [24], but trials with larger sample size are critically requested [3].

There is emerging evidence that fatty acid metabolism and homeostasis are altered in autism due to genetic defects, dietary insufficiency and abnormality in the fatty acid metabolizing enzymes [25–27]. It is well known that alterations of fatty acid metabolism can affect the normal brain function especially during the development. A direct relationship between impaired fatty acid metabolism at various sites and pathophysiology of autism was repeatedly documented.

PLA 2 is an important enzyme that maintains the membrane phospholipids. It catalyzes the release of AA, a precursor of key lipid mediators such as PGs from the *sn-2* position of phospholipids [28, 29], and it has been shown to play a critical role in neuronal plasticity [30]. Activation of PLA 2 with the excitatory neurotransmitter glutamate usually resulted in a remarkable increase of AA with concomitant impairment of membrane phospholipids [31]. Additionally, both DHA and AA can be released in the presence increased levels inflammatory cytokines [32]. ω-3 PUFA supplementation appears to provide a promising neuroprotective treatment strategy related to the reduction of neuro-progression mediated by excitotoxicity and oxidative damage (PLA_2 and PUFA supplementation in UHR individuals) [33].

COX-2 has been widely studied as important enzyme that plays critical role in the body. COX-2 is highly expressed in tissues that under stress of inflammation or neurotoxicity. In study done by Boudrault et al. [34], COX-2 was shown to be modulated by ω-3 PUFA in mice brains beside its ability to control ω-6 PUFA level. These results suggest a potential mechanism by which ω-3 PUFA mediates its biological effects on inflammation or neurotransmission. ω3 PUFA suppresses the production of interleukin 1 (IL-1β) by suppressing the IL-1β mRNA, as well as the expression of Cox2 (cytooxygenase) mRNA that is induced by IL-1β [10].

LOX is a group of iron-containing dioxygenases that catalyze the addition of oxygen to AA, DHA and other PUFAs [35]. LOXs have different isoforms according to the type of tissue where they are located. 5-LOX has been shown to play important roles in human pathology by virtue of its central role in leukotriene biosynthesis. Leukotrienes have attracted much attention because of their powerful biological effects in vitro and in vivo. These lipid mediators are active in the low level and elicit a cellular proinflammatory and immune modulatory responses. 5-LOX and leukotrienes have been proved to play role in the pathogenesis of many human acute and chronic inflammatory diseases such as asthma, rheumatoid arthritis, inflammatory bowel disease, psoriasis, dermatitis, nephritis, atherosclerosis, autism and cancer [36–39]. The anti-inflammatory properties of ω3 PUFAs, especially EPA, are due to competition with AA as a substrate for 5-lipoxygenase. The eicosanoids are considered a link between PUFA, inflammation and immunity. In addition, ω3 PUFAs have effect on reduce leukotrienes level [10]. From molecular genetic studies of the Icelandic population, variant 5-LOX genotypes were found to be associated with increased atherosclerosis, and dietary ω6 PUFAs promoted, whereas marine ω3 PUFAs inhibited, this effect [40].

PGE 2 is a signaling molecule that diffuses rapidly through the membranes and exerts its diverse effects in the brain via four G-protein coupled EP receptors: EP1, EP2, EP3 and EP4 [41, 42]. The role of PGE 2 signaling in early brain development including formation of dendritic spines and neuronal plasticity is also documented [43, 44]. Tamiji and Crawford [45] reported that expression of the four G-protein coupled EP receptors was found to be significantly increases in the mouse brain during early neurogenesis (11–15 embryonic day). This might indicate that the PGE 2 signaling pathway may have an important role during early brain development. Early brain pathology demonstrates abnormality of certain brain regions in autism [46–48]. Among these regions are cerebellum, medulla and pons which start to develop at the early stages of the neurogenesis (embryonic day 12), in addition to thalamus, hypothalamus, hippocampus and entorhinal cortex that begin developing at around day 15 [49]. A direct involvement of COX-2/PGE 2 signaling pathway in the development of these structures still remains to be ascertained.

The first reaction of mitochondrial fatty acid β-oxidation (FAO) in mitochondria is catalyzed by acyl-CoA dehydrogenase. Four different dehydrogenases participate in the complete degradation of fatty acids in mitochondria. They are flavin adenine dinucleotide (FAD)-containing enzymes which are structurally and functionally related only differ in their substrate specificities. These are, short-chain acyl-CoA dehydrogenase (SCAD), medium-chain acyl-CoA dehydrogenase (MCAD), long-chain acyl-CoA dehydrogenase (LCAD) and very long-chain acyl-CoA dehydrogenase (VLCAD), reflect the acyl chain lengths of their preferred substrates. Deficiency of long-chain acyl-CoA dehydrogenase (LCAD), as one of these dehydrogenases, is suspected to have a link with the development of autism [50].

Fatty acid β-oxidation is the major pathway to produce ATP and reducing power from different chain lengths of fatty acids [51, 52]. Transport of fatty acids from the cytoplasm into mitochondria is rate limiting step of FAO, and it requires carnitine as acyl carrier and carnitine palmitoyltransferase I (CPT1), which catalyzes the first regulatory reaction in this process. Trimethyllysine hydroxylase (TMLHE) is a second enzyme that catalyzes the first step of carnitine biosynthesis [53]. It is very interesting that several studies had reported that mutation of TMLHE is present in human population with high rate [54]. There is great evidence demonstrating an association between impaired FAO and autism [50, 55, 56]. Individuals with autism show altered levels of blood or plasma carnitine and acyl-carnitine, as a phenotype related to impaired long chain FAO. On the other hand, FAO-deficient children exhibit autistic features such as developmental delay [57]. Recently, Xie et al. [58] reported that efficient FAO is critically needed for the maintenance of neuronal stem cell (NSC) homeostasis in the mammalian embryonic neocortex. They suggested that linkage of NSC homeostatic mechanisms with inborn errors of metabolism (IEM) of developmental brain disorders has clinical implications. An increased risk of autism was found to be associated with TMLHE deficiencies [54, 59]. They also recorded that enhanced oxidative stress was observed in NSC mitochondria with impaired FAO activity, suggesting that impairment of NSC self-renewal occurs due to oxidative stress as an accepted etiological mechanism in autistic children [26, 27, 60].

Another evidence for fatty acids metabolic disturbances as one potential etiological mechanism in autism is the remarkable increase of adipic and suberic acids, as two dicarboxylic acids produced by the omega (ω)-oxidation pathway, a minor catabolic pathway for medium-chain fatty acids that becomes more important when β-oxidation is impaired [51, 61] (**Figure 3**). Based on the previously discussed association between impaired FAO and autism [50, 56, 62, 63], it was suggested that altered β-oxidation can increase the activity of ω fatty acid oxidation, thus leading to increased production of adipic and suberic acid [58, 61]. There is a strong body of evidence between mitochondrial dysfunction and PUFAs transport and metabolism in autism. For this reason, shifting from β to ω-oxidation pathway considering as an emergency pathway that protect cell from deleterious effects of mitochondrial enzyme

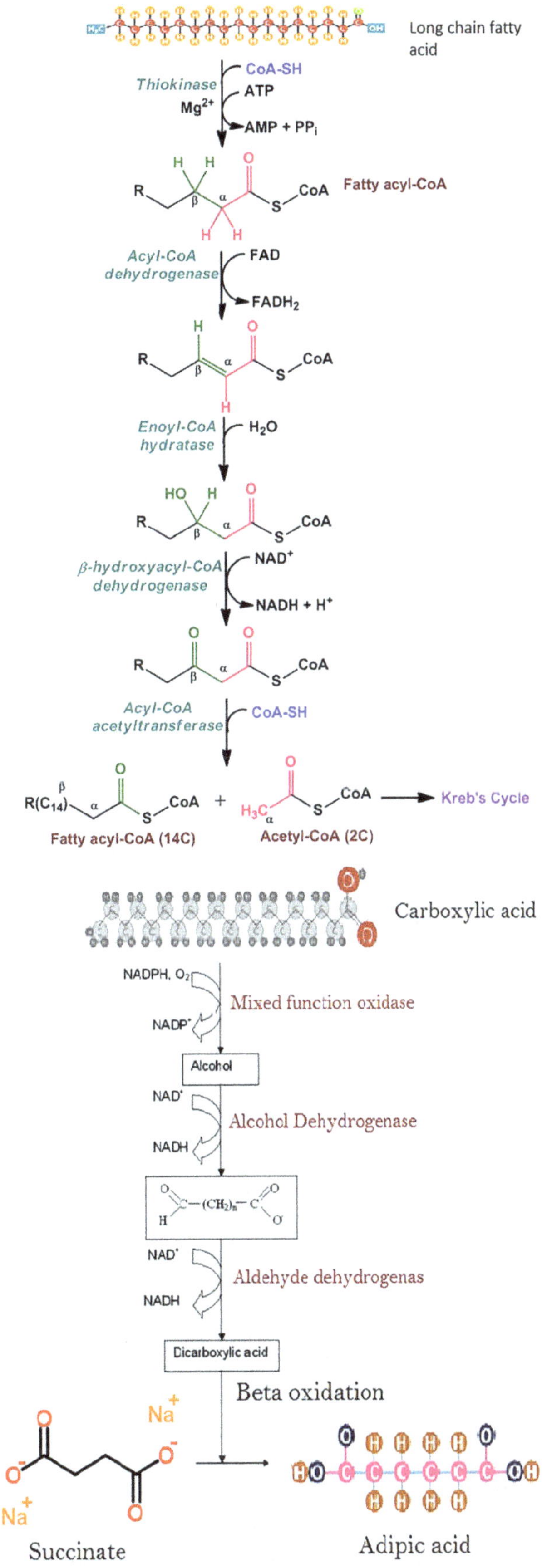

Figure 3. Beta oxidation (up) and omega oxidation [44] of fatty acids.

dysfunctions. So, researchers those days are looking for biomarkers that help to understand the activity of this pathway. The attention has focused in adipic and suberic acid measurements and their correlation with other important determiners that defined in autism.

Increased level of adipic acid has shown to inhibit the activity of both L-glutamate decarboxylase [64] and GABA transaminase [65], leading to impaired glutamate/GABA ratio that might induce glutamate excitotoxicity, as consistent autistic feature in animal model and individuals with autism, through the overstimulation of glutamate receptors [66–69].

5. Fatty acids and brain neurochemistry

5.1. Serotonin

The reported impaired profile of PUFAs and their related lipid mediators in autistic children can be related to their abnormal neurotransmitter physiology. In animals studies, feeding on essential fatty acids diet resulted in serotonin depletion in the frontal cortex of pre-adolescent but not in post-pubescent rats, suggesting a role of n-3 DHA and n-6 AA in neurotransmitter synthesis or turnover [70]. Based on this lower n-3, DHA can be related to the absence of age-dependent changes of brain serotonin synthesis in autistic children and hyperserotonemia as biomarker of clinical severity of autism [71].

5.2. Gamma-amino butyric acid (GABA)

Takeuchi et al. reported that ω-3 DHA deficiency is related to the altered GABAergic activity in autistic patients [72]. This might be through the prevention GABAA receptor blocking repeatedly reported in this disorder [73]. This provides an important link between PUFAs and pathogenesis of autism [74, 75]. A second mechanism of interaction between PUFAs and GABA neurotransmission is through the actions of phospholipase A2 (PLA2) a membrane phospholipid hydrolyzing enzyme. PLA2 is thought to inhibit GABAA receptor function by reducing chloride flux in the cerebral cortex [76]. Based on this, ω-6 AA usually induces neuronal excitability through the activation of PLA2 or phospholipase C (PLC) and inhibition of GABAA receptors [77]. This can be easily related to the imbalanced GABAergic/glutamatergic in autistic patients [66].

5.3. Glutamate

Multiple early studies demonstrated that activation of postsynaptic glutamate receptors by glutamate induces release of AA from membrane phospholipids either directly, by activation of phospholipase A2, or indirectly from degradation of diacylglycerol [78, 79]. On the other hand, AA has been shown to increase glutamate release from synaptosomes [80, 81] through the stimulation of the inositol phospholipid metabolism or activation of protein kinase C.

Elevation of AA can be easily related to glutamate excitotoxicity and glutathione depletion as etiological mechanisms of autism. Recently, elevation of PLA2 was recorded in plasma of autistic patients compared to healthy controls [66, 82]. This enzyme is involved in the selective release of AA from phospholipids such as PC, PS and PE [36, 83]. Higuchi et al. [84] proved that AA is involved and related to glutamate-induced glutathione depletion and the subsequent cell death through the accumulation of hydroperoxy eicosatetraenoic acids (HPETE) as AA reactive oxygen species (ROS) or hydroperoxides. This can be supported by the recent record of Gebremedhin [85] which reported that astrocytes of neonatal rat brain express message and protein for cytochrome P450 4A ω-hydroxylase CYP4A2/3 and synthesize 20-HETE when incubated with AA and this usually enhanced through the activation of metabotropic glutamate receptors.

5.4. Dopamine

Omega-3 intake has shown therapeutic effects through dopamine neurotransmission in major depression. The antidepressant efficacy of ω-3 supplementation may raise the possibility that they may have specific value for major depressive disorder with a dopaminergic system deficit [86]. This finding may have important implications for therapeutic strategies involving augmentation of standard antidepressant medications with fish oil. ω-3 has beneficial effect as detoxification agent that remove bad effect of reactive oxygen species in Parkinson disease [87]. PUFAs have been specially associated with dopamine activity in frontal lobe of brain. In adolescents, dietary n-3 PUFA deficiency produced a modality selective and task-dependent impairment in cognitive and motivated behavior distinct from the deficits observed in adults. This deficiency affected expression of dopamine-related proteins. Adolescent behavior and dopamine availability are uniquely sensitive to dietary omega-3 fatty acid deficiency [88].

6. PUFAs and BDNF interact with each other

Brain-derived neurotrophic factor (BDNF) showed alteration levels in sample of autistic patients, and it is involved in the regulation of neuronal development and plasticity and has a role in learning and memory. In first several years, serum BDNF concentrations increased in healthy children and then slightly decreased after reaching the adult level. In the patients with autism, mean levels were significantly lower in children compared with healthy adults [89]. Many researches [90–92] indicated that BDNF plays a critical role in the diagnosis of autism. PUFAs and BDNF interact with each other since PUFAs are known to augment the levels of BDNF in the brain [93]. PGE2 derived from AA, induced release of BDNF from glial cells and astrocytes through a G-protein-coupled receptor and then affect on the whole signaling pathway inside cell [94]. PGE2 contributes to BDNF upregulation in neurons following nerve injury in animal models, which facilitates the synthesis of BDNF in primary sensory neurons to initiate repair of damaged neurons and neuronal regeneration [95]. Other PUFA metabolites especially lipoxin A4 (LXA4), resolvins and protectins interact with BDNF. These interactions provide anti-inflammatory effect when the body needs it [96]. Deficiency in ω-3 PUFA intake is linked to decreased BDNF content, and low BDNF levels have been described after prenatal stress [97] (**Figure 4**). Glucocorticoids have been related to such an effect, since corticosterone is able to down-regulate both mRNA and protein BDNF [98]. Over-expressing of glucocorticoids showed an increased anxiety-like behavior [99]. Larrieu and colleagues have clarified that n-3 PUFA deficiency can influence neuronal cortical morphology and depressive-like

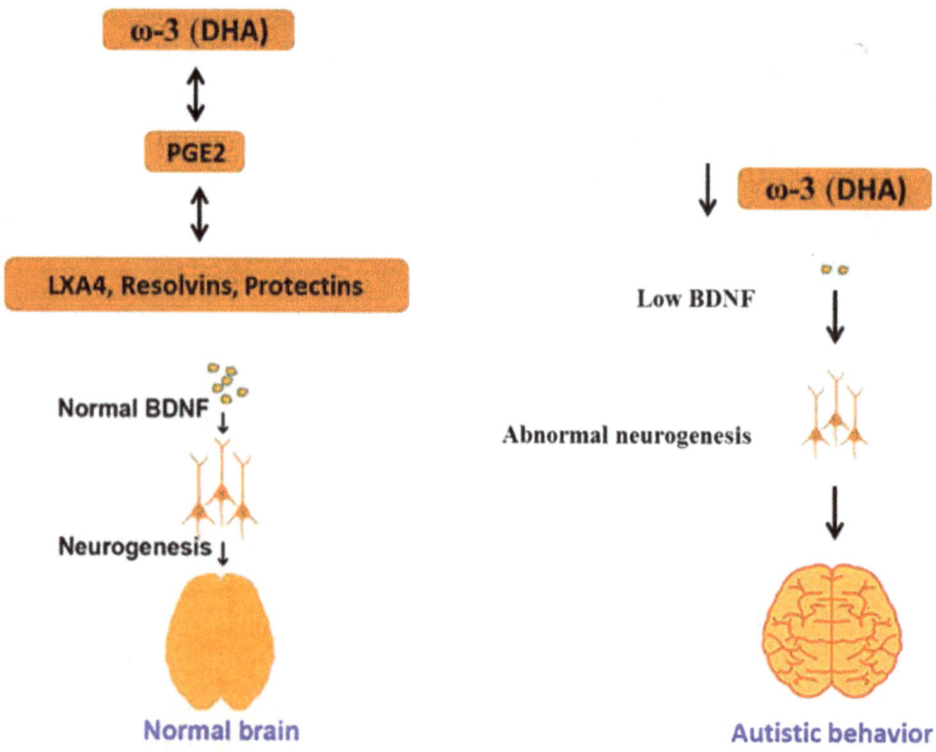

Figure 4. Interaction between PUFAs metabolites and BDNF.

behavior through corticosterone secretion. Furthermore, they showed that diet with low ω-3 induces a phenotype of social deficits and emotional behavior which is observed in autistic patients [100].

7. Amelioration of impaired lipid metabolism as treatment strategy of autism

It is well accepted that imbalances in ω-3 and ω-6 fatty acids are one of the etiological mechanisms in autism and are directly related to the abnormal behavioral severity of these patients. Interestingly, omega-3 and omega-6 fatty acids supplementation resulted in increased level of these fatty acids in the blood, reduced the elevated AA:DHA ratio ameliorates some behavioral deficits such as eye contact, hyperactivity, concentration and motor skills in autistic patients [101]. This can find support in the more recent study of Yui et al. [102, 103] which proved that large doses of AA added to DHA may improve impaired social interaction in individuals with autism, and Amminger et al. [104] who suggest that the use of pure omega-3 PUFAs (without any AA) may be beneficial in autism.

In a recent report of Klein and Kemper [105], supplementation with ω-3 fatty acids is more effective than risperidone as pharmacological drug with side effects. ω-3 fatty acids demonstrate many ameliorating effects presented as more social interaction, less irritability and more flexibility [106]. Due to the lack of evidence of effectiveness from large randomized clinical trials, the safety, and low cost of ω-3 fatty acids, clinicians can encourage families' use of supplemental ω-3, but more frequent and completely blind trials are requested to move ω-3 fatty acids from tolerated to recommended supplement for the treatment of autistic patients [105].

Due to the strong interaction between diet and the gut microbiota, it has been suggested that the role of dietary changes in influencing brain biochemistry and behavior may be mediated through changes in gut microbiota composition and function [107]. In addition to improving brain function, n-3 PUFA can be used as treatment strategy of autistic patients through its beneficial impact on restoring healthy gut-microbiota by inducing bifidobacteria, and lactobacillus growth, and inhibiting enterobacteria growth with subsequent anti-inflammatory responses [16].

Mediterranean diet as good source of ω-3 usually recommended as a healthy diet [108]. It consists mainly of cereals, vegetables, nuts and fruits, with moderate amount of fish and poultry and low amount of red meat. Polyphenols as major ingredients of olive oil, a common component of Mediterranean diet is known to promo its protective effect by modulating different signaling cascades among which is nuclear factor-kappaB (NF-kB), pro-inflammatory response and oxidative stress response as three etiological mechanisms repeatedly recorded in autism [109].

Moreover, carnitine supplements, as a compound normally required for fatty acids metabolism, and significantly reduced in some children with autism [55], it was effective in improving the remarkably reduced DHA and very long-chain fatty acid level of autistic subjects [110]. Unlike autistic children, ω-3 supplementation showed no beneficial effect on severe autistic adults [24, 111].

Author details

Afaf El-Ansary[1,2]* and Hanan Qasem[2]

*Address all correspondence to: afafkelansary@gmail.com

1 Central Laboratory, King Saud University, Riyadh, Saudi Arabia

2 Autism Research and Treatment Center, King Saud University, Riyadh, Saudi Arabia

References

[1] Knowles EE, Huynh K, Meikle PJ, Göring HH, Olvera RL, Mathias SR, Duggirala R, Almasy L, Blangero J, Curran JE, Glahn DC. The lipidome in major depressive disorder: Shared genetic influence for ether-phosphatidylcholines, a plasma-based phenotype related to inflammation, and disease risk. Eur Psychiatry. 2017.**43**:44-50.

[2] Quehenberger O., Dennis E.A. The human plasma lipidome. New England Journal of Medicine. 2011;**365**(19):1812-1823

[3] Schuchardt J.P., et al., Significance of long-chain polyunsaturated fatty acids (PUFAs) for the development and behaviour of children. European Journal of Pediatrics. 2010;**169**(2):149-164

[4] Tamiji J., Crawford D.A. The neurobiology of lipid metabolism in autism spectrum disorders. Neuro-Signals. 2010;**18**(2):98-112.

[5] Carlson S.E., Colombo J. *Docosahexaenoic Acid and Arachidonic Acid Nutrition in Early Development*. Advances in Pediatrics, 2016. **63**(1): 453-71.

[6] Bazinet R.P., Laye S. Polyunsaturated fatty acids and their metabolites in brain function and disease. Nature Reviews Neuroscience. 2014;**15**(12):771-785

[7] Rapoport S.I., Rao J.S., Igarashi M. Brain metabolism of nutritionally essential polyunsaturated fatty acids depends on both the diet and the liver. Prostaglandins, Leukotrienes and Essential Fatty Acids. 2007;**77**(5):251-261

[8] Sturm V., et al., DBS in the basolateral amygdala improves symptoms of autism and related self-injurious behavior: a case report and hypothesis on the pathogenesis of the disorder. Frontiers in Human Neuroscience. 2012;**6**:341

[9] Feller S.E., Gawrisch K., MacKerell A.D. Polyunsaturated fatty acids in lipid bilayers: intrinsic and environmental contributions to their unique physical properties. Journal of the American Chemical Society. 2002;**124**(2):318-326

[10] Simopoulos A.P. Omega-3 fatty acids in inflammation and autoimmune diseases. Journal of the American College of Nutrition. 2002;**21**(6):495-505

[11] Yehuda S. Omega-6/Omega-3 Ratio and Brain-Related Functions, in Omega-6/Omega-3 Essential Fatty Acid Ratio: The Scientific Evidence 2004, Karger Publishers. pp. 37-56

[12] Backhed F., et al., Dynamics and stabilization of the human gut microbiome during the first year of life. Cell Host & Microbe. 2015;**17**(5):690-703

[13] El-Ansary A., Shaker G.H., Rizk M.Z. *Role of gut-brain axis in the aetiology of neurodevelopmental disorders with reference to autism*. J Clinic Toxicol 2013. S6: 005. doi:10.4172/2161-0495.S6-005.

[14] Ghosh S., et al., Diets rich in n-6 PUFA induce intestinal microbial dysbiosis in aged mice. The British Journal of Nutrition. 2013;**110**(3):515-523

[15] Klaus J., et al., Small intestinal bacterial overgrowth mimicking acute flare as a pitfall in patients with Crohn's Disease. BMC Gastroenterology. 2009;**9**(1):1

[16] Kaliannan K., et al., A host-microbiome interaction mediates the opposing effects of omega-6 and omega-3 fatty acids on metabolic endotoxemia. Scientific Reports. 2015;**5**. doi: 10.1038/srep11276.

[17] Bennett M., Gilroy D. Lipid mediators in inflammation. Microbiology Spectrum. 2016;**4**(6). doi: 10.1128/microbiolspec.MCHD-0035-2016.

[18] de Jong A.J., et al., Fatty acids, lipid mediators, and T-cell function. Frontiers in Immunology. 2014;5:483

[19] King E., Coxon R. Determination of bilirubin with precipitation of the plasma proteins. Journal of Clinical Pathology. 1950;**3**(3):248

[20] Kyrkanides S., et al., Cyclooxygenase-2 modulates brain inflammation-related gene expression in central nervous system radiation injury. Brain Research. Molecular Brain Research. 2002;**104**(2):159-169

[21] Smith H.S. Arachidonic acid pathways in nociception. The Journal of Supportive Oncology. 2006;**4**(6):277-287

[22] Morales I., et al., *Neuroinflammation in the pathogenesis of Alzheimer disease. A rational framework for the search of novel therapeutic approaches*. 2014. **8**:1-9

[23] El-Ansary A.K., Bacha A.G., Al-Ayahdi L.Y. Impaired plasma phospholipids and relative amounts of essential polyunsaturated fatty acids in autistic patients from Saudi Arabia. Lipids in Health and Disease. 2011;**10**:63

[24] James S., Montgomery P., Williams K. *Omega-3 fatty acids supplementation for autism spectrum disorders (ASD)*. The Cochrane Library, 2011. (11):CD007992.

[25] Ming X., et al., Increased excretion of a lipid peroxidation biomarker in autism. Prostaglandins, Leukotrienes and Essential Fatty Acids. 2005;**73**(5):379-384

[26] Chauhan A., et al., Oxidative stress in autism: Increased lipid peroxidation and reduced serum levels of ceruloplasmin and transferrin-the antioxidant proteins. Life Sciences. 2004;**75**(21):2539-2549

[27] James S.J., et al., Metabolic biomarkers of increased oxidative stress and impaired methylation capacity in children with autism. The American Journal of Clinical Nutrition. 2004;**80**(6):1611-1617

[28] Murakami M., Kudo I. Phospholipase A2. Journal of Biochemistry. 2002;**131**(3):285-292

[29] Green J.T., Orr S.K., Bazinet R.P. The emerging role of group VI calcium-independent phospholipase A2 in releasing docosahexaenoic acid from brain phospholipids. Journal of Lipid Research. 2008;**49**(5):939-944

[30] Massicotte G., Modification of glutamate receptors by phospholipase A2: its role in adaptive neural plasticity. Cellular and Molecular Life Sciences CMLS. 2000;**57**(11):1542-1550

[31] Marin P., et al., Nicotine-induced inhibition of neuronal phospholipase A2. Journal of Pharmacology and Experimental Therapeutics. 1997;**280**(3):1277-1283

[32] Farooqui A.A., Horrocks L.A., Farooqui T. Modulation of inflammation in brain: a matter of fat. Journal of Neurochemistry. 2007;**101**(3):577-599

[33] Smesny, S., et al., N-3 fatty acid supplementation influences associations between membrane fatty acids and phospholipase A 2 activity in patients at risk to develop psychosis. Schizophrenia Research. 2010;**117**(2):487-488

[34] Boudrault C., et al., Cyclooxygenase-2 and n-6 PUFA are lower and DHA is higher in the cortex of fat-1 mice. Neurochemistry International. 2010;**56**(4):585-589

[35] Shimizu T., Wolfe L.S. Arachidonic acid cascade and signal transduction. Journal of Neurochemistry. 1990;**55**(1):1-15

[36] Qasem H., Al-Ayadhi L., El-Ansary A. Cysteinyl leukotriene correlated with 8-isoprostane levels as predictive biomarkers for sensory dysfunction in autism. Lipids in Health and Disease. 2016;**15**(1):130

[37] El-Ansary A., Al-Ayadhi L. Lipid mediators in plasma of autism spectrum disorders. Lipids in Health and Disease. 2012;**11**(1):1

[38] Czapski G.A., et al., The lipoxygenases: Their regulation and implication in Alzheimer's disease. Neurochemical Research. 2016;**41**(1-2):243-257

[39] Dahln S.-E., et al., Leukotrienes are potent constrictors of human bronchi. Nature. 1980;**288**(5790):484-486

[40] Dwyer J.H., et al., Arachidonate 5-lipoxygenase promoter genotype, dietary arachidonic acid, and atherosclerosis. New England Journal of Medicine. 2004;**350**(1):29-37

[41] Ashwood P., Van de Water J. A review of autism and the immune response. Clinical and Developmental Immunology. 2004;**11**(2):165-174

[42] Abramovitz M., et al., The utilization of recombinant prostanoid receptors to determine the affinities and selectivities of prostaglandins and related analogs. Biochimica et Biophysica Acta (BBA)-Molecular and Cell Biology of Lipids. 2000;1483(2):285-293

[43] Chen C., Magee J.C., Bazan N.G. Cyclooxygenase-2 regulates prostaglandin E2 signaling in hippocampal long-term synaptic plasticity. Journal of Neurophysiology. 2002;**87**(6):2851-2857

[44] Burks S.R., Wright C.L., McCarthy M.M. Exploration of prostanoid receptor subtype regulating estradiol and prostaglandin E 2 induction of spinophilin in developing preoptic area neurons. Neuroscience. 2007;**146**(3):1117-1127

[45] Tamiji J., Crawford D.A. Prostaglandin E 2 and misoprostol induce neurite retraction in Neuro-2a cells. Biochemical and Biophysical Research Communications. 2010;**398**(3): 450-456

[46] Courchesne E. Brain development in autism: early overgrowth followed by premature arrest of growth. Mental Retardation and Developmental Disabilities Research Reviews. 2004;**10**(2):106-111

[47] Courchesne E., et al., Mapping early brain development in autism. Neuron. 2007;**56**(2): 399-413

[48] Amaral D.G., Schumann C.M., Nordahl C.W. Neuroanatomy of autism. Trends in Neurosciences. 2008;**31**(3):137-145

[49] Rice D., Barone S. Jr. Critical periods of vulnerability for the developing nervous system: evidence from humans and animal models. Environmental Health Perspectives. 2000;**108**(Suppl 3):511

[50] Clark-Taylor T., Clark-Taylor B.E. Is autism a disorder of fatty acid metabolism? Possible dysfunction of mitochondrial β-oxidation by long chain acyl-CoA dehydrogenase. Medical Hypotheses. 2004;**62**(6):970-975

[51] Houten S.M., Wanders R.J. A general introduction to the biochemistry of mitochondrial fatty acid β-oxidation. Journal of Inherited Metabolic Disease. 2010;**33**(5):469-477

[52] Ito K., Suda T. Metabolic requirements for the maintenance of self-renewing stem cells. Nature Reviews Molecular Cell Biology. 2014;**15**(4):243-256

[53] Strijbis K., Vaz F.d.r.M, Distel B. Enzymology of the carnitine biosynthesis pathway. IUBMB Life. 2010;**62**(5):357-362

[54] Celestino-Soper P.B., et al., A common X-linked inborn error of carnitine biosynthesis may be a risk factor for nondysmorphic autism. Proceedings of the National Academy of Sciences. 2012;**109**(21):7974-7981

[55] Filipek P.A., et al., Relative carnitine deficiency in autism. Journal of Autism and Developmental Disorders. 2004;**34**(6):615-623

[56] Rossignol D., Frye R. Mitochondrial dysfunction in autism spectrum disorders: a systematic review and meta-analysis. Molecular Psychiatry. 2012;**17**(3):290-314

[57] Waisbren S.E., et al., Neuropsychological outcomes in fatty acid oxidation disorders: 85 cases detected by newborn screening. Developmental Disabilities Research Reviews. 2013;**17**(3):260-268

[58] Xie Z., et al., Inborn errors of long-chain fatty acid β-oxidation link neural stem cell self-renewal to autism. Cell Reports. 2016;**14**(5):991-999

[59] Pampols T. Inherited metabolic rare disease, in Rare Diseases Epidemiology. 2010, Springer. pp. 397-431

[60] Al-Gadani Y., et al., Metabolic biomarkers related to oxidative stress and antioxidant status in Saudi autistic children. Clinical Biochemistry. 2009;**42**(10):1032-1040

[61] Puig-Alcaraz C., Fuentes-Albero M., Cauli O. Relationship between adipic acid concentration and the core symptoms of autism spectrum disorders. Psychiatry Research. 2016;242:39-45

[62] Guevara-Campos J., Gonzlez-Guevara L., Cauli O. Autism and intellectual disability associated with mitochondrial disease and hyperlactacidemia. International Journal of Molecular Sciences, 2015;**16**(2):3870-3884

[63] Marin S.E., Saneto R.P. Neuropsychiatric features in primary mitochondrial disease. Neurologic Clinics. 2016;**34**(1):247-294

[64] Roberts E. γ-Aminobutyric acid and nervous system function—a perspective. Biochemical Pharmacology. 1974;**23**(19):2637-2649

[65] Schousboe A., Wu J.Ä., Roberts E. Subunit structure and kinetic properties of 4-aminobutyrate-2-ketoglutarate transaminase purified from mouse brain. Journal of Neurochemistry. 1974;**23**(6):1189-1195

[66] El-Ansary A., Al-Ayadhi L. GABAergic/glutamatergic imbalance relative to excessive neuroinflammation in autism spectrum disorders. Journal of Neuroinflammation. 2014;**11**(1):1

[67] Alabdali A., Al-Ayadhi L., El-Ansary A. Association of social and cognitive impairment and biomarkers in autism spectrum disorders. Journal of Neuroinflammation. 2014;**11**(1):1

[68] Fatemi S.H., et al., Consensus paper: pathological role of the cerebellum in autism. The Cerebellum. 2012;**11**(3):777-807

[69] Sgadl P., et al., Loss of GABAergic neurons in the hippocampus and cerebral cortex of Engrailed-2 null mutant mice: implications for autism spectrum disorders. Experimental Neurology. 2013;247:496-505

[70] Innis S.M., de la Presa Owens S. Dietary fatty acid composition in pregnancy alters neurite membrane fatty acids and dopamine in newborn rat brain. The Journal of Nutrition. 2001;**131**(1):118-122

[71] Jin C., et al., Chemical synthesis and structural elucidation of a new serotonin metabolite:(4R)-2-[(5′-hydroxy-1′-H-indol-3′-yl) methyl] thiazolidine-4-carboxylic acid. Tetrahedron Letters. 2006;**47**(6):943-946

[72] Takeuchi T., Iwanaga M., Harada E. Possible regulatory mechanism of DHA-induced anti-stress reaction in rats. Brain Research. 2003;**964**(1):136-143

[73] D'Hulst C., Atack J.R., Kooy R.F. The complexity of the GABA A receptor shapes unique pharmacological profiles. Drug Discovery Today. 2009;**14**(17):866-875

[74] Itomura M., et al., Factors influencing EPA+ DHA levels in red blood cells in Japan. In Vivo. 2008;**22**(1):131-135

[75] Pardo C.A., Eberhart C.G. The neurobiology of autism. Brain Pathology. 2007;17(4):434-447

[76] Roseth S., Fykse E.M., Fonnum F. Uptake of L-glutamate into synaptic vesicles: competitive inhibition by dyes with biphenyl and amino-and sulphonic acid-substituted naphthyl groups. Biochemical Pharmacology. 1998;**56**(9):1243-1249

[77] Schwartz R.D., Yu X. Inhibition of GABA-gated chloride channel function by arachidonic acid. Brain Research. 1992;**585**(1):405-410

[78] Aramori I. Nakanishi S. Signal transduction and pharmacological characteristics of a metabotropic glutamate receptor, mGluRl, in transfected CHO cells. Neuron. 1992;**8**(4):757-765

[79] Lazarewicz J.W., Wroblewski J., Salinska E. NMDA receptor-mediated arachidonic acid release in neurons: role in signal transduction and pathological aspects, in Neurobiology of Essential Fatty Acids. 1992, Springer. Mol Neurobiol. 2017. **54**(3):2090-2106. 73-89

[80] Lynch M., Voss K. Arachidonic acid increases inositol phospholipid metabolism and glutamate release in synaptosomes prepared from hippocampal tissue. Journal of Neurochemistry. 1990;**55**(1):215-221

[81] Herrero I., Miras-Portugal M.T., Sanchez-Prieto J. *Positive feedback of glutamate exocytosis by metabotropic presynaptic receptor stimulation*. 1992. **360**:163-6.

[82] Al-Yafee Y.A., et al., Novel metabolic biomarkers related to sulfur-dependent detoxification pathways in autistic patients of Saudi Arabia. BMC Neurology. 2011;**11**(1):1

[83] Murakami M., et al., Different functional aspects of the group II subfamily (types IIA and V) and type X secretory phospholipase A2s in regulating arachidonic acid release and prostaglandin generation implications of cyclooxygenase-2 induction and phospholipid scramblase-mediated cellular membrane perturbation. Journal of Biological Chemistry. 1999;274(44):31435-31444

[84] Higuchi Y., et al., Arachidonic acid promotes glutamate-induced cell death associated with necrosis by 12-lipoxygenase activation in glioma cells. Life Sciences. 2007;**80**(20): 1856-1864

[85] Gebremedhin D., et al., *Expression of CYP 4A œâ-hydroxylase and formation of 20-hydroxye-icosatetreanoic acid (20-HETE) in cultured rat brain astrocytes*. Prostaglandins & other lipid mediators, 2016. **124**:16-26.

[86] Jazayeri S., et al., Comparison of therapeutic effects of omega-3 fatty acid eicosapentaenoic acid and fluoxetine, separately and in combination, in major depressive disorder. The Australian and New Zealand Journal of Psychiatry. 2008;**42**(3):192-198

[87] Delattre A.M., et al., Maternal omega-3 supplement improves dopaminergic system in pre-and postnatal inflammation-induced neurotoxicity in Parkinson's Disease Model. Molecular Neurobiology. 2016:1-17.

[88] Kumar, A., *Significance of Lipid Profile Assay As a Diagnostic and Prognostic Tool*2011: Internal Medical Publishing, 1-14.

[89] Katoh-Semba R., et al., Age-related changes in BDNF protein levels in human serum: differences between autism cases and normal controls. International Journal of Developmental Neuroscience. 2007;**25**(6):367-372

[90] Sheikh A.M., et al., BDNF-Akt-Bcl2 antiapoptotic signaling pathway is compromised in the brain of autistic subjects. Journal of Neuroscience Research. 2010;**88**(12):2641-2647

[91] Chapleau C.A., et al., Modulation of dendritic spine development and plasticity by BDNF and vesicular trafficking: fundamental roles in neurodevelopmental disorders associated with mental retardation and autism. Journal of Neurodevelopmental Disorders. 2009;**1**(3):185

[92] Raznahan A., et al., Serotonin transporter genotype and neuroanatomy in autism spectrum disorders. Psychiatric Genetics. 2009;**19**(3):147-150

[93] Bousquet M., et al., Modulation of brain-derived neurotrophic factor as a potential neuroprotective mechanism of action of omega-3 fatty acids in a parkinsonian animal model. Progress in Neuro-Psychopharmacology and Biological Psychiatry. 2009;**33**(8): 1401-1408

[94] Hutchinson A.J., et al., Activation of EP 2 prostanoid receptors in human glial cell lines stimulates the secretion of BDNF. Neurochemistry International. 2009;**54**(7):439-446

[95] Ma W., St-Jacques B., Cruz Duarte P. Targeting pain mediators induced by injured nerve-derived COX2 and PGE2 to treat neuropathic pain. Expert Opinion on Therapeutic Targets. 2012;**16**(6):527-540

[96] Vetrivel U., et al., Agonistic effect of polyunsaturated fatty acids (PUFAs) and its metabolites on brain-derived neurotrophic factor (BDNF) through molecular docking simulation. Lipids in Health and Disease. 2012;**11**(1):1

[97] Madore C., et al., Nutritional n-3 PUFAs deficiency during perinatal periods alters brain innate immune system and neuronal plasticity-associated genes. Brain, Behavior, and Immunity. 2014;**41**:22-31

[98] Schaaf M., De Kloet E., Vreugdenhil E. Corticosterone effects on bdnf expression in the hippocampus implications for memory formation. Stress. 2000;**3**(3):201-208

[99] Sarrazin N., et al., Transcriptional effects of glucocorticoid receptors in the dentate gyrus increase anxiety-related behaviors. PloS One. 2009;**4**(11):e7704

[100] Larrieu T., et al., Nutritional omega-3 modulates neuronal morphology in the prefrontal cortex along with depression-related behaviour through corticosterone secretion. Translational Psychiatry. 2014;**4**(9):e437

[101] Meguid N.A., et al., Role of polyunsaturated fatty acids in the management of Egyptian children with autism. Clinical Biochemistry. 2008;**41**(13):1044-1048

[102] Yui K., et al., Effects of large doses of arachidonic acid added to docosahexaenoic acid on social impairment in individuals with autism spectrum disorders: a double-blind, placebo-controlled, randomized trial. Journal of Clinical Psychopharmacology. 2012;**32**(2):200-206

[103] Yui K., et al., [Therapeutic effects of larger doses of arachidonic acid added to DHA on social impairment and its relation to alterations of polyunsaturated fatty acids in individuals with autism spectrum disorders]. Nihon shinkei seishin yakurigaku zasshi= Japanese Journal of Psychopharmacology. 2011;**31**(3):117-124

[104] Amminger G.P., et al., Omega-3 fatty acids supplementation in children with autism: a double-blind randomized, placebo-controlled pilot study. Biological Psychiatry. 2007;**61**(4):551-553

[105] Klein N., Kemper K.J. Integrative approaches to caring for children with autism. Current Problems in Pediatric and Adolescent Health Care. 2016;**46**(6):195-201

[106] Mehl-Madrona L., et al., Micronutrients versus standard medication management in autism: a naturalistic case–control study. Journal of Child and Adolescent Psychopharmacology. 2010;**20**(2):95-103

[107] Oriach C.S., et al., Food for thought: The role of nutrition in the microbiota-gut–brain axis. Clinical Nutrition Experimental. 2016;6:25-38

[108] Simopoulos A.P. Evolutionary Aspects of the Dietary Omega-6/Omega-3 Fatty Acid Ratio: Medical Implications, in Evolutionary Thinking in Medicine. 2016, Springer. pp. 119-134

[109] Lazic M., et al., Reduced dietary omega-6 to omega-3 fatty acid ratio and 12/15-lipoxygenase deficiency are protective against chronic high fat diet-induced steatohepatitis. PloS One. 2014;**9**(9):e107658

[110] Pastural É., et al., Novel plasma phospholipid biomarkers of autism: mitochondrial dysfunction as a putative causative mechanism. Prostaglandins, Leukotrienes and Essential Fatty Acids. 2009;**81**(4):253-264

[111] Politi A., et al., Silica-on-silicon waveguide quantum circuits. Science. 2008;**320**(5876): 646-649

11

Importance of Fatty Acids in Physiopathology of Human Body

Katalin Nagy and Ioana-Daria Tiuca

Abstract

Fatty acids are important components of the human body, having biological, structural and functional roles. Besides their role as source of energy, they act as main constituents of cellular membranes. In this case, as part of the membrane phospholipids, they assure the fluidity, flexibility, permeability of the membrane and also assure the passive transport through the membrane and are interconnected with other proteins in intra and intercellular way. Among these fatty acids, omega-3 and omega-6 polyunsaturated fatty acids (PUFAs) seem to be the most important, due to their multiple biological roles, such as influencing the inflammatory cascade, reducing the oxidative stress, presenting neuroprotection and cardiovascular protection. Fatty acid levels have been shown to be altered in different diseases, which is why they have been used to identify potential biomarkers for several pathologies, such as polycystic ovary syndrome (PCOS). Consequently, this chapter synthesizes the most important physiological and pathological implications of fatty acids in human body functioning.

Keywords: cell membrane, physiology, pathology, (anti-)inflammatory effect, neuroprotection, cardiovascular disease

1. Introduction

Fatty acids (FAs) are part of the lipid class, widespread in the nature, food and organisms, being an important constituent of the membrane cell. They have important biological functions, structural and functional roles, and they represent an important source of energy. Their metabolism produces a huge quantity of adenosine triphosphate (ATP). The β-oxidation of the fatty acids is a well-known process, mostly used by the heart and the muscular tissue to obtain energy.

The human body can synthesize many of these fatty acids, except some essential polyunsaturated fatty acids (PUFAs): the linoleic acid (LA) and the α-linolenic acid (ALA). These two are spread especially in different vegetable oils, but their metabolites are found mainly in the fish oil. The linoleic acid is the most abundant fatty acid in nature, and it is the precursor of other omega-6 fatty acids. The omega-3 fatty acids are synthesized from α-linolenic acid. The human body cannot synthesize fatty acids with odd number of carbon atoms chain; however, there were studies in which this type of fatty acids were identified in a low concentration in plasma [1].

Once ingested, short chain PUFAs are converted to long-chain fatty acids. These are critical for mammalian cells in order to perform various biological functions, such as sustaining the structural integrity of cellular membranes and serving as signaling molecules. They are highly enriched in the adipose tissues, for example in the brain, where they participate in the development and maintenance of the central nervous system during both embryonic and adult stages [2].

The fatty acids can be identified and quantified using various analytical methods, but the most widely used technique is the gas chromatography (GC). Its main advantages are selectivity, sensibility and efficiency. One of the disadvantages of this technique is that before the main analysis, a derivatization of fatty acids is necessary to obtain the methyl esters and this way to increase their volatility. Other analytical techniques used for the detection of fatty acids mentioned in the literature were the high performance liquid chromatography (HPLC) or the capillary electrophoresis (CE) [3, 4].

The aim of this chapter was to present some of the alterations in plasma and other biological fluids fatty acid profile in different diseases, some potential biomarkers in each case and to highlight the fatty acids importance in the proper functioning of the human organism.

2. Physiology of fatty acids

Fatty acids are widely spread through the whole human organism, and they can be found under different forms: free circulating fatty acids or esterified, taking form of:

- triacylglycerols (or triglycerides), when esterified with glycerol,
- phospholipids, when esterified with phosphoric acid,
- glycolipids, when combined with glucose or other saccharides,
- sphingolipids, etc.

The great importance of fatty acids resides in the fact that they are main constituents of the human cell. The type of fatty acid, saturated or unsaturated, long-chained or short-chained can influence the physiology of the cell, as it will be described below.

2.1. Fatty acids in the human cell

Every human cell is formed by a membrane, the cytoplasm and the nucleus. The membrane, the barrier which not only protects the cell from the outside world but also which has the role in transporting nutrients inside and outside of the cell, is formed of a bilayer of lipids with role in the passive transport and proteins with role in the active transport (**Figure 1**) [5].

This double layer of lipids, which assures the main structure of the cell membrane, is formed of two layers of phospholipids. In this case, the phosphoric acid is esterified with a diacylglycerol (R1, R2), which can contain the same or different fatty acid residues, and another residue (R3) directly connected on phosphoric acid is another type of molecule (**Figure 2**). The phospholipid molecule gets, in this way, an amphiphilic character, which means it is in the same time hydrophilic, due to its phosphoric "head", and hydrophobic, due to the fatty acid "tails". In the cell membrane, the phospholipids are oriented in the bilayer with the hydrophilic head toward the outside of the layer, whereas hydrophobic tails remain on the inside of the bilayer, as seen in **Figure 1**.

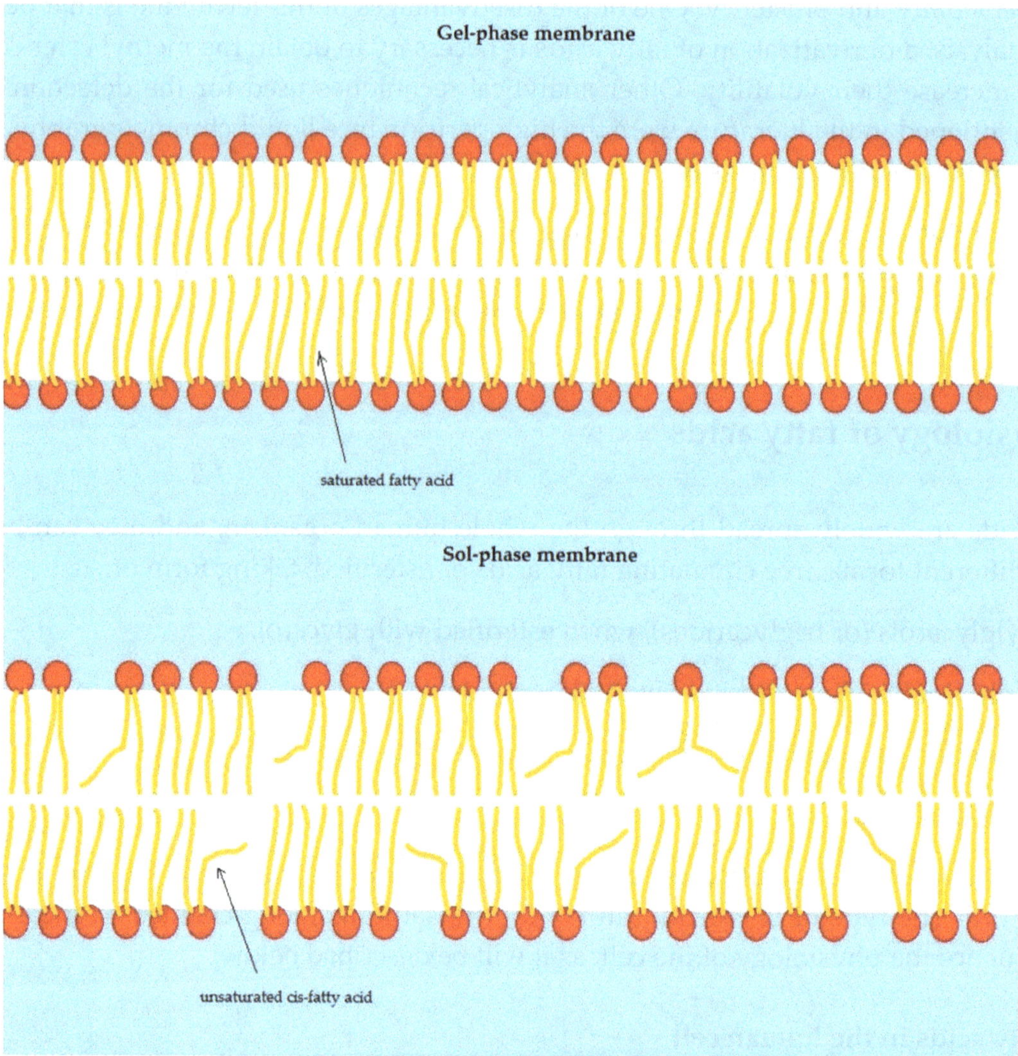

Figure 1. Different fluidity of cell membranes depending on the saturation of fatty acids: saturated fatty acids forming viscous membrane (up), unsaturated *cis* fatty acids forming fluid membrane (down).

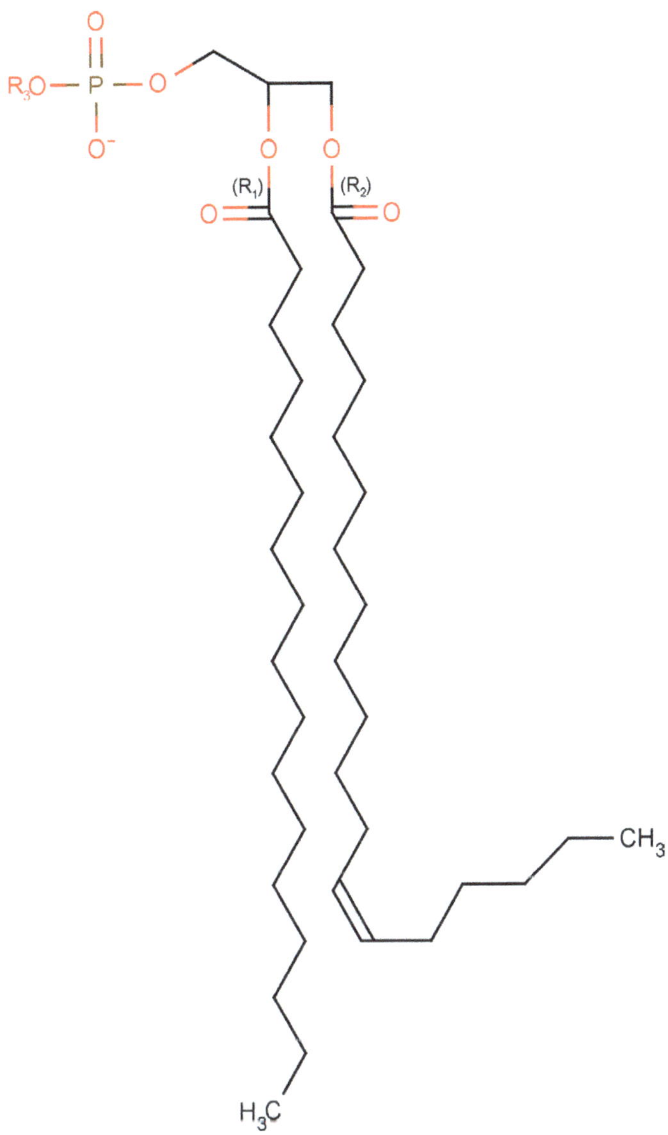

Figure 2. General structure of a phospholipid.

The type of fatty acids found in the structure of the cell membrane can affect its fluidity, its stability and its functions. First, the saturation [6] of fatty acids influences the fluidity of membranes. Thus, if the membrane is composed mostly of saturated fatty acids, which have a straight rigid chain, the phospholipids will form a more rigid bilayer, whereas a membrane formed by many *cis*-unsaturated fatty acids will be more flexible (**Figure 1**). This fact can explain the beneficial effect of polyunsaturated fatty acids on arterial and venous walls, increasing their flexibility and having positive effects in cardiovascular diseases.

One might think that a high content of polyunsaturated fatty acids in cell membranes could expose them to oxidative stress and, consequently, to lipid oxidation and peroxidation [7], but Clinics A [8] has revealed the theory of "triple cell membrane synergy", which states that fatty acids in phospholipids in membranes are surrounded by protective antioxidants. It seems that this hyperfluidity of membranes is obtained in bilayers rich in docosahexaenoic acid (DHA) [6] and is necessary in cells with high and rapid activity, such as rhodopsin disks or axons. In case of rhodopsin disks, the high fluidity of the membrane (~50% DHA) seems to contribute to the fast transport of rhodopsin on the two sides of the membrane, facilitating the initiation of the visual cascade. In case of axons, keeping the hyperfluidity theory of highly rich DHA membranes, this helps increase their permeability and avoid forming gel-phase islands, which could happen in case of lesser unsaturated fatty acids or even saturated and which would lead to a drastic change of the electrochemical behavior of the axon. In the same time, the DHA-rich membrane assures the differentials in Na^+ (outside of the cell) and K^+ (inside of the cell) necessary for signal transmission.

Moreover, studies have revealed [9] that the docosahexaenoic acid found in the structure of the membrane bilayer can influence the activity of the Na^+-K^+-ATPase pump in the same membrane; the DHA content was significantly correlated with the pump activity in heart and kidney tissues, but not in brain tissues. However, in brain tissues, both DHA concentrations and Na^+-K^+-ATPase were found to be the highest, indicating that DHA content increases in tissues with high energy needs. The authors [10] have mentioned other ionic transport proteins, which are modulated by DHA concentrations, such as voltage-gated K^+-channels, voltage-sensitive-Na^+ channels in myocardial cells or L-type Ca^{2+} channels.

In the same time, membrane protein activity can be affected by external fatty acids released by other cells activity or by exogenous source (diet) [11]. Studies have shown that syntaxin-3 activity is dependent on the presence of omega-6 arachidonic acid. Syntaxin-3, a cell membrane protein responsible for neurite outgrowth, is activated by forming a complex with the synaptosomal-associated protein of 25 kDa (SNAP25). (SNAP25). The formation of this complex has been found to be dependant on the presence of arachidonic acid, at a half-maximal effective concentration of ~100 μM. Other omega-3, such as docosahexaenoic and linoleic acid, omega-6 and linolenic acid have shown the same capacity to activate membrane syntaxin-3. This can explain the specific relevance of omega-3 and omega-6 fatty acids in good functioning of brain activity, neuronal regeneration, neurite outgrowth and their beneficial effects in degenerative neurological diseases. DHA content seems to be connected with its anti-cancer effect, influencing the activity of sphingomyelinase. The activity of protein kinase C (PKC) has been reported to be increased at high levels of DHA in membranes, where DHA is incorporated in complexes with phosphatidylethanolamine and lesser with phosphatidylcholine [10]. Other proteins influenced by DHA presence and/or concentrations reported are phospholipases A2 and C, cytochrome P450SCC or the insulin receptor.

2.2. Omega-3 and omega-6 fatty acids

The omega-3 and omega-6 fatty acids are long-chain polyunsaturated fatty acids with the first double bond located at the third, respectively, the sixth carbon atom related to the methyl end, having a *cis* configuration. The omega-3 and omega-6 PUFA families are essential fatty acids in humans, because they cannot be synthesized *de novo*. The omega-6 fatty acids are the predominant PUFAs in all diets, with the linoleic acid as their representative. The α-linolenic fatty acid is an omega-3 PUFA, which is the precursor of other omega-3 long-chain polyunsaturated fatty acids (LC-PUFAs) [12].

For their metabolism, both omega-3 and omega-6 fatty acids use the same pathway, including the same enzymes. Linoleic acid is converted into arachidonic acid through the steps presented in **Figure 3**. This fatty acid is the most important omega-6 polyunsaturated fatty acid. The arachidonic acid (AA) can be also released from cell membranes through the action of phospholipase A_2 and serves as precursor for the synthesis of the biologically active eicosanoids. These eicosanoids are the prostaglandins (PG), thromboxanes and leukotrienes. There are three types of PG: PG1, PG2 and PG3. The first one has many beneficial effects, it reduces inflammation and helps the body to recover from injury by reducing swelling and redness. PG2 has the exact opposite effect of PG1. This increases inflammation, vasoconstriction and blood clotting. PG3 has a mixture of functions in the body, from which the most important one is represented by the property to reduce inflammation caused by PG2. Dihomo-γ-linolenic acid (DGLA), an intermediate metabolite of the omega-6 pathway, can be converted to either the anti-inflammatory PG1 or into the arachidonic acid, a precursor of PG2. This transformation requires the enzyme Δ5-desaturase, whose activity can be limited. Also, the activity of the Δ6-desaturase can be compromised during inflammatory conditions. Both of these enzyme activities are influenced by diet and environmental factors. In diets high in omega-3 fatty acids, most of the enzyme Δ5-desaturase will be used in the omega-3 pathway, so DGLA will be converted into an anti-inflammatory PG1. Contrariwise, a diet low in omega-3 fatty acids will convert DGLA into AA, and this way the inflammation will increase. A balance of omega-3 and omega-6 fatty acid is therefore essential for a proper health [13].

Competition between the omega-6 and omega-3 fatty acids occurs in prostaglandin formation. Eicosapentaenoic acid (EPA) competes with arachidonic acid for prostaglandin and leukotriene synthesis at the cyclooxygenase and lipoxygenase level. Metabolism of the arachidonic acid by the cyclooxygenase enzyme gives rise to the 2-series prostaglandins and thromboxanes and by the 5-lipooxygenase (LOX) pathway hydroxy, hydroperoxy derivatives and the 4-series of leukotrienes are formed (**Figure 4**) [13].

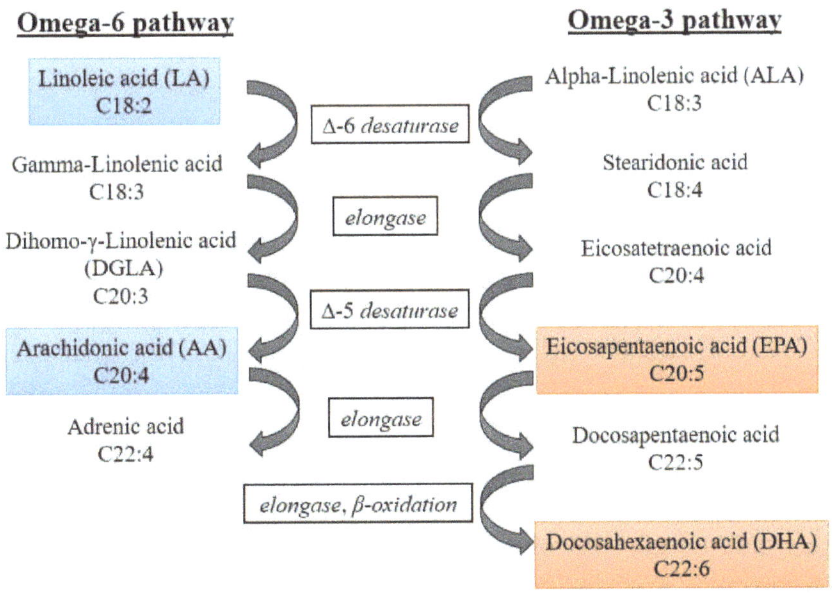

Figure 3. The omega-3 and omega-6 metabolism pathways.

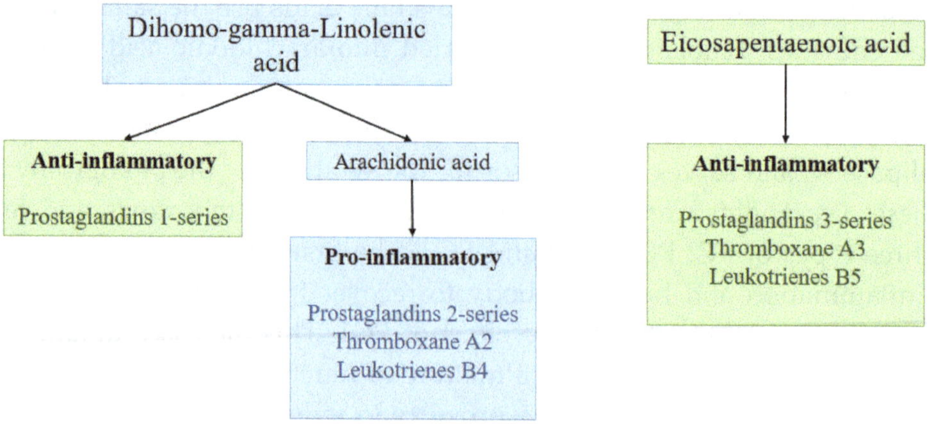

Figure 4. Pro- and anti-inflammatory metabolites of omega-3 and omega-6 fatty acids.

3. Fatty acids with pathological implications

The inflammatory process appears when the human body tries to fight infection and/or to repair damaged tissue. Most of the time this process is protective, but sometimes it can be transformed into chronic inflammation, which can lead to the development or progression of some chronic diseases, including rheumatoid arthritis, coronary vascular disease, cancer and neurological diseases [14]. Fatty acids from diet can influence people's health condition, and they can deteriorate or ameliorate the evolution of some diseases. Scientists try to identify biomarkers for different diseases in order to be able to observe their evolution, to distinguish the ones with similar symptoms and be able to give a precise diagnose. According to Biomarkers Definitions Working Group, 2001 [15], these biomarkers are defined as "Measurable characteristics that reflect biological function or dysfunction, response to a therapeutic measure, or indication of the natural progression of a disease." They are not only useful for disease risk determination but are also extremely useful in establishing a diagnosis. An ideal diagnostic biomarker would reliably reflect *in vivo* pathology with high sensitivity and specificity, while a screening biomarker would combine at least moderate sensitivity with high specificity and low cost.

3.1. Cardiovascular diseases (CVDs)

It became well known over the years that the omega-3 fatty acids have a high impact on health and play an important role in cardiovascular disease prevention. According to the American Heart Association, heart and blood vessel diseases are related to a process called atherosclerosis, which can cause a heart attack or stroke. The role of inflammation in the atherosclerosis process is well known, as well as the fact that omega-3 fatty acids can modify inflammatory cascades favorably, which may be an important factor in their protective role. Their beneficial effects regarding the cardiovascular diseases are mediated by their anti-arrhythmic, lipid lowering, anti-thrombotic and anti-inflammatory properties.

Based on large randomized control trials and *in vitro* molecular experiments, different hypotheses have been proposed for the mechanism of action of these fatty acids regarding the cardiovascular diseases. Several animal studies have demonstrated that omega-3 PUFAs have beneficial effects in the cardiovascular system, including anti-thrombotic, endothelial relaxation and anti-fibrotic effects [16]. One of the most important, the anti-atherosclerotic effect of the fatty acids, can be explained through their anti-inflammatory effect and their influence on oxidative stress, endothelial dysfunction and homeostasis. Atherosclerosis is characterized not only by inflammation but also by an endothelial dysfunction. This is caused by an epoxide hydrolase that converts the epoxyeicosatrienoic acids (EETs) to dihydroxyeicosatrienoic acids (DHETs). Decreasing this enzyme activity and increasing the EETs/DHETs ratio can have a beneficial effect on the endothelium function [17]. The epoxyeicosatrienoic acids generated from AA induce vasodilatation, stimulate angiogenesis and protect heart from ischemia. CYP450 monooxygenase also converts EPA and DHA to epoxyeicosatetraenoic acids (EpETEs) and epoxydocosapentaenoic acid (EpDPAs), which have similar properties to EETs [16].

Based on clinical trials, it has been observed that omega-3 fatty acids can reduce the hepatic triglyceride synthesis and increase the clearance of circulating ones [17]. According to this, they can have an important role in the management of the type III hyperlipidemia.

Besides these roles, the omega-3 fatty acids have an important anti-arrhythmic effect because they stabilize the partially depolarized ischemic myocytes by shortening the action potential duration and by slowing down the impulse conduction [17]. Due to their length and their increased number of double bonds, they can influence the function of different membrane proteins and can modulate the sodium channel function in cardiomyocytes, leading to antiarrhythmic effect. EPA and DHA also modulate the activity of L-type calcium channels, leading to a reduction in free cytosolic calcium ion, which stabilizes myocyte electrical excitability [16].

However, the best studied effect of the omega-3 and omega-6 fatty acids is the inflammatory one. The anti-inflammatory properties of omega-3 PUFAs have been conventionally attributed to their capability to interact with the main inflammatory signaling pathways and to their suppressive effect on inflammatory cytokine production [18]. The fatty acid metabolites are represented by different types of mediators with both inflammatory and anti-inflammatory properties. There are two types of prostaglandins, one derived from arachidonic acid, with inflammatory, platelet aggregation and vasoconstriction effect, and another one derived from EPA with the exact opposite effect. Besides this, EPA and DHA are the precursors of lipoxins, resolvins and protectins, which also regulate vascular tone and blood pressure [17]. Resolvin E-series are synthesized from EPA through the conversion of 18-hydroxyeicosapentaenoic acid (18-HEPE), and protectins, resolvin D-series and maresins are DHA-derived mediators (**Figure 5**) [16]. An increased level of omega-3 PUFA was associated with a decreased circulating concentration of inflammatory cytokines, such as tumor necrosis factor (TNF), interleukin IL-1β and IL-6.

EPA and DHA downregulate the expression of inflammation-related genes through the nuclear peroxisome proliferator-activated receptor (PPARα/γ) [16]. This nuclear receptor has been linked to *in vivo* lipid metabolism, considering that its activation stimulates β-oxidation and decreases the circulating levels of triglycerides and free FAs, which prevents adipocyte hypertrophy and hyperplasia [19].

Besides all of these mechanisms, omega-3 fatty acids increase the endothelial nitric oxide production, which result in a vasodilatory response [17]. A summary of all these cardiovascular effects of the omega-3 fatty acids is presented in **Figure 6**.

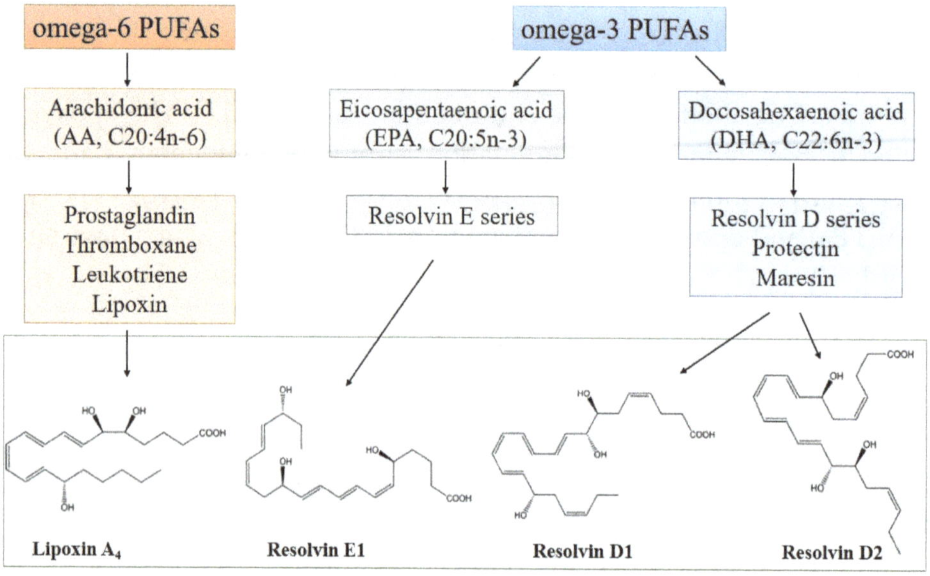

Figure 5. New lipid mediators: resolvins and lipoxins synthesis.

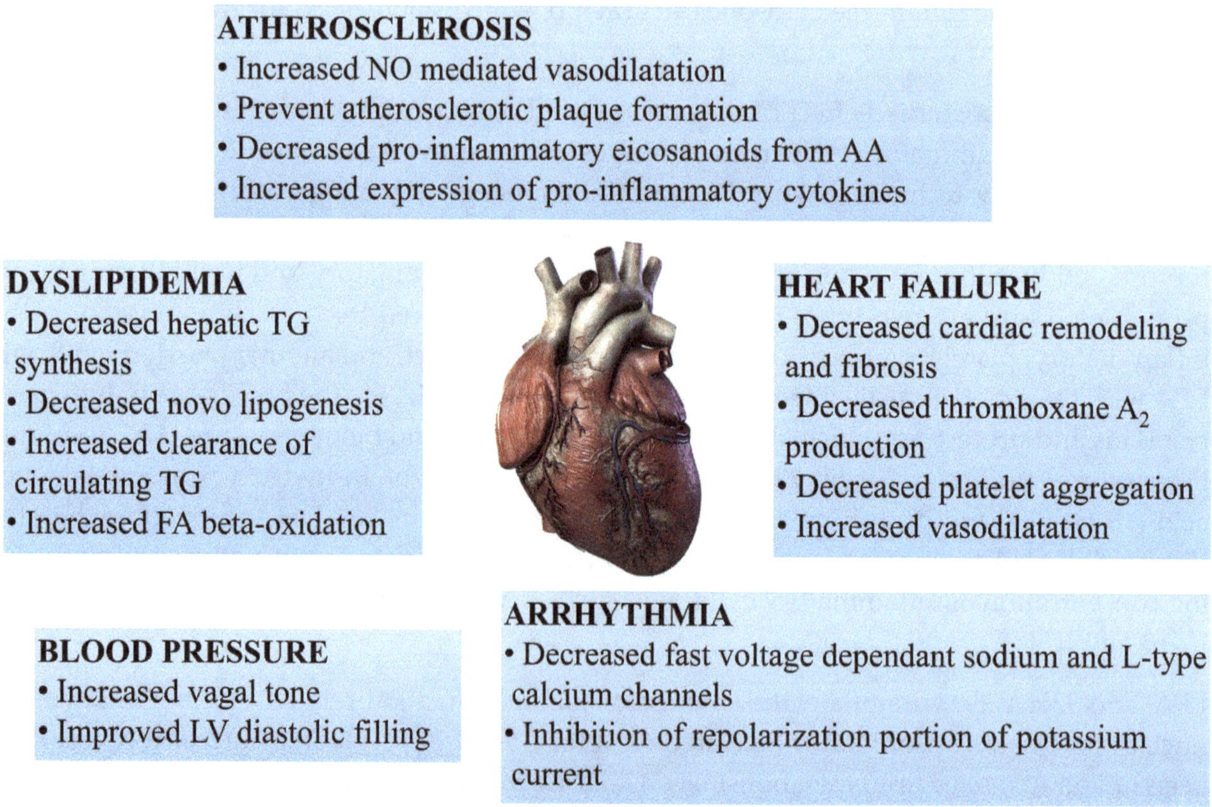

Figure 6. The cardiovascular effects of omega-3 polyunsaturated fatty acids.

Many studies have shown the importance of the omega-3 index. This was developed by Harris and von Schacky and it is defined as the percentage of EPA+DHA content in the red blood cell membranes. An omega-3 index of less than 4% indicates a low cardioprotection. A low omega-3 index is also associated with an increased risk of ventricular fibrillation during acute ischemic phase of myocardial infarction and sudden cardiac death [17], whereas levels higher than 8% confer cardioprotection. Smoking, a major risk factor for the development of the cardiovascular diseases, and higher body weight were associated with lower EPA+DHA levels. Also an inverse association was found between the levels of triglycerides and very low-density lipoprotein (VLDL) and the omega 3-index.

Contrary to the beneficial effects of the omega-3 fatty acids, consuming large amounts of omega-6 FAs increases the plasma concentrations of eicosanoids derived from AA metabolism, specifically prostaglandins, thromboxane, leukotrienes, hydroxylated FAs and lipoxins. These bioactive products have an important inflammatory, thrombosis and atherosclerosis properties and contribute to the development of allergic and inflammatory disorders and excessive cell proliferation [19]. Some studies highlight the importance of omega-6 PUFA/ omega-3 PUFA ratio, because a high ratio promotes the pathogenesis of many diseases, including cardiovascular diseases, but not only. It is also associated with inflammatory markers, including C-reactive protein and IL-6. A low omega-6/omega-3 ratio has beneficial effect on patients with asthma and suppresses inflammation in patients with rheumatoid arthritis.

3.2. Rheumatoid arthritis (RA)

Rheumatoid arthritis (RA) is an autoimmune disease that causes chronic inflammation of the joints and progressive joint destruction. The exact mechanism why the body's immune system attacks the joints is still unclear, but many studies have shown that beside the immunologic etiology there are alterations of different metabolic pathways. The plasma metabolic changes can clarify the pathological mechanism [20]. Several studies have shown an association between synovial inflammation and increased free fatty acid concentration in plasma, which demonstrates that the fat metabolism is accelerated in rheumatoid arthritis [21]. Many of these are related to inflammation and might be considered a marker of arthritic inflammation in different stages [20]. One of these fatty acids, stearic acid, was found in higher levels in patients with established rheumatoid arthritis than in patients in early stages [22].

All of these metabolic changes can be evidenced using a GC-MS method. Fatty acids are extracted from different biological samples using an organic solvent or a solid phase extraction. This step is followed by a derivatization process, when the fatty acid methyl esters are obtained. The main analysis of these compounds can be performed using both polar and non-polar capillary phases, a gradient temperature and helium as a carrier gas. The identification and separation of free fatty acids can also be done with a HPLC system, using different mobile phases in isocratic/gradient elution.

The protective effect of docosahexaenoic acid, an omega-3 polyunsaturated fatty acid, is recognized in many types of chronic inflammatory conditions, because this fatty acid can be metabolized into bioactive lipid mediators with anti-inflammatory activities, as described above. A high concentration of omega-3 PUFA is correlated with a reduced number of morning stiffness, swollen joints, pain or disease activity. It has also been shown that these fatty acids can reduce the incidence and severity of collagen-induced arthritis. In addition to their anti-inflammatory properties, they inhibit the formation of reactive oxygen species and the AA-mediated induction of tumor necrosis factor receptor type I (TNFRI). They have immune modulatory effect and can affect both T cell function and B cell function [21]. Not only PUFAs but also some monounsaturated fatty acids, such as oleic acid, have a benefic effect, due to their anti-inflammatory properties [20]. RA patients, with an elevated inflammation have a decreased level of all these fatty acids: oleic, palmitic, EPA, DHA acid, compared to healthy controls. Differences were unrelated to age, gender or body mass index (BMI) [22]. Besides their anti-inflammatory activity, omega-3 PUFAs incorporated into phospholipids can result into other lipid mediators (resolvins, protectins) with an increased anti-inflammatory activity and an altered cytokine gene expression [14]. However, the high levels of the omega-6 fatty acids, mainly the arachidonic acid, were positively correlated with synovitis. This happened because AA is the main substrate for the synthesis of the pro-inflammatory mediators, such as cytokines and eicosanoids.

The phosphatidylcholine/lysophosphatidylcholine ratio was lower in serum of RA patients compared to healthy individuals, which can result in higher levels of free fatty acids. Other lipid mediators, such as oxylipins, were also detected in plasma. The most prominent eicosanoid of the LOX pathway found in plasma was 5-hydroxyeicosatetraenoic acid (5-HETE) [21]. Another biomarker in inflammation can be considered the omega-3 index. The inflammatory mediators, C-reactive protein, monocytes and neutrophils, are inversely correlated to DHA, omega-3 index and total omega-3 PUFA [14].

3.3. Neuropsychiatric diseases

Discovering new biomarkers in the field of psychiatry has a huge importance, because they could clarify the etiology of psychiatric problems, confirm the diagnosis of disorders with similar symptoms and predict the course of the disorder and determine how to treat it [15]. Neurodegenerative diseases are caused by several factors, including genetic mutation, membrane damage, mitochondrial dysfunction and a protein or lipid metabolism alteration.

PUFAs are selectively concentrated in synaptic neuronal membranes and regulate vascular and immune functions that affect the central nervous system. Moreover, they have important functions in neurotransmitter signaling. The brain is the most lipid-enriched organ, containing several major lipid classes, including fatty acids. Omega-3 and omega-6 fatty acids constitute 30–35% of total brain fatty acids and have beneficial effects on cognitive function. During brain development, especially in the embryonic stage, polyunsaturated fatty acids are critical for cell proliferation and neuronal differentiation, and their deprivation results in apoptosis. Deregulation of fatty acids is also involved in the pathogenesis of numerous brain disorders, such as neurodegenerative diseases, mental retardation, stroke and trauma [2].

The Alzheimer disease (AD) is a chronic neurodegenerative disease, which usually affects the elderly people and causes dementia. The most common early symptom is difficulty in remembering recent events, but difficulty in speech, disorganized thinking and memory loss is also common. Different morphological modifications of the brain were observed, such as the extracellular amyloid beta (Aβ) depositions and the tau protein abnormalities, the neurofibrillary tangles formation inside the nerve cell bodies. Currently accepted biomarkers of AD include levels of brain chemicals related to amyloid or tau protein and imaging-derived estimates of the size and metabolic activity of specific brain regions. Besides these modifications of the protein metabolism, the lipid metabolism is altered as well, characterized by a decreased level of the omega-3 fatty acids.

Molecular alterations of the fatty acids that persist from preclinical stages through the dementia phase may serve as biomarkers that could aid the early diagnosis of AD. Different stages of AD may have a different gene expression for fatty acid synthesis [2]. In this disease, scientists have revealed alterations in lipid metabolism pathways and in lipid carrier proteins, such as ApoE. Alterations of the lipid metabolism were observed not only in case of patients diagnosed with Alzheimer but also in case of those with other cognitive alterations. In each case, the biggest difference reported between the healthy volunteers and the patients was represented by the level of docosahexaenoic acid. However, other fatty acids presented an altered profile as well. In different studies, a low concentration of palmitic acid (C16:0), oleic acid (C18:1n-9) and some omega-3 fatty acids has been shown, such as α-linolenic acid (C18:3n-3), eicosapentaenoic acid (EPA, C20:5n-3) and docosapentaenoic acid (C22:5n-3) [23].

The monounsaturated fatty acids, mainly the oleic acid, inhibit the production of Aβ and amyloid plaque formation both *in vitro* and *in vivo*. In contrast, arachidonic acid increases Aβ production and the formation of amyloid plaques [2]. Fatty acids contribute to the modulation of the structure and function of biological membranes, including elasticity, membrane organization and ion permeability, and may therefore facilitate brain glucose uptake, neurotransmission and neuronal function.

The docosahexaenoic acid (C22:6) is indispensable for the neuronal myelination, and it is an important precursor for the very long chain fatty acid synthesis (C24:6, C26:6, C28:6, C30:6, C32:6, C34:6), found in the brain. It is also involved in neurogenesis, neurotransmission and protects the brain from the oxidative stress. It has an important role in maintaining the integrity of the basal membrane and as a phospholipid ester maintains the flexibility of the cellular membrane, helping the synaptic transmission, and it can also adjust the speed of the signal transmission. DHA can influence the brain development because it can regulate the gene expression, monoaminergic neurotransmission or protection against apoptotic cell death [24]. During pregnancy, DHA accumulates in human neonatal brain tissue at an accelerated rate during the third trimester in association with rapid changes in cortical structural maturation. A deficit of this fatty acid in the stage of the brain development can lead to the cognitive performance alterations [24].

The polyunsaturated fatty acids, besides their role of maintaining the integrity of the neuronal cell membrane, are involved in the synthesis of eicosapentaenoic acid from which the synthesis of the 3-series prostaglandin and 5-leukotriene begins. EPA has neuroprotective, anti-oxidant and anti-inflammatory properties [24].

This fatty acid inhibits the synthesis of prostaglandins derived from the omega-3 fatty acids, such as PGE2 and PGF2α, which confers an anti-inflammatory property. In case of deficiency of DHA and EPA, the cell permeability modifies and mitochondrial dysfunctions and inflammation appear, and along with the oxidative stress, it plays an important role in the progression of the disease. DHA and EPA can play a role in alleviating oxidative stress and reducing the risk of neurodegenerative diseases [25]. The novel series of lipid mediators (resolvins, protectins and maresins) have revealed their protective and beneficial effect in neurological diseases, due to their anti-inflammatory and pro-resolving properties.

A simultaneous deficiency of LA and ALA creates serious problems in fatty acid composition of the brain. ALA deficiency alters the course of brain development and perturbed the composition of brain cell membranes, neurons, oligodendrocytes and astrocytes as well as subcellular components such as myelin, nerve endings and mitochondria.

Aging is characterized by a diminution in PPARα expression in various tissues, representing a key target in the prevention of diseases associated with old age. A decrease in the activity of PPARα-regulated genes involved in β-oxidation is accompanied by changes in the composition of FAs in brain. This leads to an increased level of the very long chain saturated fatty acids (SFAs) (C20:0, C22:0, C24:0), monounsaturated fatty acids (MUFAs) (C16:1, C18:1, C20:1, C22:1, C24:1), and decreasing the LC-PUFAs AA and DHA, which are related to the progression of brain aging. It was suggested that PPARα and their endogenous ligands have a role in neuroprotection against oxidative stress, which is key in neurodegenerative diseases, contributing to a normal brain aging. These considerations suggest that endogenous and exogenous PPARα agonists could be useful as a prevention measure for neurodegenerative diseases and ischemic injury, especially in the elderly and/or in patients with high cardiovascular risk [19].

Several other neurological disorders present altered neuronal and plasma fatty acid composition, such as depression, bipolar disorder, schizophrenia and attention deficit hyperactivity disorder.

Depression is accompanied by activation of the inflammatory response system indicated by an increased production of inflammatory cytokines and oxidative biomarker. Cytokine production is accompanied by increased oxidative stress leading to elevated production of reactive oxygen species (ROS) and nitric oxide (NO) or decreased anti-oxidant defense, such as superoxide dismutase (SOD) and glutathione peroxidase. Epidemiological studies also showed that low intake and blood levels of omega-3 PUFAs are associated with an increased risk for being diagnosed with major depressive disorder. Erythrocyte levels of C16:0, C18:0, EPA and the omega-3 index were significantly lower in the case of patients diagnosed with major depression than in the controls, whereas erythrocyte levels of C16:1, C18:3n6, C18:3n3, C18:1t and C18:2t were significantly higher [26]. Different studies suggest that omega-3 fatty acid status influences the development of central serotonin systems. A deficit of the omega-3 fatty acids leads to impaired serotonin release and behavioral signs of depression and aggression. Patients with major depressive disorder present a DHA deficiency compared to healthy controls [27]. The plasma and erythrocyte phospholipid levels of these people showed significant and positive correlation between the ratio of AA/EPA and severity of depression and suicidal behavior.

The patients diagnosed with bipolar disorder presented higher plasma concentrations of all saturated fatty acids than the controls. Lignoceric acid was over 50% higher in the patient group than in the healthy volunteers. In this disorder too, the most important differences between the two groups were represented by the significant decreases in DHA levels and strong increases in levels of EPA and ALA [28].

The presently available data of the literature suggest that the metabolism of PUFAs is altered in patients with schizophrenia, both in the acute and chronic stages of the disease. Altered neuronal membrane structure and metabolism might contribute to some of the symptoms of schizophrenia. A change in membrane lipid composition in neuronal cells can affect neurotransmission and this way can affect the behavior in schizophrenia. Studies showed no difference between the schizophrenia patients and control subjects in the contribution of omega-3 fatty acids to the lipid composition of the phospholipid fraction. However, the values of total omega-6 PUFAs and docosapentaenoic acid are shown to be significantly lower in case of patients with schizophrenia than in case of the control subjects. Membrane lipids seem to fluctuate in different disease phases. This may be related to changes in neuroinflammatory and oxidative processes, which are reported to contribute to disease progression and underlie symptom severity. The healthy group presents stable PUFA levels compared to the patients group. PUFAs are not only important components of neuronal cell membranes but also play an important role in regulation of inflammation through the formation of eicosanoids. Inflammation and oxidative stress may play a role in disease progression through lipid peroxidation and cholesterol oxidation, leading to neuronal cell death [29]. PUFA deficiency will also impair dopaminergic and glutamatergic neurotransmission, which are linked to negative symptoms.

Based on evidence that erythrocyte EPA+DHA composition of total fatty acid lower than 4% increases risk for sudden cardiac death, and evidence that the majority of patients with psychiatric disorders present an omega-3 index lower than 4%, major recurrent psychiatric disorders may increase risk of cardiovascular disease, a principle cause of excess premature mortality in patients with mood and psychotic disorders [27].

3.4. Other pathologies

Inflammation plays a role in the etiology of many types of cancer. It was reported that high concentration of serum long-chain omega-3 fatty acid phospholipids, EPA and DHA, in particular, was associated with the increased risk of high-grade prostate cancer. However, high concentrations of *trans*-fatty acids, which are known to produce inflammation, are associated with a reduced risk of prostate cancer [30]. Other studies have been realized on colorectal cancer tissues from which different fatty acids have been separated. The analysis revealed high concentrations of saturated fatty acids and low levels of monounsaturated fatty acids. Compared to healthy tissues as controls, the malign ones presented a low omega-3/omega-6 PUFA ratio [31]. Likewise, the dysfunction of the lipid metabolism from the hepatocellular carcinoma can lead to an altered plasma lipid profile. The main differences were evidenced in case of C18:2n-6, C20:4n-6, C16:0 and C18:1n-9. These fatty acids can be considered potential biomarkers in case of hepatocellular carcinoma [32].

Obesity is a metabolic disease, and it increases the risk to develop diabetes, non-alcoholic fatty liver disease or other cardiovascular diseases. Compared to healthy volunteers, plasma levels of fatty acids are increased in the obese patients. The biomarker identified in this disease is the increased level of the unsaturated fatty acids, especially the palmitoleic acid (C16:1) and dihomo-gamma-linolenic acid (C20:3) [33]. Studies show that 40% of the people diagnosed with diabetes will develop a kidney or a cardiovascular disease. A high level of omega-3/omega-6 ratio is associated with a low risk to develop a kidney failure. Besides obesity, the polycystic ovary syndrome (PCOS) can lead to insulin resistance. In this case, two biomarkers were identified: nervonic acid (C24:1) for the presence of PCOS and dihomo-gamma-linolenic acid for insulin resistance [4]. The fatty acid lipolysis in the fatty tissue is increased as well, but the β-oxidation of the fatty acids is decreased. This way the cell permeability and the inflammatory cell infiltration are increased. This may explain why arachidonic acid is considered a biomarker in the plasma of diabetic patients, with or without different stages of nephropathy.

The fatty acid profile is modified in case of an infectious disease, too. In the incipient stage of the dengue fever, a decreased level of C14:0, C16:0, C18:0, C20:4-n6 and C22:6-n3 was observed [4].

Due to the pro-inflammatory and immunoactive properties of the arachidonic acid, a high concentration of this fatty acid and a high level of the AA/EPA ratio are associated with the sickle cell disease and cystic fibrosis. The last one is a genetic disease in which the fatty acid profile is altered. The LA and DHA concentration decreases and increases the concentration of palmitoleic acid and Δ5,8,11-eicosatrienoic acid. In case of cystic fibrosis, the AA concentration increases only at the highest LNA:ALA ratio [4].

The pathobiology of sickle cell disease is initiated by episodic vascular occlusion in which the adherence of circulating blood cells to vascular endothelium is modulated by polyunsaturated fatty acids. Patients diagnosed with this disease have altered red cell and PUFA composition. This is characterized by an increased AA, decreased DHA and EPA levels, and this may play a role in abnormal blood-endothelial cell interactions [34].

Long chain fatty acids are found in a low concentration in plasma, being difficult to identify them. They are synthesized from the short chain fatty acids with the elongase enzyme and are degraded in peroxisome by the β-oxidase enzyme. In the peroxisomal disorders, such as the Zellweger or the adrenoleukodystrophy, the very long chain fatty acids accumulate in the plasma, which leads to the intracellular calcium accumulation and decreases the mitochondrial respiration, which leads to the cellular death of the oligodendrocyte and astrocytes. Important biomarkers in this disease are C26:0, C26:1 and C26:2 [35].

The chronic obstructive pulmonary disease (COPD) is a chronic inflammatory disorder of the airways in which the airways narrow and swell and produce extra mucus. This can make breathing difficult and trigger coughing, wheezing and shortness of breath. Asthma is thought to be caused by a combination of genetic and environmental factors. Once it is installed, the inflammation starts and involves various cell types and mediators. Smoking is one of the major risk factors for the development of COPD, although other risk factors, such as air pollution and genetic factors, exist. In both smokers and COPD patients, a decreased level of omega-3 PUFAs was reported. The C20:5 and the C22:6 were the most significantly decreased, whereas the monounsaturated fatty acid, C16:1, was increased in COPD compared to non-smokers [36]. Many epidemiological studies showed the protective role of DHA in allergic diseases because it suppresses airway eosinophilic inflammation. A new monoglyceride DHA derivative and EPA derivative showed their protective effects on airway inflammation and inflammatory cytokine production. Patients with severe asthma present a selective dysregulation of the 15-lypooxigenase pathway, the reason why a 5-lipoxygenase-dependent metabolite of arachidonic acid, 5-HETE, was similar in patients and healthy subjects [37]. **Table 1** summarizes the biological effects of the most important fatty acids.

Classification	Fatty acid	Biological effect	Diseases
SCFA	Propionic acid	Immunosuppressive effect, improves tissue insulin sensitivity	Prevention of obesity and diabetes type 2
	Butyric acid	Inhibits angiogenesis, antimicrobial effects	Prevention of colorectal cancer, irritable bowel syndrome
MCFA		Thermogenesis, anti-steatosis, weight control	Treating hyperlipidemias, prevention of obesity
LCFA	Oleic acid	Anti-inflammatory effect, inhibit the production of Aβ and amyloid plaque formation	Benefic in rheumatoid arthritis, Alzheimer disease
	α-Linolenic acid (omega-3)	Anti-steatosis, anti-inflammatory	Benefic in Alzheimer disease, CVD prevention, increases prostate cancer risk
	EPA+DHA (omega-3)	Anti-inflammatory, anti-arrhythmic, anti-atherosclerotic	Prevention of obesity, CVDs, beneficial effect in Alzheimer disease, brain development, rheumatoid arthritis, type 2 diabetes
	Arachidonic acid (omega-6)	Inflammatory, platelet aggregation, vasoconstriction, immuno-active properties	Promotes synovitis, Alzheimer disease, diabetes, sickle-cell disease, cystic fibrosis

Table 1. Summary table of biological effects of some FA and their implications in some pathologies.

A significant consumption of omega-3 PUFAs results in a decreased level of arachidonic acid in the membranes of inflammatory cells. This will lead to a decreased level of pro-inflammatory eicosanoids. There is a large amount of literature based on studies investigating the effects of omega-3 PUFAs on inflammation and immune function. The most studies are investigating the fish oil effect on human health. The omega-6/omega-3 fatty acids ratio from modern diets is ranged from 15:1 to 17:1 although it should be 1:1 [8]. The cardioprotective effect of the omega-3 fatty acid was well studied, that is why The American Heart Association recommends 3 g/day EPA+DHA for reducing elevated triglyceride levels [38]. Many studies have been published on the effect of omega-3 PUFAs on brain structure and function. Most of them indicate an increased functional activation in children, although not all of them found any effect of omega-3 fatty acids on cognition [24].

4. Conclusions

In conclusion, this chapter demonstrates the importance of fatty acids in human health, both regarding on the physiology of human body, but also influencing the appearance and/or evolution of different diseases, which are seemingly not related. In this way, especially omega-3 and omega-6 fatty acids become a common point to these pathologies, such as cardiovascular, neurologic, oncologic or endocrinologic diseases, due to their mechanism of action at a cellular level. We have proved the protective effect of DHA and EPA on different types of tissues; however, other types of fatty acids are not to be ignored.

Aknowledgements

The authors of this chapter gratefully acknowledge the financial support of the University of Medicine and Pharmacy "I. Hațieganu", through the internal grant no. 1300/45/13.01.2017.

Author details

Katalin Nagy and Ioana-Daria Tiuca*

*Address all correspondence to: tiuca.daria@umfcluj.ro

University of Medicine and Pharmacy "I. Hațieganu", Cluj-Napoca, Romania

References

[1] Tiuca ID, Nagy K, Oprean R. Development and optimization of a gas-chromatographic separation method of fatty acids in human serum. World J Pharm Sci. 2015;3(8):1713-9.

[2] Liu Q, Zhang J. Lipid metabolism in Alzheimer's disease. Neurosci Bull. 2014;30(2):331-45. DOI: 10.1007/s12264-013-1410-3

[3] Salimon J, Omar TA, Salih N. An accurate and reliable method for identification and quantification of fatty acids and trans fatty acids in food fats samples using gas chromatography. Arab J Chem. 2013; DOI: 10.1016/j.arabjc.2013.07.016

[4] Tiuca I, Nagy K, Oprean R. Recent developments in fatty acids profile determination in biological samples—a review. Rev Rom Med Lab. 2015;23(4):371-84. DOI: 10.1515/rrlm-2015-0035

[5] Hackney C, Furness D. Introducing the cell: the unit of life. In: Orchard G, Nation B, editors. Cell Structure & Function. Glasgow: Oxford University Press; 2015. pp. 1-45.

[6] Valentine RC, Valentine DL. Omega-3 fatty acids in cellular membranes: a unified concept. Prog Lipid Res. 2004;43(5):383-402. DOI: 10.1016/j.plipres.2004.05.004

[7] Pamplona R, Barja G, Portero-Otin M. Membrane fatty acid unsaturation, protection against oxidative stress, and maximum life span a homeoviscous-longevity adaptation?. New York Acad Sci. 2002;959:475-90. DOI: 10.1111/j.1749-6632.2002tb02118.x

[8] Kidd PM. Omega-3 DHA and EPA for cognition, behavior, and mood: clinical findings and structural-functional synergies with cell membrane phospholipids. Altern Med Rev. 2007;12(3):207-27.

[9] Turner N, Else PL, Hulbert AJ. Docosahexaenoic acid (DHA) content of membranes determines molecular activity of the sodium pump: implications for disease states and metabolism. Naturwissenschaften. 2003;90(11):521-3. DOI: 10.1007/s00114-003-0470-z

[10] Stillwell W, Wassall SR. Docosahexaenoic acid: membrane properties of a unique fatty acid. Chem Phys Lipids. 2003;126(1):1-27. DOI: 10.1016/S0009-3084(03)00101-4

[11] Darios F, Davletov B. Omega-3 and omega-6 fatty acids stimulate cell membrane expansion by acting on syntaxin 3. Nature. 2006;440(7085):813-7. DOI: 10.1038/nature04598

[12] Lorente-Cebrián S, Costa AGV, Navas-Carretero S, Zabala M, Martínez JA, Moreno-Aliaga MJ. Role of omega-3 fatty acids in obesity, metabolic syndrome, and cardiovascular diseases: a review of the evidence. J Physiol Biochem. 2013;69(3):633-51. DOI: 10.1007/s13105-013-0265-4

[13] Calder PC. Polyunsaturated fatty acids and inflammation. Biochem Soc Trans. 2005;33(Pt 2):423-7. DOI: 10.1051/ocl.2004.0038

[14] Olliver M, Veysey M, Lucock M, Niblett S, King K, MacDonald-Wicks L, et al. Erythrocyte omega-3 polyunsaturated fatty acid levels are associated with biomarkers of inflammation in older Australians. J Nutr Intermed Metab. 2016;5:61-9. DOI: 10.1016/j.jnim.2016.03.002

[15] Kalia M, Costa E Silva J. Biomarkers of psychiatric diseases: current status and future prospects. Metabolism. 2015;64(3):S11-5. DOI: 10.1016/j.metabol.2014.10.026

[16] Endo J, Arita M. Cardioprotective mechanism of omega-3 polyunsaturated fatty acids. J Cardiol. 2016;67(1):22-7. DOI: 10.1016/j.jjcc.2015.08.002

[17] Peter S, Jacob JJ. Role of omega-3 fatty acids in cardiovascular disorders. In: Hegde MV, Zanwar AA, Adekar SP, editors. Omega-3 Fatty Acids: Keys to Nutritional Health. Cham: Springer International Publishing; 2016. pp. 513-30. DOI: 10.1007/978-3-319-40458-5_37

[18] Lorente-Cebrián S, Costa AGV, Navas-Carretero S, Zabala M, Laiglesia LM, Martínez JA, et al. An update on the role of omega-3 fatty acids on inflammatory and degenerative diseases. J Physiol Biochem. 2015;71(2):341-9. DOI: 10.1007/s13105-015-0395-y

[19] Echeverria F, Ortiz M, Valenzuela R, Videla LA. Long-chain polyunsaturated fatty acids regulation of PPARs, signaling: relationship to tissue development and aging. Prostaglandins Leukot Essent Fat Acids. 2016;114(October):28-34. DOI: 10.1016/j.plefa.2016.10.001

[20] Zhou J, Chen J, Hu C, Xie Z, Li H, Wei S, et al. Exploration of the serum metabolite signature in patients with rheumatoid arthritis using gas chromatography-mass spectrometry. J Pharm Biomed Anal. 2016;127:60-7. DOI: 10.1016/j.jpba.2016.02.004

[21] Brouwers H, von Hegedus J, Toes R, Kloppenburg M, Ioan-Facsinay A. Lipid mediators of inflammation in rheumatoid arthritis and osteoarthritis. Best Pract Res Clin Rheumatol. 2015;29(6):741-55. DOI: 10.1016/j.berh.2016.02.003

[22] Rodriguez-Carrio J, Alperi-Lopez M, Lopez P, Ballina-Garcia FJ, Suirez A. Non-esterified fatty acids profiling in rheumatoid arthritis: associations with clinical features and Th1 response. PLoS One. 2016;11(8):1-17. DOI: 10.1371/journal.pone.0159573

[23] Olazarán J, Gil-De-Gómez L, Rodríguez-Martín A, Valentí-Soler M, Frades-Payo B, Marín-Muñoz J, et al. A blood-based, 7-metabolite signature for the early diagnosis of Alzheimer's disease. J Alzheimer's Dis. 2015;45(4):1157-73. DOI: 10.3233/JAD-142925

[24] Bos DJ, van Montfort SJT, Oranje B, Durston S, Smeets PAM. Effects of omega-3 polyunsaturated fatty acids on human brain morphology and function: what is the evidence?. Eur Neuropsychopharmacol. 2016;26(3):546-61. DOI: 10.1016/j.euroneuro.2015.12.031

[25] Katyare SS, Mali AV. Omega-3 fatty acids and mitochondrial functions. In: Hegde MV, Zanwar AA, Adekar SP, editors. Omega-3 Fatty Acids: Keys to Nutritional Health. Cham: Springer International Publishing; 2016. pp. 229-33. DOI: 10.1007/978-3-319-40458-5_17

[26] Baek D, Park Y. Association between erythrocyte n-3 polyunsaturated fatty acids and biomarkers of inflammation and oxidative stress in patients with and without depression. Prostaglandins Leukot Essent Fat Acids. 2013;89(5):291-6. DOI: 10.1016/j.plefa.2013.09.008

[27] Messamore E, McNamara RK. Detection and treatment of omega-3 fatty acid deficiency in psychiatric practice: rationale and implementation. Lipids Health Dis. 2016;15(1):25. DOI: 10.1186/s12944-016-0196-5

[28] Erika S, Reidera A, Singha G, Gelenberga AJ, Rapoportd SI. Low unesterified:esterified eicosapentaenoic acid (EPA) plasma concentration ratio is associated with bipolar disorder episodes, and omega-3 plasma concentrations are altered by treatment. Bipolar Disord. 2016;8(5):583-92. DOI: 10.1111/bdi.12337

[29] Solberg DK, Bentsen H, Refsum H, Andreassen OA. Lipid profiles in schizophrenia associated with clinical traits: a five year follow-up study. BMC Psychiatry. 2016;16(1):299. DOI: 10.1186/s12888-016-1006-3

[30] Brasky TM, Darke AK, Song X, Tangen CM, Goodman PJ, Thompson IM, et al. Plasma phospholipid fatty acids and prostate cancer risk in the select trial. J Natl Cancer Inst. 2013;105(15):1132-41. DOI: 10.1093/jnci/djt174

[31] Mal M, Koh PK, Cheah PY, Chan ECY. Metabotyping of human colorectal cancer using two-dimensional gas chromatography mass spectrometry. Anal Bioanal Chem. 2012;403(2):483-93. DOI: 10.1007/s00216-012-5870-5

[32] Qiu J-F, Zhang K-L, Zhang X-J, Hu Y-J, Li P, Shang C-Z, et al. Abnormalities in plasma phospholipid fatty acid profiles of patients with hepatocellular carcinoma. Lipids. 2015;50(10):977-85. DOI: 10.1007/s11745-015-4060-6

[33] Ni Y, Zhao L, Yu H, Ma X, Bao Y, Rajani C, et al. Circulating unsaturated fatty acids delineate the metabolic status of obese individuals. EBioMedicine. 2015;2(10):1513-22. DOI: 10.1016/j.ebiom.2015.09.004

[34] Aslan M, Celmeli G, Özcan F, Kupesiz A. LC-MS/MS analysis of plasma polyunsaturated fatty acids in patients with homozygous sickle cell disease. Clin Exp Med. 2015;15(3):397-403. DOI: 10.1007/s10238-014-0293-6

[35] Hama K, Nagai T, Nishizawa C, Ikeda K, Morita M, Satoh N, et al. Molecular species of phospholipids with very long chain fatty acids in skin fibroblasts of Zellweger syndrome. Lipids. 2013 Dec;48(12):1253-67. DOI: 10.1007/s11745-013-3848-5

[36] Titz B, Luettich K, Leroy P, Boue S, Vuillaume G, Vihervaara T, et al. Alterations in serum polyunsaturated fatty acids and eicosanoids in patients with mild to moderate chronic obstructive pulmonary disease (COPD). Int J Mol Sci. 2016;17(9):1-21. DOI: 10.3390/ijms17091583

[37] Miyata J, Arita M. Role of omega-3 fatty acids and their metabolites in asthma and allergic diseases. Allergol Int. 2015;64(1):27-34. DOI: 10.1016/j.alit.2014.08.003

[38] McNamara RK, Strimpfel J, Jandacek R, Rider T, Tso P, Welge JA, et al. Detection and treatment of long-chain omega-3 fatty acid deficiency in adolescents with SSRI-resistant major depressive disorder. PharmaNutrition. 2014;2(2):38-46. DOI: 10.1016/j.phanu.2014.02.002

12

Fatty Acid's Profiles of Aquatic Organisms: Revealing the Impacts of Environmental and Anthropogenic Stressors

Ana M.M. Gonçalves, João C. Marques and

Fernando Gonçalves

Abstract

There is a great concern about the impacts of climate changes namely due to salinity seawater and temperature alterations in aquatic organisms with the estuarine and coastal environments being the major affected areas. The intensive usage of chemicals in an indiscriminate way in agriculture practices, achieving, in some cases, values above the limits of contamination authorized by the European legislation, also drastically affects the surrounded estuarine areas with profound consequences to the water quality and the aquatic communities. It is known that stressors affect organisms' physiological conditions with recent works concerning alterations in the fatty acid (FA) profiles associated with environmental and contamination events that become more frequent. FA plays a key role in immune and physiological functions and is associated with the prevention of some diseases, shown to be good bio-indicators to assess the organisms' impacts under stress conditions. Thus, this chapter proposes to address natural (salinity and temperature) and chemical (herbicide and metal) stressors' impacts in the FA profiles of *Thalassiosira weissflogii* and *Cerastoderma edule* and infers about the effects on organisms' physiological processes and along the food web. Consequences in food resources and to healthier and nutritious food consumption with benefits to human beings are also assessed.

Keywords: fatty acids, bio-indicator, stressors, climate changes, salinity, temperature, pollutants, herbicide, metal, aquatic organisms, food quality, estuaries

1. Introduction

Fatty acids (FAs) are essential molecules with a crucial role in the maintenance of physiological functions of many organisms. These carboxylic acids provide fuel for the brain and the tissues at the metabolic level and are a major constituent in the cellular wall as part of phospholipids. Once transferred across the food chain, FAs perform the connection between primary producers and secondary consumers [1]. Greatly abundant in the brain tissues, these molecules represent almost half of the brain weight [2]. Fatty acid nomenclature is represented by X:YωZ, where X represents the number of carbon atoms in the chain, Y represents the number of double bonds and Z gives information about the position of the first double bond counting from the methyl group [3]. There are two major groups of FA: saturated fatty acids (SFAs) and unsaturated fatty acids (UFAs), differing between them by the existence of double bonds in UFA. The position of the double bonds is determined by the desaturase enzyme activity, which performs the double bond in different positions accordingly to the type of enzyme present in different organisms (e.g. Δ3 in animals, Δ6 in animals and plants and Δ15 in algae and plants with chlorophyll) [3]. Saturated FAs are metabolized mainly as a source of energy and also as lipid storage. UFA can be identified by the number of double bonds: Monounsaturated FA (MUFA) have one double bond and can be synthesized *de novo* almost by all organisms [4]. Polyunsaturated FAs (PUFAs) are fatty acids with two or more double bonds that play an essential role in the brain development [5] and many physiological functions such as down regulating inflammation, cellular signalling [6] and regulation of transcription factors [7]. Essential fatty acids (EFAs) are some PUFAs that play major important functions in physiological and biochemical processes and that must be acquired externally, through dietary input, once the majority of the animals cannot synthesize them *de novo* [8], due to the lack of the desaturase enzyme [9]. Although some animals are able to synthetize EFA from linolenic precursors such as α-linolenic acid C18:3 (n-3), C18:4 (n-3) and C18:5 (n-3), found almost exclusively in plants [3] via elongation and desaturation, the rate this conversion succeeds is residual to supply the necessary amount of EFA required for an optimal growth and development. Thus, EFAs must be obtained by direct feeding on phytoplanktonic, plants or bacteria species [10], or by ingesting lipid emulsions with high content of EFAs [11]. Since EFA are later transferred along the food web, aquatic species like fish and other organisms from higher trophic levels are an important food sources of such molecules [4]. EFAs are represented mainly by some PUFAs. Highly unsaturated fatty acids (HUFAs) are a subset of PUFAs with a chain of 20 carbon atoms or more and with three or more double bonds that play an important role in cellular growth, with special relevance in tissue growth, energy storage, neural development and also reproductive fitness [4, 12]. The physiological activities of EFAs in animals are mainly represented by eicosapentaenoic acid (EPA—20:5n3) and docosahexaenoic acid (DHA—22:6n3), two of the most important HUFAs that are synthesized *de novo* by phytoplankton and bio-accumulated by animals. The arachidonic acid (ARA—20:4n-6) is also a representative EFA with functions as a precursor of animal hormones such as prostaglandins and leukotrienes amongst others [3]. EPA and DHA play a major role in brain development and maintenance of brain structure and function [13]. EPA intake influences many physiological processes such as reproduction, immunity efficiency and osmoregulation [3]. DHA is important for the health and developing of neurons and for neurotransmission, with strong influence on cognition and behavior, and it is also proved to be important in the protection against oxidative stress [14]. Fatty acids are considered to be an accurate tool in trophic interaction studies [15], mainly due to their importance in the health/stability of the ecosystem, and because they are transferred conservatively to higher trophic levels along the trophic food web [4, 16]. Furthermore, FA profiles can reflect structural changes in species' biochemical composition in response to stressors [17–20]. Lipid components are also very sensitive to environmental changes, which make them an efficient assessment tool to monitor toxicological effects on the marine biota as bio-indicators of ecosystem health [19].

Environmental pollution worldwide is an undesirable by-product of the increased demand for natural resources in the modern civilization. However, since the advent of human societies, there have always been foci of environmental contamination, though nothing on the scale we see today. Practically, whole environment suffers from some degree of contamination in concentrations above those expected for the region. The pollutants that damage the ecosystem are the pollutants from industry and mining that release toxic substances such as metals and organic pollutants. Some pesticides and mainly metals (e.g. Cd, Cr, Pb, Hg, Ni, Cu) are non-degradable and therefore accumulate in nature, where they continue to affect ecosystem's function over the course of decades or even centuries. These chemicals can distress several biological organization levels affecting flora and fauna aquatic organisms, interfering with the metabolic and physiological processes and thus compromising the structure and physicochemical properties of the membrane, damage cells, tissues and organs. Long-term effects may lead to higher mortality among population, changing the diversity and structure of the communities. Furthermore, due to global climate changes, environmental conditions are expected to change considerably in several areas. In some regions, it is expected, seasonal differences become more notorious than they used to be exposing the organisms to a wide physiological stress. These changes not only act as additional stress factors but may also considerably modify the toxicity of pollutants in aquatic ecosystems.

Estuaries are coastal ecosystems, which are biologically highly productive and having great importance in the ecological and at socioeconomic contexts, providing exceptional natural resources and services to human beings, mainly to local populations. Some of these systems are located near farmlands, industrial and residential areas being under anthropogenic pressures that affect water quality and the aquatic communities. Estuarine systems are very useful model systems to study the ecological and evolutionary responses of organisms to highly variable, discontinuous habitats due to the extreme daily variations that occur in these transitional areas exposing the organisms to a widely physiological stress [21]. The transition between the freshwater and marine environments creates a gradient of physical and chemical conditions that determine the amount and distribution of the species and communities that live at these ecosystems [22], with salinity being one of the major controlling factor of species' distribution in estuarine systems. Aquatic organisms from these ecosystems are exposed to physical and chemical environmental conditions that vary greatly, on both seasonally and daily basis. Because planktonic species are strongly influenced by climatic factors, and particularly sensitive to changes in hydrological conditions [23], the increase in frequency of flooding episodes has a strong impact on macrobenthic communities and is proven to be result into a significant decline in the diversity of suspension-feeder species, such as microcrustaceans (e.g. copepods) [24]. The temperature of the water is also linked with changes in salinity, and although some studies have tested the two parameters separately, both should also be studied in a bi-dimensional approach, to best simulate natural conditions [25]. According to Kinne [26] and Williams and Geddes [27], temperature can alter the physiological tolerance of an organism to salinity changes, and in turn, salinity can influence the impact of environmental temperature on the same organism. Salinity is indeed of major importance in the distribution of aquatic organisms because the ability of osmoregulation affects ecological tolerances and the type of ecosystem (marine, coastal, estuarine, freshwater) [28].

Since past decades, extreme weather episodes are frequent worldwide, and Portugal is not an exception. At the Mondego estuary, located in the western coast of Portugal, near Figueira da Foz city, episodes of drought and flood have been registered and are well documented in the literature revealing ecological impacts on aquatic communities [24, 29–31]. This temperate estuarine system surrounded by agriculture fields, the commercial port, beach and industries, and with a high marine exploitation of resources, suffers high anthropogenic pressures similar to many other estuarine systems mainly from the Mediterranean region. Rice and corn fields are the main agriculture production in the Mondego valley, being Viper and Primextra® Gold TZ the most used pesticides in agriculture practices, respectively, according to information from the cooperatives of the region. Furthermore, copper is one of the main constituents of pesticides formulations, with application in agricultural activities. It is an essential metal, with vital importance in low concentrations to organisms, acting as a co-factor of many enzymes, i.e. it is a component of superoxide dismutase, an enzyme defending living organisms against reactive oxygen species [32]. Still becomes toxic at high concentrations affecting several biochemical and metabolic processes such as FA metabolism, cell division, photosynthesis, respiration and synthesis of carbohydrates, pigments and chlorophyll [33, 34]. In 1998, and similarly to other estuarine systems near intensive agriculture practices with wide usage of pollutants, a pesticide-monitoring program was implemented in Mondego estuary to recover the system [35].

Stressors affect organisms' growth and biochemical processes and also their performance and healthy status. To compensate extreme conditions, or at least conditions that are far from the optimal, some organisms developed strategic and adaptive mechanisms to compensate physiological requirements. Nevertheless, it is apparent to have the occurrence of significant losses and large alterations on the FA contents. Some studies allow to recognize and assess the response of specific markers in order to identify and validate precise bio-indicators that are able to capture the impact of disturbances resulted at extreme conditions or at the presence of stressors, which might be used as early warning signals of stressing conditions and, eventually, be associated to strategic prevention of stress-associated diseases. Therefore, it is crucial to determine and assess lethal effects and physiological responses of aquatic organisms under the influence of environmental (e.g. salinity, temperature) and chemical (e.g. pollutants) stressors in order to predict the impacts on communities and thus on aquatic ecosystems and food quality. Still, there are some characteristics that must be taken into account when choosing the species to be tested, such as (1) the species' sensitiveness to the studied parameter/substance; (2) well-known nutrient requirements; (3) low genetic and phenotypic variability among strains/organisms; and (4) fast and cost efficient maintenance in the laboratory [36]. The marine phytoplankton species, *Thalassiosira weissflogii,* is often used as a bio-indicator in several studies, becoming increasingly relevant in environmental monitoring studies [36]. The marine diatom value is closely related to its many applications in 1) ecotoxicological studies; 2) biodiesel production; 3) prey for zooplankton (rotifers, copepods, brine shrimp) and 4) production of metabolites such as lipids, proteins, carbohydrates, pigments and vitamins [37]. *Cerastoderma edule* plays a key role between primary producers and consumers. They live in intertidal shallow areas, presenting a suspension-feeder behavior [38]. This bivalve species lives worldwide occurring from the Northern Norway to the North Africa, on the east coast of the Atlantic and in Murmansk in the Arctic [39]. Due to its sessile life style, easy sampling collection, maintenance, handling and sensitivity to chemicals, *C. edule* is widely used as standard species in ecotoxicological bioassays [19, 20]. Due to its high ability to filtrate and accumulate large amount of pollutants, this species is also used as bio-indicator in ecological studies [39–43]. Furthermore, *C. edule* is very much appreciated as food source mainly by local populations, which highlights its importance to the socioeconomic sector.

In this chapter, it is proposed to determine and assess the effects of environmental (salinity and temperature) and chemical (Primextra® Gold TZ and Copper) stressors, individually and combined, in the fatty acid profiles of a marine phytoplankton species (*T. weissflogii*) and an estuarine bivalve species (*C. edule*). The impacts of global stressors to the quality of aquatic food resources and thus to a healthy and nutritive food consumption are also assessed.

2. Material and methods

2.1. Study area and sampling procedure

The Mondego estuary is a small mesotidal system in the West Atlantic coast of Portugal (40°08′N, 8°50′W). The estuary is divided into two arms, north and south (**Figure 1**). The northern arm is characterized by a salt-wedge during low tide, which changes to a partially mixed water column at high tide. It is characterized by a partially mixed water column at low tide and a well-mixed one at high tide at spring tides [44]. The southern arm is shallower and its water circulation is mostly dependent on tides and on freshwater input from a small tributary system, the Pranto River. Freshwater discharges of this river are controlled by a sluice according to the water needs by the rice fields of Mondego valley [29].

C. edule was sampled at the south arm of the estuary (**Figure 1**). Organisms were transported from the field in cold boxes with brackish water.

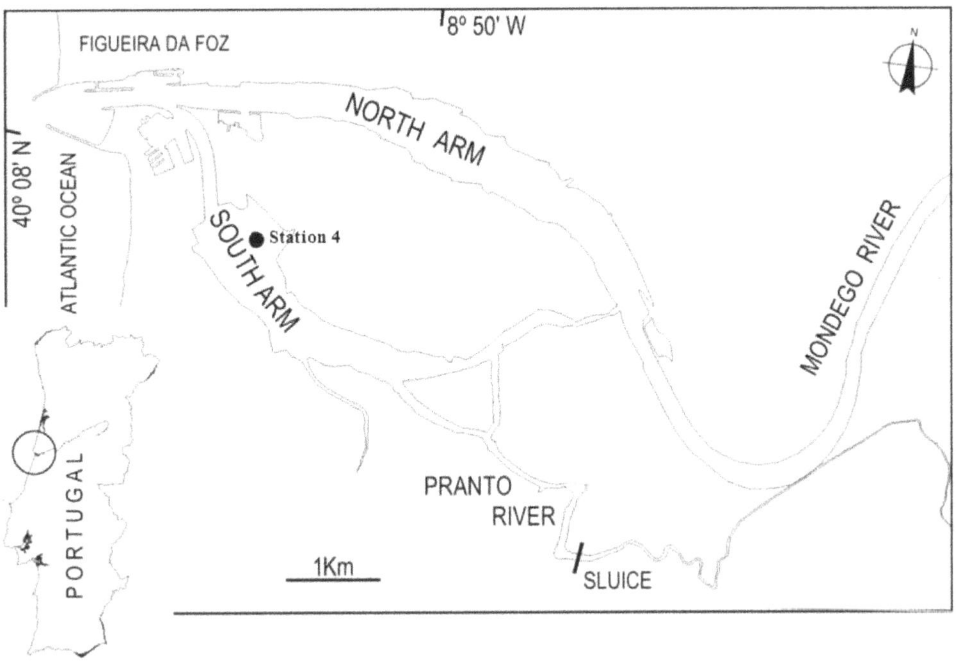

Figure 1. The Mondego estuary location and the sampling site within the estuary.

2.2. Laboratory procedures and bioassays

2.2.1. Microalga species

2.2.1.1. Culture maintenance and laboratory bioassays

T. weissflogii was obtain from the Scottish Marine Institute, Dunbeg, PA37 1QA, UK (strain number 1085/18). The microalga was maintained under laboratory conditions by renewing the alga medium once a week, maintaining a cell density of 2×10^4 cells/mL. f/2 medium was prepared accordingly to Guillard and Reyther [45] with water collected from the Mondego estuary with a salinity of 30 psu, previously filtered with Whatman glass microfiber filters with 1.2 μm pores and stored at 4°C. By the experiments with copper, the medium was prepared without EDTA, adapted after Rippingale and Payne [46]. A renew of algae culture was done weekly. All assays and organism cultures were maintained under artificial light with photoperiod of $16h^L$:$8h^D$. Different cultures were maintained in 15, 20 and 25 ± 2°C, respectively.

Before the beginning of bioassays, an inoculum of *T. weissflogii* was harvested from the bulk culture and incubated for 3 days in a chamber with photoperiod ($16h^L$: $8h^D$) at 20°C [18]. This procedure was repeated from other two temperatures (15 and 25°C) to the salinity experiments where a set of different temperatures (15, 20 and 25°C) were assessed to a range of salinity concentrations. The cellular concentration was then adjusted to 10^4 cells/mL after cellular density determination at a Neubauer haemocytometer, and the microalgae was then exposed to a range of salt and chemical concentrations, respectively. Eight salinity concentrations (0, 5, 10, 15, 20, 25, 33) plus the control were performed at three distinct temperatures (15, 20 and 25°C). The experiments with the both toxicants were conducted to a range of concentrations from 0.200 to 0.800 mg/L of copper(II) sulphate pentahydrate and 0.005 to 0.040 mg/L of the herbicide Primextra® Gold TZ, plus the control. Glass beakers were used to saline and herbicide experiments, whereas plastic beakers were used in the experiments with the metal. Three replicas per treatment were prepared in each bioassay. The experi-ments were conducted under photoperiod ($16h^L$: $8h^D$), with a duration of 96 h. At the end of the experiment, the cellular density was determined to each treatment using a Neubauer haemocytometer.

2.2.1.2. Microcosm bioassays

Microcosm bioassays had a duration of 7 days and has been conducted under the same conditions of the laboratory bioassays previously described. Three treatments corresponding to 96h-EC_{10}, EC_{20} and EC_{50} calculated from the bioassays described in the subsection above plus the control treatment were employed whenever possible due to the duration of the microcosm experiments. According to the results of the microalga growth obtained at the previous experimental bioassays, the control treatments of the salinity microcosm bioassays were performed to a salinity concentration of 33, 33 and 25 to the temperature of 15, 20 and 25°C, respectively. Although at 20°C the microalga presented a similar growth at the saline treatments of 30 (CTL) and 33, and being considered the optimal growth to *T. weissflogii* at the salinity of 30, the control treatment to the microcosm bioassay was performed to a salinity concentration of 33 in order to compare the results with the microcosm experiments conducted at 15°C. Three replicas per treatment were conducted. At the end of the microcosm bioassays, a final concentration of 7.2×10^6 cells/mL was measured in each Erlenmeyer flask and then filtered through a GF/F Whatman filter and frozen at −80°C for FA analysis.

2.2.2. Bivalve species

2.2.2.1. Culture maintenance and laboratory bioassays

In the lab, organisms were divided in aquaria with aeration and filtrated sea water at the salinity of 20. On 10 selected organisms, collected in the field and not under any laboratorial process, a set of measurements (shell length, total weight, tissue weight and foot weight) were assessed to determine the condition indices. After the measurements, the muscle (foot) of each organism was removed and stored at −80°C for fatty acid analysis. The remaining organisms collected in the field were maintained in the aquaria, under photoperiod conditions ($12h^L$:$12h^D$) and control temperature (20 ± 2°C), without food, during a depuration period of 48 h, previously to the experiments.

Salinity bioassay was performed on organisms exposed to a range of saline concentrations from 0 to 35 plus the control. The salinity concentrations were obtained from successive dilutions of filtrated seawater at the salinity of 35 in distilled water. Experiments with the contaminants were conducted on individuals under six concentrations of copper(II) sulphate pentahydrate ranging from 0.6 to 2.1 mg/L and a set of eight concentrations from 0.5 to 60 mg/L of the herbicide Primextra® Gold TZ plus the control, respectively. The test medium was used as negative control. Bioassays were conducted under control temperature (20 ± 2°C), $12h^L$:$12h^D$photoperiod, with filtrated sea water medium at the salinity of 20, during 120 h. Tests were carried out in glass (to saline and herbicide experiments) and plastic (to the metal) vials, 10 per treatment, containing a final test volume of 1000 mL per replicate. Organisms were fed daily with a commercial mixture of microalgae and rotifers. Organisms were transferred to newly prepared test solutions every next day. Bivalves were checked daily for mortality and behavioural conditions (to evaluate the conditions of the valves, organism behavior during feeding and the activity of the siphon). After the exposure period, all survival organisms were dissected, measured the weight and the body length and evaluated the condition indices of each individual. After the measurements, the muscle tissue (foot) was stored at −80°C for further fatty acid analysis.

2.2.3. *Fatty acid analysis*

The sample extraction for the FA analysis was obtained accordingly to the method described in Gonçalves et al. [19], substituting BF3-methanol by H_2S0_4, due to reported deficiency in PUFA detection [47]. A differentiated phase was extracted, and an internal standard (fatty acid Methylnonadecanoate (C19) Fluka 74208) was added to the quantification of FA. The samples were later analyzed using a gas chromatograph with a mass spectrometer and an HP88 column (60 m × 25 mm × 0.20 μm). It was conducted in splitless mode, with a 1 μL injector per run. The column temperature was set to increase from 75 to 230°C at a rate of 2°C/min. The carrier gas was helium at a flow rate of 1.3 mL/min. The results of the GC analyses were obtained, and fatty acid methyl esters (FAMEs) were identified by comparison of their retention times with those of individual purified standards. FAMEs can also be quantified by determining the area of the peaks of each fatty acid with the help of calibration factors [47].

2.2.4. *Statistical analysis*

The cellular density of *T. weissflogii* measured with a Neubauer Haemocytometer, at the end of the bioassays, was used to estimate the concentrations which induced x% growth inhibition (EC_x values, with x = 10, 20, 50) and the corresponding 95% confidence intervals by non-linear regression, using the least-squares method to fit the data to the logistic equation.

The LC10, LC20 and LC50 values with corresponding 95% confidence intervals for *C. edule* were determined using Probit analysis [48].

To determine significant differences between treatments, one-way analysis of variance (ANOVA) was performed, followed by Dunnett's multiple comparison test to identify significant differences between salinity treatments and the control treatment, considering a level of significance of 0.05.

The FA profiles were assessed by determining total (mg/ind) or relative (%) FA concentrations.

One-way analysis of similarity (ANOSIM) was applied to determine differences in FA profiles of each species across the different treatments.

2.2.5. *Fatty acid trophic markers*

FA ratios of bacteria, algae or animal were assessed at the extracts of lipids of *C. edule*. The FA ratios determined and respective food sources are described in **Table 1**.

Marker	Source	Reference
DHA/EPA	Dinoflagellates/diatoms, carnivory	[16, 64, 67]
EPA	Diatoms	[16, 64–65]
DHA	Carnivory, dinoflagellates	[16, 64, 66–67, 70]
18:1n9	Carnivory	[66, 70]
18:2n6	Carnivory	[66, 70]
Σiso and anteiso C15 and C17	Bacteria	[68, 69]

Table 1. Dietary and trophic fatty acid markers used in the present study.

3. Results

3.1. Bioassays

3.1.1. *Thalassiosira weissflogii*

After 96 h of exposure to a range of salinity concentrations and three different temperatures, significant statistical differences were observed between the control and the lower salinity concentrations with an exception to the bioassay conducted at 20°C where a significant statistical difference was also observed to the highest salinity treatment (**Figure 2**). At the lowest temperature (15°C), the microalga presented the lowest growth than at higher temperatures. Still, to salinities near the optimal value of salinity to this microalga (30), the growth was high at all temperatures tested, not observing statistical significant differences between those concentrations and the control (**Figure 2**).

Considering the exposure to both contaminants (the herbicide and the metal), *T. weissflogii* showed to be more sensitive to Primextra®Gold TZ than to copper (**Figure 3**).

A significant growth inhibition was detected after the exposure to both contaminants with the herbicide revealing to be more toxic than the metal. In fact, all treatments of the herbicide showed statistical significant differences with the control, whereas to copper, only the three higher concentrations presented statistical significant differences with the control.

3.1.2. *Cerastoderma edule*

Considering the optimality of salinity for the activity of *C. edule* is 20–25, the species revealed to be mostly affected by low salinities (LC_{50} = 11.01 (10.66–11.54) mg/L) with 100% of mortality at salinity concentrations below 10. Although the growth inhibition of *T. weissflogii* to lower salinity concentrations, the microalga demonstrated to be more tolerant than *C. edule*.

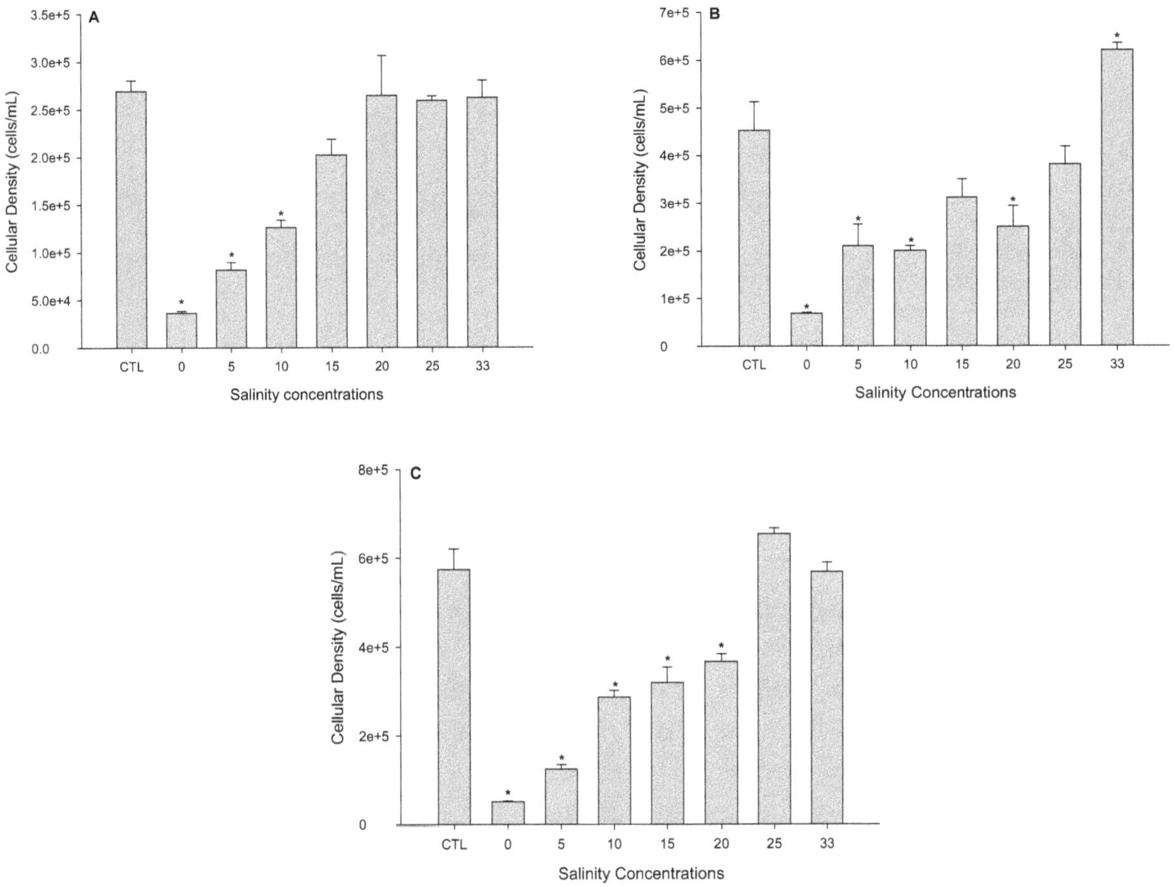

Figure 2. Cell density of *T. weissflogii* at (a) 15, (b) 20 and (c) 25°C after 96 h of exposure to salinity treatments, where CTL refers to the negative control treatment. Symbol '*' indicates a significant (P < 0.05) difference of the treatments compared to the CTL.

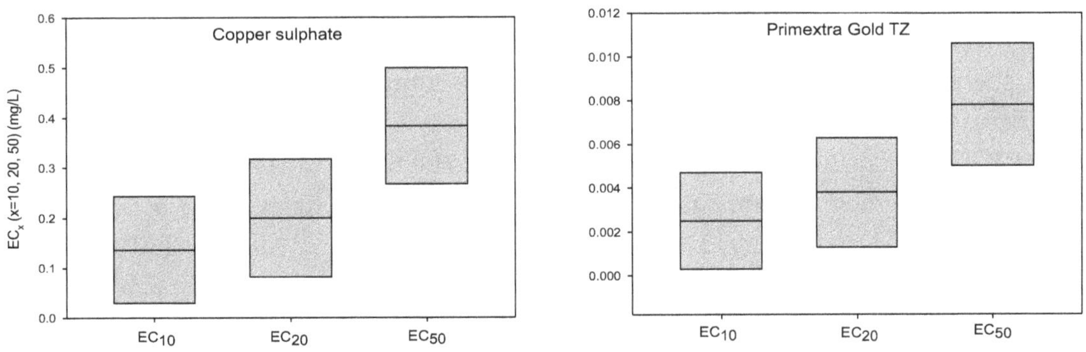

Figure 3. Selected EC_x (x = 10, 20, 50) values (mg/L^{-1}) estimated for *Thalassiosira weissflogii*. The central band of each box denotes the EC_X (x=10, 20, 50) value and the bottom and top of the box represent the lower and upper 95% confidence limits.

In **Table 2**, there are the lethal concentration (LC) values determined to *C. edule* exposed to the herbicide and the metal. The results clearly revealed that estuarine bivalve is more sensitive to the metal than to the herbicide, showing an opposite pattern of the one observed with *T. weissflogii*.

	Copper (II) sulfate pentahydrate (mg/L)	Primextra® Gold TZ (mg/L)
LC_{10}	0.341 (0.000; 0.571)	21.298 (11.008; 25.222)
LC_{20}	0.504 (0.083; 0.698)	23.868 (16.271; 27.422)
LC_{50}	0.818 (0.595; 0.987)	28.784 (24.731; 33.238)

Table 2. Lethal concentration (LC) values to *C. edule* exposed to copper sulfate and Primextra® Gold TZ. In brackets are indicated the 95% confidence limits.

3.2. Fatty acid profiles

3.2.1. Thalassiosira weissflogii

3.2.1.1. Salinity experiments

The fatty acid composition (mg/individual) of *T. weissflogii* exposed to three salinity treatments (10.46, 15.9, 18.5) plus the control is summarized in **Table 3** (15°C), **Table 4** (20°C) and **Table 5** (25°C). After the saline exposure to 15°C, the FA composition of *T. weissflogii* is varied in general with a significant decrease in MUFA and PUFA, especially at the salinity treatments of 10.46 and 15.9. The amount of SFAs in the control treatment is very low compared to other treatments. The saline treatment of 18.5 registered a drastic increase in the abundance of SFA compared to all treatments, with the control registering the lowest amount of saturated FA. MUFA was the dominant FA group in the FA profile of the marine diatom, with a decrease at the saline treatment of C2 (15.9), followed by C1 (18.5) and then C3 (10.46). In fact, the control at 15°C was the treatment that registered the highest amount in MUFA from the three temperatures tested. A dominance of LC-MUFAs is notorious, although C20:1n9(cis11) was absent at the control and at the concentration C1 (18.5). PUFA also displayed a high abundance at the control and at C1 (18.5), with a decrease at the concentrations C2 (15.9) and C3 (10.46). In fact the total amount of PUFA at C1 treatment was almost twofold of the quantity registered at the control. Important precursors of LC-PUFA such as C18:2n6c were presented only in the control and in the salinity concentration of 15.9, whereas C18:3n3 was presented in all treatments, with a residual quantity (closely zero) at the salinity treatment of 10.46. This FA group showed to be the most sensitive to salinity. HUFAs were presented at highest quantity at the extreme salinity treatments (C1 = 18.5; C3 = 10.46) with the control and the C2 treatment (15.9) registering the lowest values. All EFAs (ARA, EPA, DHA) had a similar pattern after saline exposure: their level remained very close in each concentration apart from EPA that registered the lowest value of the HUFA determined at all treatments. DHA was the most abundant EFA in the control treatment and at the salinity treatment of 18.5.

At 20°C, the FA profile of *T. weissflogii* exposed to the three saline treatments (15, 30.5 and 32) plus the control reported an increase in SFA content from the control to the lowest salinity concentration (15), although with relatively low diversity, had been notorious for higher amount of SFA of longer chain (**Table 4**). A significant rise was observed at the PUFA and HUFA contents at the lowest salinity treatment (15). An opposite trend was identified at the total amount of MUFA to all tested treatments when compared to the control. The salinity

	CTL	±Std. error	C1 (18.5)	±Std. error	C2 (15.9)	±Std. error	C3 (10.46)	±Std. error
C14:0			54.9151	21.8215				
C15:0								
C16:0	32.4438	7.8688			32.6365	13.7101		
C17:0								
C18:0								
C20:0			808.2884	321.1812			24.9692	10.8120
C21:0			803.2505	162.62326	467.5087	87.8701	195.8909	38.09074
C22:0								
C23:0			1207.2149	479.6986				
C24:0	75.8923	18.4066	395.4571	157.1387				
Total SFA	108.3361	26.2754	3269.1270	1142.4632	500.14525	101.5802	220.8601	48.9027
C14:1	52.5108	12.7357						
C15:1								
C16:1	34777.3500	1912.2145	15868.4131	3200.9993	9562.9247	3989.8029		
C17:1	49.0293	2.9871	1262.8013	339.8028	10046.7004	4146.2759	427.3974	99.6523
C18:1n9t	34.7431	1.4553	359.6901	93.3756	36.7104	8.8740	2.1760	0.9422
C18:1n9c	48.2467	0.3976	417.0501	117.3960	45.6232	4.9456	215.0315	84.0745
C20:1n9(cis-11)					28.5713	12.0023	255.5463	110.6548
C22:1n9	82.6656	20.0494	570.6772	172.9728	84.4090	10.3481	484.0046	109.6170
C24:1n9	316.5629	39.3603	425.4274	169.0477	139.9360	58.7849	116.9618	29.7734
Total MUFA	35361.1083	1989.1999	18904.0593	4093.5942	19944.8750	8231.0337	1501.1175	434.7142
C18:2n6t	212.8497	26.0495	400.9721	79.7362	19.2169	8.0727	6.9028	2.9890
C18:2n6c	21.1455	5.12853			73.1763	30.7402		

	CTL	±Std. error	C1 (18.5)	±Std. error	C2 (15.9)	±Std. error	C3 (10.46)	±Std. error
C20:2cis(11-14)	4.7466	1.15123	309.2416	110.9660	42.2756	6.8694	23.8036	10.3073
C22:2cis(13-16)			23.3495	9.2782			33.5178	14.5136
C18:3	111.1073	26.9475			22.6410	9.5111	32.1885	8.5154
C18:3n3	83.7401	20.3000	83.5544	33.2012	27.8042	11.6801		
C20:3n3			4462.2590	1751.2907	99.5302	37.7730	95.8512	23.9086
Total PUFA	433.5892	79.5867	817.1177	233.1816	185.1140	66.8735	96.4128	9.0813
C20:4n6 (ARA)	32.0602	7.7756	693.0573	265.71303	55.3162	7.4896	141.8455	27.7393
C20:5n3 (EPA)	42.1809	10.2304	175.5987	57.4274	56.2248	23.6191	64.45161	27.9084
C22:6n3 (DHA)	110.8065	13.1428	787.9480	313.0988			110.4819	47.8401
Total HUFA	185.0476	31.1488	1656.604	636.23923	111.541	31.1087	316.77901	103.4878
N	19		21		18		17	

Table 3. Abundance of fatty acids (saturated fatty acids—SFA, monounsaturated fatty acids—MUFA, polyunsaturated fatty acids—PUFA and highly unsaturated fatty acids—HUFA, in mg/ind) in the profile of *T. weissflogii* after exposure to salinity treatments (10.46; 15.9; 18.5) at 15°C.

	CTL	±Std. error	C1 (32)	±Std. error	C2 (30.5)	±Sth. error	C3 (15)	±Std. error
C14:0							524.4776	198.2339
C15:0								
C16:0					185.5345	44.1380	1299.3503	491.1083
C17:0								
C18:0								
C20:0			10.7682				171.8261	64.9442
C21:0	43.9605	11.4958	51.9012	11.4958			143.5637	33.9657
C22:0	31.5296	15.7648	45.9347	15.7648	8.3243	2.9431	6.1552	2.3265
C23:0	67.8728	21.6031	45.6180	21.6031			35.3845	13.3741
C24:0			10.0689		12.5133	4.4241	60.0374	22.6920
Total SFA	143.3628	48.8636	164.2909	48.8636	206.3721	51.5052	2240.7950	826.6445
C14:1								
C15:1								
C16:1	32041.6830	480.9976	10230.4003	480.9976	28290.2142	1943.4874		
C17:1	89.0514	13.4534	27.[illegible]444	13.4534	8.4810	2.9985	21.0489	7.9558
C18:1n9t					58.9215	13.1667004	197.3032	74.5736
C18:1n9c	386.1857	112.0796	76.3323	112.0796	25.4705	1.9659	390.8133	131.4833
C20:1n9(cis-11)			95.0662				1704.3170	629.4505
C22:1n9	176.2199	7.9133	278.4564	7.9133	78.0943	2.8059	989.9304	324.5449
C24:1n9	14.1459	7.0729	238.0316	7.0729			156.5543	59.1720
Total MUFA	32707.2860	621.5169	10945.4313	621.5169	28461.1817	1964.4245	3459.9672	1227.1800
C18:2n6t	201.34907	58.6126	67.9254	58.61256	32.2866	1.7356	198.9214	59.3448
C18:2n6c								

	CTL	±Std. error	C1 (32)	±Std. error	C2 (30.5)	±Sth. error	C3 (15)	±Std. error
C22:2cis(13-16)			155.7379		31.9686	2.1246	116.8187	39.0555
C18:3					5.7043728	2.01680		
C18:3n3			79.5902				11.1572	4.2170
C20:2cis(11-14)	30.5477	15.2738	32.7481	15.2738	1.2155	0.4298	354.8936	131.3944
C20:3n3	181.6203	49.4586	108.7435	49.4586	47.9214	8.4011	20.5600	4.1581
C20:4n6	147.5854	21.7236	122.3678	20.5766	46.5139	5.8593	1731.6336	848.9185
Total PUFA	561.1024	145.0686	567.1130	143.9215	165.6103	20.5672	2433.9846	1087.0884
C20:4n6 (ARA)	147.5854	19.15840	122.3678	19.1584	46.5137	5.2815	1731.6336	632.7465
C20:5n3 (EPA)			182.5960		41.0818	6.0011	1318.1238	452.1425
C22:6n3 (DHA)					11.1902	3.9563	65.5965	24.7931
Total HUFA	147.5854	19.15840	304.9637	19.1584	98.7860	15.2390	3115.35390	1109.6821
N	15		22		20		25	

Table 4. Abundance of fatty acids (saturated fatty acids—SFA, monounsaturated fatty acids—MUFA, polyunsaturated fatty acids—PUFA and highly unsaturated fatty acids—HUFA, in mg/ind) in the profile of *T. weissflogii* after exposure to salinity treatments (15; 30.5; 32) at 20°C.

	CTL	±Std. error	C1 (23.76)	±Std. error	C2 (22.59)	±Std. error	C3 (14.79)	±Std. error
C14:0			1170.3476	361.1771				
C15:0	15.8093	6.2820						
C16:0			265.4131	81.9083	91.3424	55.9356		
C17:0								
C18:0			56.2356	17.3547				
C20:0	14.7009	2.9211	51.8891	16.0133				
C21:0	34.1393	9.4712	58.1041	17.9313	52.5980	32.2000	15.8881	11.2346
C22:0	133.0046	21.8387	301.2867	92.9791				
C23:0	324.9469	59.3290	716.0168	156.4173	16.1034	4.9800		
C24:0	144.2169	32.1972						
Total SFA	666.8180	132.0393	2619.2930	743.78106	160.0438	93.1342	15.8881	11.23460
C14:1								
C15:1								
C16:1	3920.3316	1557.7818			10178.9007	3313.9257		
C17:1								
C18:1n9t			58.3992	18.0224				
C18:1n9c			106.4004	15.3800	37.1824	22.76945		
C20:1n9(cis-11)	23.9182	4.8087	56.05451	17.2988				
C22:1n9	283.4383	26.872	1218.6964	295.9116	31.2481	9.9562	77.1268	11.6267
C24:1n9	136.0564	41.155	440.7829	105.7048				
Total MUFA	4363.7445	1877.394	1880.3334	518.1971	3415.7770	5184.6099	77.1268	14.7068
C18:2n6t	89.4747	26.9962	261.2764	33.1211				
C18:2n6c								

	CTL	±Std. error	C1 (23.76)	±Std. error	C2 (22.59)	±Std. error	C3 (14.79)	±Std. error
C20:2cis(11-14)	36.7421	10.5830	163.4404	33.0000				
C22:2cis(13-16)	78.7990	20.2553	383.4063	79.1263				
C18:3	9.3324	3.7083	39.3623	12.1475				
C18:3n3	38.6596	7.7462	106.3339	32.8154				
C20:3n3	252.3527	53.0700	1185.6409	323.0043				
Total PUFA	505.3605	122.3691	2139.4602	513.3264				
C20:4n6 (ARA)	344.5002	55.6143	640.6545	135.8612	12.2297	7.4891	18.2719	4.7866
C20:5n3 (EPA)	185.2005	7.6715	384.7258	88.5744				
C22:6n3 (DHA)	91.9278	5.9549	513.8760	103.9311				
Total HUFA	621.6286	69.2407	1539.2573	328.3666	12.2297	7.4891	18.2719	4.7866
N	19		21		7		3	

Table 5. Abundance of fatty acids (saturated fatty acids—SFA, monounsaturated fatty acids—MUFA, polyunsaturated fatty acids—PUFA and highly unsaturated fatty acids—HUFA, in mg/ind) in the profile of *T. weissflogii* after exposure to salinity treatments (14.79; 22.59; 23.76) at 25°C.

concentration C2 (30.5) registered a sharp decrease in the total amount of MUFA, PUFA and HUFA when compared to the control. MUFA constituted the most representative group with the highest amount of FA in each treatment.

Comparing to the salinity experiments conducted at 15°C, the control treatment at 20°C presented the highest content of SFA and PUFA, with the control at 15°C registered a higher amount of total MUFA and HUFA. Comparing the salinity treatments tested, the highest quantity of SFA, PUFA and HUFA was observed at the salinity concentrations C1 (18.5) and C3 (15) to the experiments performed at the temperature of 15 and 20°C, respectively. The total MUFA registered the highest values at the middle concentrations (C2 = 15.9; C2 = 30.5) of the experiments conducted at 15 and 20°C correspondingly.

At the FA profiles of *T. weissflogii* exposed to the salinity treatments at 25°C, a sharp decreased of total SFA from the control to the other treatments was observed with the exception of salinity concentration of 23.76, where a great increase was observed (**Table 5**). In general, MUFA constituted the most abundant group of FA, mainly represented by longer chain MUFAs, except at the highest salinity treatment (23.76). Still this salinity concentration registered the highest quantity of PUFA from all salinity experiments conducted at distinct temperatures (15, 20 and 25°C). PUFA content was only observed in the control and at salinity concentration C3 (23.76). C18:3n3 was present in the higher salinity treatment and in low amounts at the control, which may indicate that these LC-PUFA precursors were desaturated and elongated in the synthesis of HUFA. PUFA was characterized by a great diversity, still absent at the lower salinity treatments (14.79 and 22.59). HUFA registered the highest amount in the highest saline concentration (23.76) followed by the control treatment, decreased drastically at the other salinity treatments. Also, the control that registered the highest amount of HUFA was the one performed at 25°C. EFAs are mainly represented by ARA that were identified at all treatments. In the negative control, ARA is the most abundant EFA followed by EPA and lastly by DHA. Indeed, EPA and DHA were absent in the two lowest salinity treatments.

The FA content of *T. weissflogii* at the different salinity concentrations plus the control showed sharp changes among all treatments at 25°C than at the other temperatures (15 and 20°C). Furthermore, a great decrease in FA diversity was observed in C2 and C3 salinity treatments at the experiment conducted at 25°C, whereas the experiments conducted at 15 and 20°C presented higher or similar diversity of FA among the treatments, which indicate a detrimental effect of salinity in the FA content of the species.

3.2.1.2. Pollutants experiments

The results obtained demonstrated the fatty acid profiles of the microalga were affected by the presence of both toxicants, mainly by the metal (**Table 6**). Although it was not detected clear differences among the treatments, moderate changes were observed at the highest herbicide concentration compared to the other treatments. Small changes at the total amount of SFA and MUFA were also registered between the control and the highest concentration (0.0078 mg/L) of Primextra.

	Primextra® Gold TZ (mg/L)				Copper (mg/L)			
	CTL	0.0025	0.0038	0.0078	CTL	0.1361	0.1995	0.3834
% total SFA	42.51	41.72	41.18	44.51	48.85	42.16	43.30	54.44
% total MUFA	15.03	14.63	16.03	16.46	15.17	17.70	16.81	14.96
% total PUFA	25.95	29.03	27.48	26.70	21.48	23.21	23.14	18.02
% total HUFA	16.48	14.62	15.30	12.32	14.49	16.93	16.75	12.59
EPA	13.69	12.16	12.71	10.32	12.33	14.53	14.33	10.59
DHA	2.79	2.46	2.59	2.00	2.16	2.40	2.42	2.00
N	18	18	18	18	18	18	18	18

Table 6. Total fatty acid and EFA (%) content in *T. weissflogii* after the exposure to the herbicide Primextra® Gold TZ and the metal copper.

A marginal decrease of SFA from the CTL to the copper concentration of 0.1995 mg/L, with a rise at the highest copper treatment was observed. An opposite trend was verified to the unsaturated fatty acids (MUFA, PUFA and HUFA). A slightly rise in the total amount of the UFA was registered from the CTL to the metal concentration of 0.1995 mg/L followed by a decrease at the highest treatment. Although significant differences were not registered at the fatty acid content between the control and the lowest copper treatments, clear differences between the control and the highest copper concentrations were verified.

The amount of HUFA in herbicide and copper treatments was related to the presence of DHA and EPA, with ARA being absent at all treatments. EPA was the dominant EFA at all treatments (**Table 6**).

3.2.2. *Cerastoderma edule*

3.2.2.1. *Salinity experiments*

A considerable increase in the main FA groups (saturated fatty acids—SFA and unsaturated fatty acids—UFA) was clearly observed in the individuals from the field to the organisms exposed to a range of salinity concentrations, unless to polyunsaturated fatty acid (PUFA) where an opposite trend was verified (**Table 7**). Individuals from the field were mostly constituted by PUFA (95.773%), presented lower amounts of SFA (3.314%), HUFA (0.708%) and MUFA (0.205%). Under a range of salinity concentrations, a lacking of SFA of short chain was clearly observed in all organisms, with only the individuals exposed to the highest salinity treatment presented slightly amounts of C6:0, C8:0 and C10:0. Omega-6 was mainly represented by γ-linolenic acid, C18:3n6 and arachidonic acid (ARA), C20:4n6, whereas omega-3 occurred mainly at the forms of docosahexaenoic acid (DHA), C22:6n3, eicosapentaenoic acid (EPA), C20:5n3 and α-Linolenic acid (ALA), C18:3n3.

	Salinity concentrations					
	Field	10	15	25	30	35
% total SFA	3.314	26.561	31.434	19.843	59.711	30.511
% total MUFA	0.205	32.280	28.336	15.043	14.606	19.847
% total PUFA	95.773	25.130	16.162	44.712	4.095	42.821
% total HUFA	0.708	16.030	24.068	20.401	21.588	6.821
ARA	0.000	0.000	3.849	5.401	9.963	3.245
EPA	0.117	14.835	15.468	7.888	7.838	2.988
DHA	0.591	1.195	4.750	7.112	3.786	0.588
N	28	21	15	15	15	22

Table 7. Total fatty acid and EFA (%) content in *C. edule* in the field and after the exposure to a range of salinity treatments.

3.2.2.2. *Pollutants experiments*

A slightly increase at all FA groups was observed in the organisms exposed to the herbicide compared with the individuals from the field (**Table 8**). An opposite pattern was found in the organisms exposed to copper where a slender decrease in SFA (at the first metal treatment), PUFA and HUFA was observed compared to the individuals from the field. Comparing the HUFA content in the organisms exposed to both pollutants, a slightly increase was observed at the herbicide treatments with a reduction at the copper concentrations. DHA was the EFA occurring in higher amount after the exposure to commercial formulation of the herbicide. In the treatments exposed to the metal, a clear pattern was not observed. In general, there was a higher diversity in FA in the individuals exposed to Primextra compared to the organisms under copper treatments, with exception to the highest copper concentration that registered the highest diversity of FA (**Table 8**).

	Primextra® Gold TZ (mg/L)								Copper (mg/L)				
	Field	CTL	0.5	2.5	5	10	20	30	Field	CTL	0.6	0.9	1.2
% total SFA	0.074	0.129	0.204	0.0744	0.155	0.130	0.165	0.098	0.016	0.001	0.005	0.052	0.077
% total MUFA	0.019	0.041	0.057	0.023	0.039	0.036	0.039	0.023	0.025	0.030	0.029	0.029	0.031
% total PUFA	0.051	0.108	0.061	0.049	0.066	0.054	0.069	0.048	0.153	0.042	0.078	0.086	0.121
% total HUFA	0.113	0.187	0.258	0.110	0.221	0.196	0.224	0.150	0.116	0.084	0.159	0.144	0.075
EPA	0.047	0.084	0.095	0.050	0.069	0.075	0.070	0.062	0.084	0.042	0.094	0.074	0.034
DHA	0.067	0.102	0.164	0.061	0.156	0.121	0.154	0.088	0.032	0.041	0.065	0.070	0.040
N	11	12	15	12	11	13	13	12	11	9	12	12	16

Table 8. Total fatty acid and EFA (%) content in *C. edule* in the field and after the exposure to a set concentrations of the herbicide Primextra® Gold TZ and the metal copper.

3.2.2.3. Fatty acid trophic markers (FATMs): Cerastoderma edule

In **Table 9**, the FA composition of the food (rotifers and microalgae) used daily to feed *C. edule* in the lab is represented. Microalgae food source presented a highest richness in PUFA whereas rotifers showed higher composition in SFA.

FATM ratios indicated an omnivorous diet with the organisms ingesting phytoplankton and zooplankton, with some of the individuals in the lab and at the field consuming higher amounts of phytoplankton (diatoms). Still some organisms from the field also showed a diet based mainly on zooplankton with few also feeding on bacteria.

	Food source	
	Microlagae	**Rotifera**
% total SFA	7.7	62.7
% total MUFA	3.0	23.0
% total PUFA	86.5	9.0
% total HUFA	3.0	5.4

Table 9. Fatty acid composition of food source used daily to feed *Cerastoderma edule* in the lab.

4. Conclusion

In general, an organism under stress conditions may change its physiological and biochemical responses as a strategic mechanism to compensate the organism's requirements [20]. In this study, the species from different trophic levels (a marine microalgae and estuarine bivalve species) exposed to a range of salinity concentrations under distinct temperature conditions and under different treatments of a metal (copper) and a herbicide (Primextra) revealed changes in its fatty acid content. The results obtained confirmed that the environmental and chemical stressors affect the fatty acid profile of aquatic species with sharp changes in the FA content of these species and reflecting then in lower quality food. It was clear that higher temperature had a great impact on the FA composition of the microalga, with the diatom not presenting several FA mainly PUFA, DHA and EPA in its profile, making the microalga more vulnerable to the effects of different salinities, mainly under salinity concentrations lower than 22.59. In the salinity treatments, an increase of SFA and MUFA was also observed to both studied species which can indicate that PUFA and HUFA are being metabolized so that the cells can obtain more energy necessary to maintain homeostatic ionic balance while maintaining basal functions of the body, such as respiration and excretion of ammonia. Since synthesizing of PUFAs and HUFAs becomes too energetically costly, the organism may not complete the elongation processes presenting higher concentrations of SFAs and MUFAs in the FA composition. The increase in saturation level can also be explained by a cellular response toward the osmotic shock, an attempt to maintain the stability of lipid membranes [49] and as such maintain the osmotic pressure from cell damaging. A low concentration of HUFA in general may also occur due to the high sensitiveness of this FA group to environmental fluctuations and to cell stress.

In literature studies, the most characteristic FA of diatoms are included in SFA and MUFA group and are characteristic of the plant domain: C14:0, C16:0, C16:1, C18:4 and C20:5, precursors for LC-PUFA such as EPA and DHA, while in turn most diatoms are reported to be very poor in C18:2 and C18:3 [50, 51]. In the present study, the absence of most of these FA at both species with a wide lacking of SFA is clearly shown. Fisher [52] stated *Thalassiosira pseudonana* Hasle & Heimdal was dominated by the FA reported above, although in different relative amounts. Fisher [52] also observed that the concentration of C16:0 varied with the culture age and that C16:1 declined drastically in the dark. The LC-PUFA content of microalgae can indeed depend not only on the species, but also on factors related to culture condition including composition of the medium, aeration, light intensity, temperature and age of culture [53], being crucial algae be maintained under its optimality of growth conditions.

In bivalve species, the reproductive success is related to the presence of great amount of lipids that are the second major constituent of bivalves' eggs [54, 55]. The maturation of the germ cell is closely related to the FA C18:3n3, C18:4n3, C20:1n9 and C20:2n6, while C14:0, C16:0, C16:1n7, C18:1n7 and C18:1n9 play a key role at the embryonic development of the bivalve eggs [54, 56]. Thus, it is crucial the presence of these FA during the breeding and thus in the FA profiles of the individuals.

The low concentration of HUFA reported in the lowest salinity treatments (22.59 and 14.79) at 25°C to *T. weissflogii* and to *C. edule* exposed to the metal may translate a metabolic response from the cells, which cannot efficiently maintain the cellular homeostatic ionic balance and cannot use energetic reservations in the elongation and synthesis of LC-PUFA such as EPA (C20:5n3) and DHA (C22:6n3). In the salinity treatments, EFAs of both studied species (the diatom and the bivalve) were mainly represented by ARA that is an important precursor of signalling molecules including prostaglandins, prostacyclins and thromboxanes [57]. DHA was the EFA that presented the lowest concentration or was absent in all salinity treatments at the highest temperature (25°C) and also under the exposure to both pollutants of the diatom species. A similar pattern was reported to the bivalve species exposed to saline and copper concentrations. This EFA is crucial in neurophysiologic processes and influences visual acuity once is abundant in retina cells, as well as a preventive role in cardiac diseases being also presented in high amount in the brain tissues [4, 16]. In fact, a high concentration of EFA such as the combination DHA + EPA is linked to the reduction of coronary heart diseases. These two FA molecules also show a key role in bipolar disorder and other neurological dysfunctions, cognitive functions and fetal development. They are associated with benefits in the treatment of rheumatoid arthritis and inflammatory bowel disease, as well as for Crohn's disease, which is related with the suppression of ARA-derived eicosanoids [4]. Once EPA and ARA play an important role in mediating immunological responses to infections and regulating ion and water flux, a low or absent content of such EFAs can translate into alterations in membrane phospholipids once ARA and DHA components can influence cellular signalling but also sharply alter many membrane physical properties such as fluidity and bilayer thickness, among others. In humans, a deficiency of DHA affects neurotransmission, membrane-bound enzyme and ion channel activities, intensity of inflammation and immunity, all of these associated with normal aging, Alzheimer disease, hyperactivity and schizophrenia [13]. These EFAs (EPA and ARA) are also associated to the further improvement of adaption of the individuals to anthropogenic and environmental stress conditions [58]. Furthermore, C18:3n3 (α LNA) is associated with the processes of elongation and desaturation at the synthesis of EPA and DHA in mammals, whereas C18:2n6, C20:4n6 and C18:3n6 are involved in the biosynthesis of PUFA of long chain [59].

The level of essential fatty acids (n-3 EFA) in algae can be highly variable, [60], including among major taxa [61], making it hard to compare between phylogenetic close species. PUFAs and HUFAs can reach their highest content during periods of rapid cell growth or bloom episodes [62] and are important components of the microalgae membranes that affect cell membrane fluidity [3] that promotes a rapid response to environmental changes, such as variations in temperature, light and pH. Thus, HUFAs are the most affected FA group in case of cellular damage due to failed osmoregulation. The reduction of EFA content, mainly in the base of the trophic food web, may have serious implications at higher levels in the food web, once the compromising of the nutritious value of the primary producers influences the uptake of EFAs and thus the fundamental processes of regulation of many species in the ecosystems. Some of these species are the food source of human beings, having the climate events in larger scale a great impact at the biochemical values of nutritional requirements of many aquatic species and thus at human health. Thus, a balanced fatty acid profile is essential, with a balanced amount of essential fatty acids (EFAs) and of other fatty acids with a key role in the regulation and functioning of the organisms.

Fatty acid proves to be a useful bio-indicator of ecological and healthy status of aquatic ecosystems, providing crucial information about the impacts of global stressors in the aquatic communities and thus in the trophic food web with severe repercussions to human beings and food quality. Recent reports predict the occurrence, at the next 100 years, of changes in salinity seawater, rise in temperature and water acidification [63]. In addition to these climate changes, an intensive agriculture production with an excessive usage of fertilizers and pesticides near coastal wetlands will have severe impacts to the aquatic communities and thus to the ecosystem. Therefore, it is of major importance and becomes a priority to determine and predict the effects of environmental and anthropogenic stressors in the aquatic systems in order to maintain the healthy status and the biodiversity and thus the food quality.

Acknowledgements

This study was supported by the Fundação para a Ciência e Tecnologia (FCT) through the strategic projects UID/MAR/04292/2013 granted to MARE and UID/AMB/50017/2013 granted to CESAM. A.M.M. Gonçalves also thanks the Fundação para a Ciência e Tecnologia (FCT) for the financial support provided through the post-doctoral grant SFRH/BPD/97210/2013 co-funded by the Human Potential Operational Programme (National Strategic Reference Framework 2007–2013), European Social Fund (EU) and the program POPH/FSE.

Author details

Ana M.M. Gonçalves[1,2*], João C. Marques[1] and Fernando Gonçalves[2]

*Address all correspondence to: anamartagoncalves@ua.pt; anamartagoncalves@gmail.com

1 IMAR (Marine and Environmental Research Centre) & MARE (Marine and Environmental Sciences Centre), Faculty of Sciences and Technology, University of Coimbra, Coimbra, Portugal

2 Department of Biology and CESAM, University of Aveiro, Aveiro, Portugal

References

[1] Richmond CE, Wethey DS, Woodin SA. Climate change and increased environmental variability: Demographic responses in an estuarine harpacticoid copepod. Ecological Modelling. 2007;**209**(2):189-202

[2] Benatti P, Peluso G, Nicolai R, Calvani M. Polyunsaturated fatty acids: Biochemical, nutritional and epigenetic properties. Journal of the American College of Nutrition. 2004;**23**(4):281-302

[3] Bret MT, Müller-Navarra DC. The role of highly unsaturated fatty acids in aquatic food web processes. Freshwater Biology. 1997;**13**:1055-1063

[4] Arts MT, Ackman RG, Holub BJ. Essential fatty acids in aquatic ecosystems: A crucial link between diet and human health and evolution. Canadian Journal of Fisheries and Aquatic Sciences. 2001;**58**(1):122-137

[5] Liu JJ, Green P, Mann JJ, Rapoport SI, Sublette ME. Pathways of polyunsaturated fatty acid utilization: Implications for brain function in neuropsychiatric health and disease. Brain Research. 2015;**1597**:220-246

[6] Stillwell W, Wassall SR. Docosahexaenoic acid: Membrane properties of a unique fatty acid. Chemistry and Physics of Lipids. 2003;**126**:1-27

[7] Calder PC. Polyunsaturated fatty acids and inflammation. Biochemical Society Trans actions. 2005;**33**(2):423-427

[8] Kattner G, Hagen W. Lipids in marine copepods: Latitudinal characteristics and perspective to global warming. In: Arts MT, Brett MT, Kainz MJ, editors. Lipids in Aquatic Ecosystems. USA (New York): Springer; 2009. pp. 257-280. DOI: 10.1007/978-0-387-89366-2

[9] Sargent JR, Bell JG, Bell MV, Henderson RJ, Tocher DJ. The metabolism of phospholipids and polyunsaturated fatty acids in fish. In: Lahlou B, Vitiello P, editors. Aquaculture: Fundamental and Applied Research Coastal and Estuarine Studies. USA (Washington, D C): American Geophysical Union; 1993. pp. 103-112

[10] Veloza AJ, Chu FLE, Tang KW. Trophic modification of essential fatty acids by heterotrophic protists and its effects on the fatty acid composition of the copepod *Acartia tonsa*. Marine Biology. 2006;**148**(4):779-788

[11] Coutteau P, Sorgeloos P. Manipulation of dietary lipids, fatty acids and vitamins in zooplankton cultures. Freshwater Biology. 1997;**38**(3):501-512

[12] Ladhar C, Ayadi H, Denis F, Tastard E, Sellami I. The effect of environmental factors on the fatty acid composition of copepods and *Artemia* in the Sfax solar saltern (Tunisia). Biochemical Systematics and Ecology. 2014;**56**:237-245

[13] Horrocks LA, Akhlaq AF. Docosahexaenoic acid in the diet: Its importance in maintenance and restoration of neural membrane function. Prostaglandins, Leukotrienes and Essential Fatty Acids. 2004;**70**(4):361-372

[14] Innis SM. Dietary (n-3) fatty acids and brain development. The Journal of Nutrition. 2007;**137**(4):855-859

[15] Kelly JR, Scheibling RE. Fatty acids as dietary tracers in benthic food webs. Marine Ecology Progress Series. 2012;**446**:1-22

[16] Dalsgaard J, John MS, Kattner G, Müller-Navarra D, Hagen W. Fatty acid trophic markers in the pelagic marine environment. Advances in Marine Biology. 2003;**46**:225-340

[17] Filimonova V, Gonçalves F, Marques JC, De Troch M, Gonçalves AMM. Fatty acid profiling as bioindicator of chemical stress in marine organisms: A review. Ecological Indicators. 2016a;**67**:657-672. DOI: 10.1016/j.ecolind.2016.03.044

[18] Filimonova V, Gonçalves F, Marques JC, De Troch M, Gonçalves AMM. Biochemical and toxicological effects of organic (herbicide Primextra® Gold TZ) and inorganic (copper) compounds on zooplankton and phytoplankton species. Aquatic Toxicology. 2016b;**177**:33-43. DOI: http://dx.doi.org/10.1016/j.aquatox.2016.05.008

[19] Gonçalves AMM, Mesquita AF, Verdelhos T, Coutinho JAP, Marques JC, Gonçalves F. Fatty acids' profiles as indicators of stress induced by of a common herbicide on two marine bivalves species: *Cerastoderma edule* (Linnaeus, 1758) and *Scrobicularia plana* (da Costa, 1778). Ecological Indicators. 2016;**63**:209-218. DOI: 10.1016/j.ecolind.2015.12.006

[20] Gonçalves AMM, Barroso DV, Serafim TL, Verdelhos T, Marques JC, Gonçalves F. The biochemical response of two commercial bivalve species to exposure to strong salinity changes illustrated by selected biomarkers. Ecological Indicators Journal. 2017;**76**:59-66. DOI: 10.1016/j.ecolind.2017.01.020

[21] Caudill CC, Bucklin A. Molecular phylogeography and evolutionary history of the estuarine copepod, *Acartia tonsa*, on the Northwest Atlantic coast. Hydrobiologia. 2004;**511**: 91-102

[22] Boesch DF. A new look at the zonation of benthos along the estuarine gradient. Ecology of Marine Benthos. 1977;**6**:245-66

[23] Beaugrand G, Ibañez F, Reid PC. Spatial, seasonal and long-term fluctuations of plankton in relation to hydroclimatic features in the English Channel, Celtic Sea and Bay of Biscay. Marine Ecology Progress Series. 2000;**200**:93-102

[24] Cardoso PG, Raffaelli D, Lillebø AI, Verdelhos T, Pardal MA. The impact of extreme flooding events and anthropogenic stressors on the macrobenthic communities' dynamics. Estuarine, Coastal and Shelf Science. 2008;**76**(3):553-565

[25] Holliday DK, Elskus AA, Roosenburg WM. Impacts of multiple stressors on growth and metabolic rate of *Malaclemys terrapin*. Environmental Toxicology and Chemistry. 2009;**28**(2):338-345

[26] Kinne O. The effects of temperature and salinity on marine and brackish water animals—I. Temperature. Oceanography and Mairne Biology, An Annual Review. 1963;**1**:1301-1340

[27] Williams WD, Geddes MC. Anostracans of Australian salt lakes, with particular reference to a comparison of *Parartemia* and *Artemia*. In: Browne RA, Sorgeloos P, Trotman CNA, editors. Artemia Biology. Boca Raton: CRC Press; 1991. pp. 351-368

[28] Rippingdale RJ, Hodgkin EP. Food availability and salinity tolerance in a brackish water copepod. Marine and Freshwater Research. 1977;**28**(1):1-7

[29] Gonçalves AMM, Pardal MA, Marques SC, Mendes S, Fernández-Gómez MJ, Galindo-Villardón MP, Azeiteiro UM. Responses of Copepoda life-history stages to climatic variability in a Southern-European temperate estuary. Zoological Studies. 2012;**51**(3):321-335

[30] Grilo TF, Cardoso PG, Dolbeth M, Bordalo MA, Pardal MA. Effects of extreme climate events on the macrobenthic communities' structure and functioning of a temperate estuary. Marine Pollution Bulletin. 2011;**62**:303-311

[31] Verdelhos T, Cardoso PG, Dolbeth M, Pardal MA. Recovery trends of *Scrobicularia plana* populations after restoration measures, affected by extreme climate events. Marine Environmental Research. 2014;**98**:39-48

[32] Barman TE. Enzyme Handbook. Supplement I. Heidelberg, Germany: Springer Verlag GmbH; 1974. pp. 504. DOI: 10.1007/978-3-662-11691-3

[33] Ritter A, Goulitquer S, Salaün JP, Tonon T, Correa JA, Potin P. Copper stress induces biosynthesis of octadecanoid and eicosanoid oxygenated derivatives in the brown algal kelp *Laminaria digitata*. New Phytologist. 2008;**180**:809-821. http://dx.doi.org/10.1111/j.1469-8137.2008.02626.x

[34] Sibi G, Anuraag TS, Bafila G. Copper stress on cellular contents and fatty acid profiles in *Chlorella* species. Online Journal of Biological Sciences. 2014;**14**:209-217. http://dx.doi.org/10.3844/ojbssp.2014.209.217

[35] Galhano V, Gomes-Laranjo J, Peixoto F. Exposure of the cyanobacterium *Nostoc muscorum* from Portuguese rice fields to Molinate (Ordram®): Effects on the antioxidant system and fatty acid profile. Aquatic Toxicology. 2011;**101**:367-376. http://dx.doi.org/10.1016/j.aquatox.2010.11.011

[36] Araújo CFC, Souza-Santos LP. Use of the microalgae *Thalassiosira weissflogii* to assess water toxicity in the Suape industrial-port complex of Pernambuco, Brazil. Ecotoxicology and Environmental Safety. 2013;**89**:212-221. http://dx.doi.org/10.1016/j.ecoenv.2012.11.032

[37] Borges L, Morón-Villarreyes JA, D'Oca MGM, Abreu PC. Effects offlocculants on lipid extraction and fatty acid composition of the microalgae *Nannochloropsis oculata* and *Thalassiosira weissflogii*. Biomass Bioenergy. 2011;**35**:4449-4454. http://dx.doi.org/10.1016/j.biombioe.2011.09.003

[38] Verdelhos T, Marques JC, Anastácio P. The impact of estuarine salinity changes on the bivalves *Scrobicularia plana* and *Cerastoderma edule*, illustrated by behavioral and mortality responses on a laboratory assay. Ecological Indicators. 2015;**52**:96-104. DOI: 10.1016/j.ecolind.2014.11.022

[39] Freitas R, Camino BM, Martins RC, Figueira E, Soares AMVM, Montaudouin X. Trematode communities in cockles (*Cerastoderma edule*) of the Ria deAveiro (Portugal): Influence of inorganic contamination. Marine Pollution Bulletin. 2014;**82**:117-126

[40] Paul-Pont I, Gonzalez P, Baudrimont M, Nili H, de Montaudouin X. Short term metallothionein inductions in the edible cockle *Cerastoderma edule* after cadmium or mercury exposure: Discrepancy between mRNA and protein responses. Aquatic Toxicology. 2010;**97**:260-267

[41] Paul-Pont I. Paillard C, Raymond N, Baudrimont M. Interactive effects of metal contamination and pathogenic organisms on the marine bivalve *Cerastoderma edule*. Marine Pollution Bulletin. 2010;**60**:515-525

[42] Cardoso PG, Grilo TF, Pereira E, Duarte AC, Pardal MA. Mercury bioaccumulation and decontamination kinetics in the edible cockle *Cerastoderma edule*. Chemosphere. 2013;**90**:1854-1859

[43] Nilin J, Pestana JLT, Ferreira NG, Loureiro S, Costa-Lotufo LV, Soares AMVM. Physiological responses of the European cockle *Cerastoderma edule* (Bivalvia: Cardidae) as indicators coastal lagoon pollution. Science of the Total Environment. 2012;**435-436**:44-52

[44] Cunha PP, Dinis J. Sedimentary dynamics of the Mondego estuary. In: Pardal MA, Marques JC, Graça MA, editors. Aquatic Ecology of the Mondego River Basin—Global Importance of Local Experience. Portugal (Coimbra): Imprensa da Universidade de Coimbra; 2002. pp. 43-62. DOI: http://dx.doi.org/10.14195/978-989-26-0336-0_4

[45] Guillard RR, Ryther JH. Studies of marine planktonic diatoms: I. *Cyclotella nana Hustedt*, and *Detonula* confervacea (CLEVE) Gran. Canadian Journal of Microbiology. 1962;**8**(2):229-239

[46] Rippingale RJ, Payne MF. Intensive cultivation of the calanoid copepod *Gladioferens imparipes*. Aquaculture. 2001;**201**:329-342. http://dx.doi.org/10.1016/S0044-8486(01)00608-1

[47] Eder K. Gas chromatographic analysis of fatty acid methyl esters. Journal of Chromato graphy. 1995;**671**:113-131

[48] Finney DJ, editors. Probit Analysis. 3rd ed. England (Cambridge): Cambridge University Press; 1971. p. 333. DOI: 10.1002/jps.2600600940

[49] Popova OB, Sanina NM, Likhatskaya GN, Bezverbnaya IP. Effects of copper and cadmium ions on the physicochemical properties of lipids of the marine bacterium *Pseudomonas putida* IB28 at different growth temperatures. Russian Journal of Marine Biology. 2008;**34**(3):179-185

[50] Verity PG, Smayda TJ. Nutritional value of *Phaeocystis pouchetii* (Prymnesiophyceae) and other phytoplankton for *Acartia* spp. (Copepoda): Ingestion, egg production and growth of nauplii. Marine Biology. 1989;**100**:161-171

[51] Chuecas L, Riley JP. Component fatty acids of the total lipids of some marine phytoplankton. Journal of the Marine Biological Association of the United Kingdom. 1969;**49**(01): 97-116

[52] Fisher NS, Schwarzenbach, RP.Fatty acid dynamics in Thalassiosira Pseudonana (Bacillariophyceae): Implications for physiological ecology. Journal of Phycology. 1978;**14**(2):143-150

[53] Dunstan GA, Volkman JK, Barrett SM, Leroi JM, Jeffrey SW. Essential polyunsaturated fatty acids from 14 species of diatom (Bacillariophyceae). Phytochemistry. 1993;**35**(1): 155-161

[54] Baptista M, Repolho T, Maulvault AL, Lopes VM, Narciso L, Marques A, Bandarra N, Rosa R. Temporal dynamics of amino and fatty acid composition in the razor clam *Ensis siliqua* (Mollusca: Bivalvia). Helgoland Marine Research. 2014;**68**:465-482

[55] Ruano F, Ramos P, Quaresma M, Bandarra NM, Fonseca IP. Evolution of fatty acid profile and Condition Index in mollusc bivalves submitted to different depuration periods. Revista Portuguesa de Ciências Veterinárias. 2012;**107**(581-582):75-84

[56] Ojea J, Pazos AJ, Martínez D, Novoa S, Sánchez JL, Abad M. Seasonal variation in weight and biochemical composition of the tissues of *Ruditapes decussatus* in relation to the gametogenic cycle. Aquaculture. 2004;**238**:451-468

[57] Schlechtriem C, Arts MT, Zellmer ID. Effect of temperature on the fatty acid composition and temporal trajectories of fatty acids in fasting *Daphnia pulex* (Crustacea, Cladocera). Lipids. 2006;**41**(4):397-400

[58] Fokina NN, Ruokolainen TR, Nemova NN, Bakhmet IN. Changes of blue mussels *Mytilus edulis* L. lipid composition under cadmium and copper toxic effect. Biological Trace Element Research. 2013;**154**:217-225. DOI: http://dx.doi.org/10.1007/s12011- 013-9727-3

[59] Givens I, Baxter S, Minihane AM, Shaw E, editors. Health Benefits of Organic Food: Effects of the Environment. Trowbridge: Cromwell Press; 2008. p. 315. ISBN: 13: 978 1 84593 459 0

[60] Morris RJ, McCartney MJ, Robinson GA. Studies of a spring phytoplankton bloom in an enclosed experimental ecosystem. I. Biochemical changes in relation to the nutrient chemistry of water. Journal of Experimental Marine Biology and Ecology. 1983;**70**(3):249-262

[61] Sargent JR, Whittle KJ. Lipids and hydrocarbons in the marine food web. In: Longhurst AR, editors. Analysis of Marine Ecosystems. London: Academic Press; 1981. pp. 491-533. ISBN 0-12-455560-8

[62] Fraser AJ, Sargent JR, Gamble JC. Lipid class and fatty acid composition of *Calanus finmarchicus* (Gunnerus), *Pseudocalanus* sp. and *Temora longicornis* Müller from a nutrient-enriched seawater enclosure. Journal of Experimental Marine Biology and Ecology. 1989;**130**(1):81-92

[63] IPCC. Summary for policymakers. In: Stocker TF, Qin D, Plattner G-K, Tignor M, Allen SK, Boschung J, Nauels A, Xia Y, Bex V, Midgley PM, editors. Climate Change 2013: The Physical Science Basis. Contribution of Working Group I to the Fifth Assessment Report of the Intergovernmental Panel on Climate Change. Cambridge, United Kingdom and New York, NY, USA: Cambridge University Press; 2013

[64] Budge SM, Parrish CC. Lipid biochemistry of plankton, settling matter and sediments in Trinity Bay, Newfoundland II. Fatty acids. Organic Geochemistry. 1998;**29**:1547-1559.

[65] Dunstan GA, Volkman JK, Barrett SM, Leroi JM, Jeffrey SW. Essential polyunsaturated fatty acids from 14 species of diatom (Bacillariphyceae). Phytochemistry. 1994;**35**: 155-161.

[66] Kharlamenko VI, Kyiashko SI, Imbs AB, Vyshkvartzev DI. Identification of food sources of invertebrates from the seagrass Zostera marina community using carbon and sulphur stable isotope ratio and fatty acid analyses. Marine Ecology Progress Series. 2001;**220**: 103-117

[67] Mansour MP, Volkman JK, Jackson AE, Blackburn SI. The fatty acid and sterol composition of five marine dinoflagellates. Journal of Phycology. 1999;**35**: 717-720

[68] Mayzaud P, Chanut JP, Ackman RG. Seasonal changes on the biochemical composition of the marine particulate matter with special reference of fatty acids and sterols. Marine Ecology Progress Series. 1989;**56**:189-204.

[69] Nadjek M, Debobbis D, Mioković D, Ivančić I. Fatty acid and phytoplankton composition of different types of mucilaginous aggregates in the Northen Adriatic. Journal of Lipid Research. 2002;**5**:600-608.

[70] Zhukova NV, Kharlamenko VI. Sources of essential fatty acids in the marine microbial loop. Aquatic Microbial Ecology. 1999;**17**:153-157

13

Fatty Acids from Microalgae: Targeting the Accumulation of Triacylglycerides

Paola Scodelaro Bilbao, Gabriela A. Salvador and Patricia I. Leonardi

Abstract

Microalgae were originally considered as sources of long-chain polyunsaturated fatty acids (PUFAs), mainly for aquaculture purposes. However, based on the fact that their fatty acids (FA), stored as triacylglycerides (TAG), can be converted into biodiesel via a transesterification reaction, several microalgal species have emerged over the last decade as promising feedstocks for biofuel production. Elucidation of microalgae FA and TAG metabolic pathways is therefore becoming a cutting-edge field for developing transgenic algal strains with improved lipid accumulation ability. Furthermore, many of the biomolecules produced by microalgae can also be exploited. In this chapter, we describe recent advances in the field of FA and TAG pathways in microalgae, focusing in particular on the enzymes involved in FA and TAG synthesis, their accumulation in lipid droplets, and their degradation. Mention is made of potentially high-value products that can be obtained from microalgae, and possible molecular targets for enhancing FA and TAG production are outlined. A summary is provided of transcriptomics, proteomics, and metabolomics of the above-mentioned pathways in microalgae. Understanding the relation between anabolic and catabolic lipid enzyme pathways will provide new insights into biodiesel production and other valuable biomolecules obtained from microalgae.

Keywords: fatty acids, triacylglycerides, lipid metabolism, microalgae

1. Introduction

Despite the drop in crude oil prices over the last few years, global efforts to develop alternative renewable energy sources continue to be driven by increasing air pollution and growing energy consumption. Extensive research is therefore being conducted in the field of biofuels

[1], which are derived from renewable biological sources. Biodiesel is the main substitute for diesel fuel and can be produced from both edible and non-edible oils. The use of edible oils has generated controversy because of the negative impact on food availability and the environment [2, 3]. As a consequence of these ethical considerations, non-food crops have emerged as a viable alternative for the production of biodiesel [4–6]. However, since non-food crops do not produce sufficient biomatter to feasibly cover the fuel requirements of the world's transport sector, attention is turning to oleaginous microalgae which are able to produce and accumulate large amounts of fatty acids (FA) in the form of triacylglycerides (TAG) that can be converted into biodiesel through a transesterification reaction [2, 3, 7]. Furthermore, some species of oleaginous microalgae can also produce high-value products such as long-chain polyunsaturated fatty acids (docosahexaenoic (DHA) and eicosapentaenoic (EPA) acids), carbohydrates (cellulose, starch), proteins, and other high-value compounds, such as pigments, antioxidants (i.e., β-carotene, astaxanthin), and vitamins, which may have commercial application in various industrial sectors [2, 3, 8, 9]. In addition to their potential as biological factories, the advantage of these photosynthetic microorganisms is that their simple growing requirements (light, $CO_{2,}$ and nutrients) offer several environmental benefits such as high solar energy conversion efficiency, utilization of saline water, CO_2 sequestration from the air and self-purification if coupled with wastewater treatment [10].

Despite the wide range of metabolites able to be synthesized by microalgae, little is known about the regulation of FA and TAG biosynthetic pathways and their storage and turnover in microalgae. In this chapter, we therefore describe recent advances in these fields and possible high-value co-products that could render the production of biodiesel from microalgae more sustainably. Recent studies on the transcriptomics, proteomics, and metabolomics of the above-mentioned pathways are also outlined. Understanding these metabolic pathways will accelerate the availability of biodiesel and other valuable biomolecules obtained from microalgae.

2. FA and TAG biosynthetic pathways in microalgae

Fatty acids are organic acids containing a carboxylic functional group with an aliphatic chain that can be saturated (SFA), monounsaturated (MUFA), or polyunsaturated (PUFA). The number of carbon atoms can vary, generating short-chain, medium-chain, or long-chain FA.

In plants, the FA biosynthetic pathway occurs in the chloroplasts (**Figure 1**).

As shown in **Figure 1**, the first step in the pathway involves the acetyl-CoA carboxylase (ACCase) which catalyzes the formation of malonyl-CoA from acetyl-CoA and bicarbonate [11]. There is evidence suggesting the presence of genes encoding this enzyme (accA and accD) in *Chlorella pyrenoidosa*. In fact, the transcription of these genes showed to be up-regulated under lipid accumulating conditions [12]. Moreover, a marked increase in the level of acetyl-CoA together with a moderate augmentation of malonyl-CoA and CoA was detected in the green microalgae *Chlorella desiccata, Dunaliella tertiolecta,* and *Chlamydomonas reinhardtii* under stress conditions, denoting increased activity of ACCase in these strains [13].

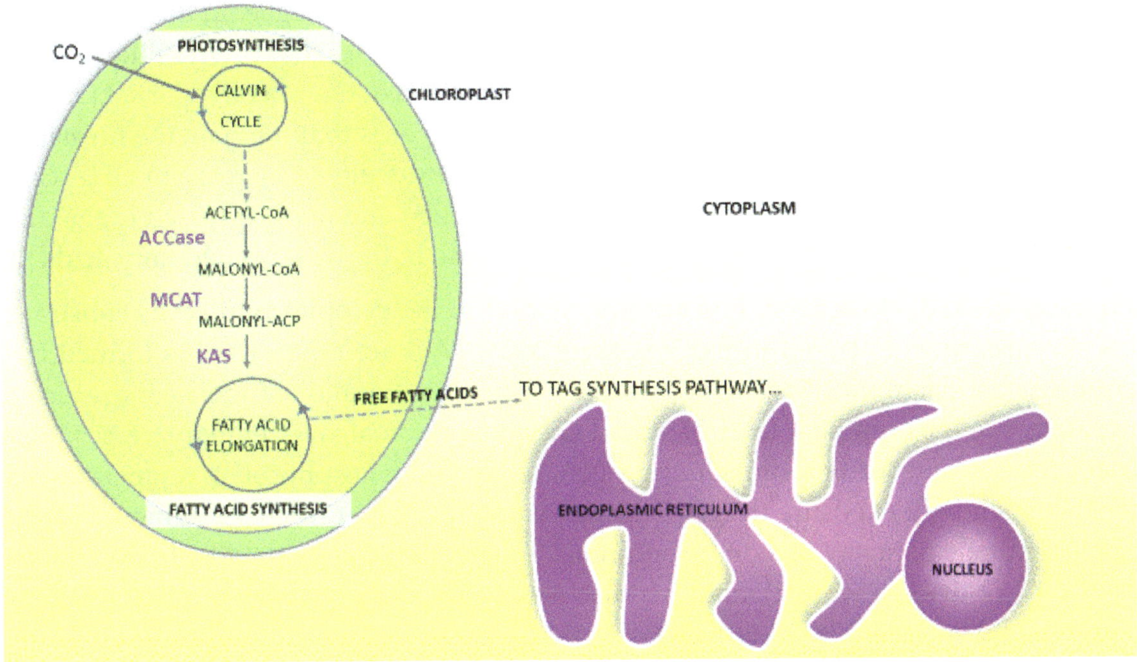

Figure 1. Simplified overview of the pathways involved in FA synthesis in plants. Enzyme abbreviations: ACCase, acetyl-CoA carboxylase; MCAT, malonyl-CoA:Acyl Carrier Protein (ACP) transacylase; KAS, ketoacyl-ACP synthases.

The next step in the FA synthesis is mediated by the malonyl-CoA:Acyl Carrier Protein (ACP) transacylase (MCAT) which transfers the malonyl group from malonyl-CoA to malonyl-ACP [11]. A putative MCAT was identified as a part of the FA biosynthetic pathway in *Nannochloropsis oceanica* [14]. In *Haematococcus pluvialis*, the genes encoding ACP were up-regulated under TAG accumulating conditions (high temperature, high salinity, and nitrogen deficiency) together with other genes involved in FA biosynthesis [15]. In addition, proteomic studies on *Neochloris oleoabundans* revealed an augmented expression of ACP, among other enzymes of the lipid synthesis, under nitrogen starvation [16].

Acyl-ACP is the carbon source or substrate for the elongation of FA. This reaction is catalyzed by enzymes known as ketoacyl-ACP synthases (KASIII, KASI, and KASII). After each condensation, a reduction, dehydration, and second reduction occur. These steps are catalyzed by enzymes known as the FAS complex: beta-ketoacyl-ACP reductase (KAR), hydroxyacyl-ACP dehydrase (HAD), and enoyl-ACP reductase (EAR), respectively [11]. Transcriptome analysis of the diatom *Chaetoceros* sp. GSL56 helped to identify putative enzymes of the FA synthesis pathway. In addition, replacement of ketoacyl-ACP synthase of *Synechococcus* 7002 with *Chaetoceros* ketoacyl-ACP synthase III induced FA synthesis [17]. In line with this, TAG accumulating conditions increased the levels of transcripts for KAS in *H. pluvialis* [15].

The *de novo* resulting FA often with 16 or 18 carbon atoms can undergo the action of elongases and desaturases that add carbon or double bonds, respectively [11]. Particularly, desaturases and elongases are being intensively studied to achieve transgenic long-chain PUFA production [18, 19].

Some reports suggest the presence of both enzyme types in microalgae. In the marine microalgae *Pavlova* sp. and *Isochrysis sp.*, two genes encoding elongases that catalyze the elongation of eicosapentaenoic acid (EPA) to docosahexaenoic acid (DHA) have been reported [20]. In the diatom *Thalassiosira pseudonana*, the genes encoding elongases that mediate the formation of DHA from EPA were successfully overexpressed, thus inducing an increase in DHA content [19]. A delta 5 desaturase was also identified, characterized and overexpressed in the diatom *Phaeodactylum tricornutum* inducing a significant increase in the unsaturated fatty acids [21] .

Upon completion of elongation, FAs are transported to the cytoplasm to act as substrates of the acyl transferases involved in the TAG synthesis. TAG are neutral lipids formed by the esterification of one molecule of glycerol with three FAs. Because of their energy-rich acyl chains, they are the dominant form of stored energy in microalgae. Cellular stresses, such as nutrient deprivation (carbon dioxide, nitrogen, silica, and phosphorous), temperature fluctuation, or high light exposure trigger their formation [22–28]. It has been demonstrated that lipid biosynthetic pathways are induced under these conditions to potentiate the lipid storage (30–60% of dry cell weight), and this mechanism is thought to play a role in microalgae adaptation and survival [24, 29–39]. It has further been reported that multiple stressors have no additive effect on lipid accumulation [24, 40].

Data relating to plant FA and TAG metabolism provided the key to identifying possible molecular targets involved in lipid synthesis and accumulation in microalgae [41]. As shown in **Figure 2**, in plants, the first step of the conventional Kennedy pathway involves the acylation of the glycerol-3-phosphate (G-3-P), catalyzed by the glycerol-3-phosphate acyltransferase (GPAT) to yield lysophosphatidic acid (LPA). GPAT is the rate-limiting step subject to many regulatory controls at the transcriptional and post-transcriptional level and to allosteric mechanisms [42, 43]. Recent studies have revealed the presence of this enzyme in microalgae. In the marine diatom *T. pseudonana*, a membrane-bound GPAT designated TpGPAT was cloned and characterized. The authors observed that G-3-P was the preferred substrate of TpGPAT [44]. A sequence for GPAT with high homology to that of plants was found in *C. reinhardtii, Volvox carteri, Ostreococcus lucimarinus, Ostreococcus tauri, Cyanidioschyzon merolae*, and *P. tricornutum*. As in *T. pseudonana*, G-3-P and fatty acyl molecules are likely to be the enzyme substrates, as suggested by the residues present in their active sites [45].

As described in **Figure 2**, lysophosphatidic acid acyltransferase (LPAAT) participates in the second step of the Kennedy pathway. This enzyme catalyzes the acylation of the LPA to yield phosphatidic acid (PA) [46]. Candidate LPAATs have been found in some algal genomes including that of *H. pluvialis* [47, 48], where it has been shown that LPAAT mRNA is induced under high irradiance stress [47]. In addition, it was recently reported that the expression of *C. reinhardtii* LPAAT (CrLPAAT1) is associated with an increase in lipid synthesis and accumulation under nitrogen starvation [48].

Phosphatidic acid phosphohydrolase (PAP) uses PA as substrate to form diacylglycerol (DAG), a precursor of TAG (**Figure 2**) [49]. In eukaryotes, PAP enzymes are the members of the evolutionarily conserved lipin protein family whose activity is related to TAG storage [50]. In the green microalga *C. reinhardtii*, PAP transcripts (named CrPAP2) are induced under stress conditions. In addition, CrPAP2 silencing slightly lowers the lipid content. Thus, in *C. reinhardtii*, as in other eukaryotes, PAP expression is related to lipid synthesis and accumulation [49].

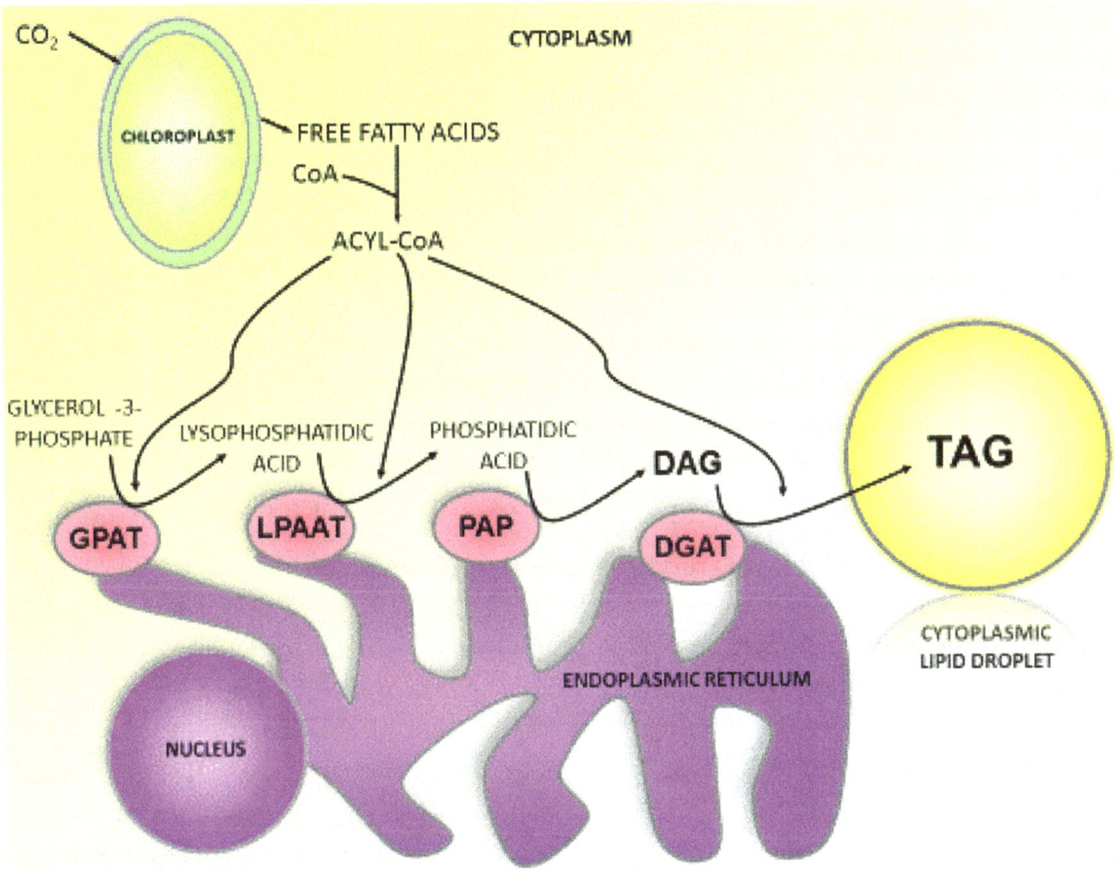

Figure 2. Simplified overview of the pathways involved in TAG synthesis in plants. Enzymes of the conventional Kennedy pathway involved in TAG synthesis and their subcellular localization in plants. Enzyme abbreviations: glycerol-3-phosphate acyltransferase (GPAT); lysophosphatidic acid acyltransferase (LPAAT); phosphatidic acid phosphohydrolase (PAP); diacylglycerol acyltransferase (DAGAT or DGAT). The same enzymes are involved in TAG synthesis in microalgae, but their intracellular localization has not yet been determined.

The last enzyme of the *de novo* TAG synthesis is acyl-CoA:diacylglycerol acyltransferase (DGAT), which catalyzes the acylation of DAG to yield TAG (**Figure 2**) [51]. This enzyme employs DAG and acyl-CoA as substrates, so the resulting TAG is formed through an acyl-CoA-dependent pathway [46] and is a key target to increase TAG synthesis and storage through genetic manipulation [52, 53]. In higher plants, three different types of DGATs participate in the formation of TAG: DGAT1, DGAT2, or DGAT3 [54]. Sequences for DGAT1 and DGAT2 isoforms were found in several algal strains [55]. Sequences for DGAT2, but not DAGT1, or DGAT3, were identified in the green microalga *O. tauri* [56]. DGAT2 was also found in *T. pseudonana* (TpDGAT2). In addition, the expression of DGAT in a TAG-null yeast mutant restored the synthesis of these neutral lipids [57]. In the oleaginous microalga *C. pyrenoidosa* grown under stress conditions, a high correlation was found between DGAT and TAG accumulation [58]. Also in *N. oceanica* IMET1, another oleaginous microalga, seven putative DGAT genes were up-regulated under nitrogen-deficient conditions, when the synthesis of TAG-neutral lipids was significantly increased [59]. In *C. reinhardtii* dgat1 and dgtt1 to dgtt5 genes encode for DGAT1 and DGAT2, respectively [60, 61]. Increased transcript expression of the genes dgat1 and dgtt1 was detected under stress conditions (less sulfur, phosphorous, iron, zinc, or nitrogen). Once more, the evidence suggests that both DGAT1 and DGAT2 could play a role in TAG synthesis as their expression is induced under TAG-accumulating conditions [62, 63]. In support of this hypothesis, overexpression of a DGAT2 isoform in the marine diatom *P. tricornutum* stimulated the synthesis of neutral lipids and their accumulation in lipid droplets [64].

As can be observed, much research has focused on the acyl-CoA-dependent reaction catalyzed by DGAT. However, the relative contribution of DGAT1 and DGAT2 isoenzymes to TAG accumulation appears to be species-dependent, so further studies should be performed to gain insight into this aspect.

TAG can be formed by the acyl-CoA-dependent pathway, detailed previously, or through acyl-CoA-independent reactions. Acyl-CoA-independent formation of TAG is mediated by the activities of two types of enzyme: the phospholipid:diacylglycerol acyltransferases (PDAT), which catalyze the formation of TAG using DAG and phosphatidylcholine (PC); and the DAG:DAG transacylases (DGTA) which utilize two molecules of DAG to form TAG and MAG [54, 65].

In fact, in *N. oceanica* IMET1, it was reported that membrane polar lipids were converted into TAG when the microalgae were grown under nitrogen deficiency [59]. In agreement with this, the gene encoding the acyltransferase PDAT1 was induced under nitrogen starvation in *C. reinhardtii*. Moreover, TAG content in the *C. reinhardtii* PDAT-null mutant was 25% lower than in the parent strain. It would thus appear that PDAT has a relevant role in TAG accumulation, stimulating the transacylation pathway in both strains [62]. Furthermore, in *C. reinhardtii* it was suggested that PDAT functions as a DGTA with acyl hydrolase activity. PDAT might, therefore, mediate membrane polar lipid turnover in a favorable environment whereas under stress conditions it may participate in phospholipid degradation contributing to TAG synthesis [66].

As already mentioned, many aspects of *C. reinhartii* lipid metabolism have already been characterized, making it the microalga of choice for current purposes [23, 67–73]. Nevertheless, *Chlamydomonas* is a non-oleaginous strain [23]. Other microalgal species with greater potential to yield biodiesel and other high-value products should therefore be more thoroughly investigated.

3. Transcriptomics, proteomics, and metabolomics

A better understanding of the mechanisms involved in TAG enrichment under stress conditions will help to maximize microalgae productivity. However, many biochemical approaches for elucidating molecular pathways depend on the availability of genomic sequence data [29]. Transcriptomics, proteomics, and metabolomics, however, are able to provide a detailed description of cell transcripts (RNA), proteins and metabolites, respectively while completely bypassing the requirement of genomic information [74, 75].

Transcriptome analysis helped to identify sequences of the enzymes involved in the biosynthesis and catabolism of FA, TAG, and starch in *D. tertiolecta*, revealing that this strain shares genetic information, at least in terms of the mentioned pathways, with closely related microalgae species such as *V. carteri* and *C. reinhardtii* [76]. The transcriptome of *N. oleoabundans* was also determined. In this case, the authors quantified the differences between nitrogen-replete and nitrogen-limiting culture conditions. Under nitrogen deficiency, *N. oleoabundans* showed higher levels of transcripts of FA and TAG synthesis pathways and inhibition of the FA β-oxidation pathway, compared to nitrogen-replete culture conditions [29]. In agreement with this finding, in *C. vulgaris*, transcriptomic [31] and proteomic [77] studies revealed an induction of the enzymes of the FA and TAG synthesis machinery under lipid enrichment conditions. Also, transcription factors associated with these metabolic pathways were augmented under the stress condition [77].

The transcriptome of *C. reinhardtii* showed that genes involved in FA and TAG metabolic pathways and in membrane remodeling were highly induced under neutral lipid accumulation conditions [78]. In this microalga, proteomic studies revealed an augmented rate of lipid synthesis machinery with a concomitant enhancement in FA and TAG; higher levels of starch than under non-stress conditions were also detected by metabolomic analyses. Metabolic pathways such as nitrogen assimilation, amino acid metabolism, oxidative phosphorylation, glycolysis, TCA cycle, and the Calvin cycle suffered adjustments during *C. reinhardtii* [79, 80].

As in *C. vulgaris*, nutrient-deprivation stress in *C. reinhardtii, D. tertiolecta*, and *N. oleoabundans* induced the expression of genes involved in FA and TAG synthesis pathways in *P. tricornutum* [81], *Chlorella protothecoides* [82], and *Tisochrysis lutea* [83].

In conclusion, these assembled transcriptomes, proteomes, and metabolomes offer valuable approaches for improving microalgal productivity, providing possible targets for molecular engineering that could enhance microalgae-derived products.

4. Molecular targets for enhancing lipid biosynthesis

Genetic strain modification to improve microalgal productivity and accelerate the industrialization of algal-derived products is a major challenge [84]. Reflecting the fact that enhancement of the FA synthesis pathway had little effect on total lipid content in some plants [85, 86], a growing body of research now focuses on overexpression of the enzymes or heterologous expression of genes involved in the TAG biosynthetic pathway. **Table 1** provides an outline of some of the genetic manipulations performed on several microalgal strains, leading to an improvement in their TAG content.

Enzymes overexpressed or heterologously expressed	Organism	Effect on lipid production (changes over control condition)	References
ME of *Phaeodactylum tricornutum*	*Chlorella pyrenoidosa*	18.7%	[87]
GPAT of *Thalassiosira pseudonana*	Yeast GPAT-deficient mutant	Restored TAG formation	[44]
GPAT of *Lobosphaera incisa*	*Chlamydomonas reinhardtii*	50%	[88]
GPAT of *Phaeodactylum tricornutum*	*Phaeodactylum tricornutum*	2-fold	[89]

Enzymes overexpressed or heterologously expressed	Organism	Effect on lipid production (changes over control condition)	References
G3PDH, GPAT, DGAT, LPAAT and PAP of *Saccharomyces cerevisiae* and *Yarrowia lipolytica*	*Chlorella minutissima*	2-fold	[90]
LPAAT of *Chlamydomonas reinhardtii*	*Chlamydomonas reinhardtii*	20%	[91]
DGAT2 of *Nannochloropsis oceanica*	*Nannochloropsis oceanica*	3.5-fold	[92]
DGAT1 and DGAT2 of *Myrmecia incisa*	*S. cerevisiae* lipid deficient mutant	Re-stored TAG formation	[93]
DGAT2 of *Neochloris oleoabundans*	*S. cerevisiae* DGAT deficient mutant	Re-stored TAG formation	[94]
DGAT 1 of *Phaeodactylum tricornutum*	*S. cerevisiae* DGAT deficient mutant	Re-stored TAG synthesis and lipid body formation	[63]
DGAT 2 of *Phaeodactylum tricornutum*	*Phaeodactylum tricornutum*	35%	[64]
Various DGAT type 2	*Chlamydomonas reinhardtii*	20–44%	[95]

Enzyme abbreviations: ME, malic enzyme; DGAT, diacylglycerol acyltransferase; G3PDH, glycerol-3-phosphate-dehydrogenase; GPAT, glycerol-3-phosphate acyltransferase; LPAAT, lysophosphatidic acid acyltransferase; PAP, phosphatidic acid phosphatase.

Table 1. Some of the genetic modifications performed on metabolic pathways related to lipid synthesis in microalgae and their effect on lipid enrichment.

5. TAG-accumulation in lipid droplets

Lipid droplets (LDs) are cell organelles that are currently the subject of in-depth study in various organisms. These lipid globules not only act as a reservoir of cell carbon and energy, they may also have a role in lipid homeostasis, signaling, trafficking, and interorganelle communications [96, 97]. As previously mentioned, under stress conditions microalgae synthesize TAG and store them as cytoplasmic LDs [22–28], which can vary in size, shape, and function depending on the cell type and the environmental conditions (**Figure 3**) [98]. In eukaryotic cells, LD structure consists of a TAG-rich hydrophobic core surrounded by surface polar glycerolipids into which proteins of the perilipin (Plin) (animal cells) or oleosin and caleosin (plants) families are embedded [99–102]. In microalgae, LD structure is conserved from eukaryotes but different LD proteins have been identified. The analysis of *C. reinhardtii* LDs recognized 16 proteins related to lipid metabolism and a major lipid droplet protein (named MLDP) was identified. MLDP silencing increased the size of the LD, without modifying LD TAG content [68]. In the green microalga, *Nannochloropsis* sp., a hydrophobic lipid droplet surface protein, named LDSP, was identified. The expression of LDSP increased concomitantly with TAG content under oil-accumulating conditions [99]. In *H. pluvialis*, seven proteins were found to be

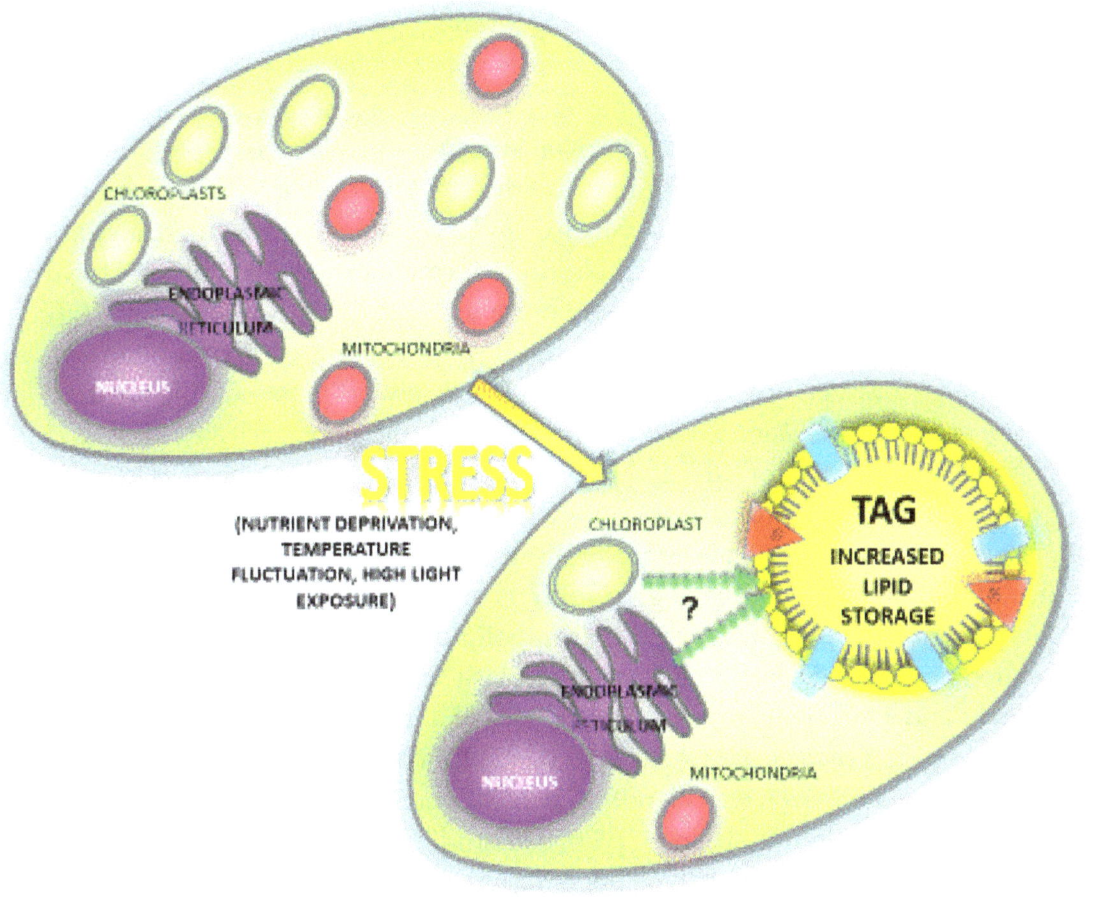

Figure 3. Schematic representation of a cytoplasmic lipid droplet (LD) from microalgae.

associated with LDs. The most abundant of these, *Haematococcus* Oil Globule Protein (HOGP), was homologous to the MLDP of *C. reinhardtii* and its expression was induced under TAG accumulating conditions [103]. LD-associated proteins may also help in the accumulation of TAG in the green microalga *Myrmecia incisa* [104]. Moreover, LDs from *C. reinhardtii* showed the presence of enzymes involved in TAG synthesis (GPAT, and PDAT) and in sterol synthesis, lipid signaling, and trafficking [69]. Further in-depth research should be able to determine the proteins associated with LDs and their role in TAG metabolism in microalgae.

In the oleaginous diatom *Fistulifera* sp., two proteins located in the oil bodies were also detected in the endoplasmic reticulum (ER), suggesting that oil bodies might originate in the ER [105]. The same authors found a signal sequence typical of ER localization in an LD protein called diatom-oleosome-associated-protein 1 (DOAP1) in *Fistulifera solaris* JPCC DA0580 [106]. Related to these findings, the induction of ER stress leads to LD formation in *C. reinhardtii* and *C. vulgaris* [107]. In addition, LDs from *C. reinhardtii* were associated not only with the ER membrane but also with the outer membrane of the chloroplasts [108]. Available data therefore suggest that in microalgae, cytoplasmic LDs are produced in the ER. However, additional studies are required to arrive at a better understanding of the mechanism of LD formation in the ER, and to determine whether chloroplasts play a role in this process.

6. TAG degradation pathways in microalgae

As previously mentioned, the economic feasibility of using microalgae as a source of FA for biodiesel depends to a great extent on improvements in the production process, one of the most significant challenges being to increase lipid yields. The selection of oleaginous strains and the search for different culture strategies to increase lipid biosynthesis constitute viable approaches; blocking the competing pathways of carbohydrate formation may be another. However, both the approaches give rise to a decrease in strain growth [22]. Lipid catabolism has largely been ignored as a relevant pathway for engineering, despite being a competing pathway to lipid biogenesis [109]. However, lipases were identified in *C. reinhardtii* [66, 72, 73] and *T. pseudonana* [110]. In the case of *C. reinhardtii,* CrLIP1 could restore the lipase activity in a *Saccharomyces cerevisiae* lipase-null strain. In addition, *C. reinhardtii* TAG content decreased with increasing expression of CrLIP1 under stress conditions, hydrolyzing mainly DAG and polar lipids [72]. In agreement with this, a galactoglycerolipid lipase was found in *C. reinhardtii.* The main substrates of the enzyme are galactoglycerolipids and the main products are FAs employed for TAG synthesis [74]. In *C. reinhardtii,* phospholipid:diacylglycerol acyltransferase (PDAT) demonstrated both transacylation and acyl hydrolase activities, and could mediate membrane lipid turnover and TAG synthesis [66]. The activity of a multifunctional lipase/phospholipase/acyltransferase of *T. pseudonana* lowered lipid content under both normal and stress conditions [110]. A single gene for PDAT was identified in *H. pluvialis,* though no functional analysis was performed for the gene in this strain [47]. Further studies are required to gain insight into the molecular mechanisms involved in TAG degradation, which could be the key to increased lipid yields in microalgae.

7. Microalgae-based biorefineries

In the context of improving the economic feasibility of microalgae-based biodiesel, a closer look should be taken at the large amounts of TAG produced in some oleaginous microalgae alongside high-value products such as carbohydrates (cellulose and starch); proteins and other high-value compounds like pigments, antioxidants (i.e., β-carotene, astaxanthin), and vitamins [2, 3, 8, 9], all of which may have commercial application in different industrial sectors. Some potentially high-value products found in microalgae are described in **Table 2**.

Product	Microalgal strain	Applications	References
Carbohydrates			
Exopolysaccharides	*Navicula cincta*	Pharmaceutics and agronomics	[111]
Starch, glucose, cellullose	*Chlorella vulgaris FSP-E*	Bioethanol production	[112, 113]
Sulfated extracellular polisaccharides	*Graesiella* sp.	Pharmaceutics	[114]
Lipids			

Product	Microalgal strain	Applications	References
Phytosterols; linoleic (C18:2n6) and alpha linolenic (C18:3n3) fatty acids	*Haematococcus pluvialis*	Human dietary supplement, nutraceutics	[115]
Phytosterols	*Dunaliella tertiolecta, D. salina*	Nutraceutics	[116]
Omega-3 long chain-PUFA	*Isochrysis, Nannochloropsis, Phaeodactylum, Pavlova*, and *Thalassiosira*	Functional food	[117]
Proteins			
Proteins	*Chlorella pyrenoidosa*	Nutraceutics	[118]
Proteins	*Chlorella vulgaris, Nannochloris bacillaris, Tetracystis* sp., *Micractinium reisseri*	Animal feed	[119]
Vitamins			
Tocopherol	*Nannochloropsis oculata, Tetraselmis suecica*	Human dietary supplement, nutraceutics	[120]
Tocopherol, pigments, phenolic compounds	*Desmodesmus* sp.	Human dietary supplement, nutraceutics	[121]
Pigments			
Astaxanthin	*Haematococcus pluvialis*	Antioxidant, cosmetics, pharmaceutics	[122]
Lutein	*Dunaliella salina*	Functional food, animal feed	[123]
Carotenoids	*Phaeodactylum tricornutum*	Cosmetics, pharmaceutics, animal feed	[124]
Carotenoids	*Dunaliella salina*	Cosmetics, pharmaceutics, animal feed	[125]
Others			
Silica	*Navicula cincta*	Abrassive products, insecticides	[111]

Table 2. Recent advances in microalgal-derived high-value products.

8. Conclusion

Oleaginous microalgae grown under stress conditions can synthesize and accumulate large quantities of FA, mainly in the form of TAG, which can then be converted into biodiesel. Although microalgae constitute a promising source of clean energy, knowledge gaps continue to abound in almost all aspects of FA and TAG metabolism for these microorganisms, including the precise identity of enzymatic machinery, the relative contributions of each enzyme and their precise regulation. Further studies are therefore required to establish the exact metabolic pathways involved in FA and TAG synthesis, accumulation, and degradation in order to develop genetic engineering strategies to obtain microalgal strains with improved capacity to convert their biomass into TAG and other valuable co-products.

Acknowledgements

The authors are grateful for research funds provided by the Consejo Nacional de Investigaciones Científicas y Técnicas de la República Argentina (CONICET); Agencia Nacional de Promoción Científica y Tecnológica, PICTs 2014-0893, 2013-0987, and 2015-0800; and the Secretaría de Ciencia y Tecnología de la Universidad Nacional del Sur, PGIs 24/B226 and 24/B196. Paola Scodelaro Bilbao, Gabriela Salvador and Patricia Leonardi are Research Members of CONICET.

Author details

Paola Scodelaro Bilbao[1,2,3]*, Gabriela A. Salvador[2,3] and Patricia I. Leonardi[1,2]

*Address all correspondence to: pscodela@criba.edu.ar

1 Laboratorio de Estudios Básicos y Biotecnológicos en Algas (LEBBA), Centro de Recursos Naturales Renovables de la Zona Semiárida (CERZOS), Consejo Nacional de Investigaciones Científicas y Técnicas (CONICET), Camino de La Carrindanga, Bahía Blanca, Argentina

2 Universidad Nacional del Sur (UNS), Departamento de Biología, Bioquímica y Farmacia, San Juan, Bahía Blanca, Argentina

3 Instituto de Investigaciones Bioquímicas de Bahía Blanca (INIBIBB), Consejo Nacional de Investigaciones Científicas y Técnicas (CONICET), Camino de La Carrindanga, Bahía Blanca, Argentina

References

[1] Misra N, Kumar Panda P, Kumar Parida B, Kanta Mishra B. Phylogenomic study of lipid genes involved in microalgal biofuel production: candidate gene mining and metabolic pathway analyses. *Evol Bioinform Online*. 2012; 8: 545-564. doi: 10.4137/EBO.S10159.

[2] Brennan L, Owende P. Biofuels from microalgae: a review of technologies for production, processing, and extractions of biofuels and co-products. *Renew Sust Energy Rev*. 2010; 14: 557-577. doi: 10.1016/j.rser.2009.10.009.

[3] Mata TM, Martinsa AA, Caetano NS. Microalgae for biodiesel production and other applications: a review. *Renew Sust Energy Rev*. 2012; 14: 217-232. doi: 10.1016/j.rser.2009.07.020.

[4] Foild N, Foild G, Sánchez M, Mittelbach M, Hackel S. J. curcas L. as a source for the production of biofuel in Nicaragua. *Bioresour Technol*. 1996; 58: 77-82.

[5] Rashid U, Anwar F, Moser BR, Knothe G. Moringa oleifera oil: a possible source of biodiesel. *Bioresour Technol*. 2008; 99: 8175-8179. doi: 10.1016/j.biortech.2008.03.066.

[6] Ahmad M, Zafar M, Azam A, Sadia H, Khan MA, Sultana S. Techno-economic aspects of biodiesel production and characterization. *Energy sources*. 2011; 6: 166-177.

[7] Yu WL, Ansari W, Schoepp NG, Hannon MJ, Mayfield SP, Burkart MD. Modifications of the metabolic pathways of lipid and triacylglycerol production in microalgae. *Microb Cell Fact*. 2011; 10: 91. doi: 10.1186/1475-2859-10-91.

[8] Williams PJ le B, Laurens LML. Microalgae as biodiesel and biomass feedstocks: review and analysis of the biochemistry, energetics and economics. *Energy Environ. Sci*. 2010; 3: 554-590. doi: 10.1039/B924978H.

[9] Borowitzka MA. High-value products from microalgae—their development and commercialization. *J Appl Phycol*. 2013; 25: 743-756. doi:10.1007/s10811-013-9983-9.

[10] Lam MK, Lee KT. Microalgae biofuels: a critical review of issues, problems and the way forward. *Biotechnol Adv*. 2012; 30: 673-690. doi: 10.1016/j.biotechadv.2011.11.008.

[11] Brown AP, Slabas AR, Rafferty JB. Fatty acid biosynthesis in plants-metabolic pathways, structure and organization. *Lipids in Photosynthesis*. 2010; 30: 11-34. doi: 10.1007/978-90-481-2863-1_2.

[12] Fan J, Cui Y, Wan M, Wang W, Li Y. Lipid accumulation and biosynthesis genes response of the oleaginous *Chlorella pyrenoidosa* under three nutrition stressors. *Biotech Biofuels*. 2014; 7: 17. doi: 10.1186/1754-6834-7-17.

[13] Avidan O, Brandis A, Rogachev I, Pick U. Enhanced acetyl-CoA production is associated with increased triglyceride accumulation in the green alga *Chlorella desiccate. J Exp Botany*. 2015; 66: 1-11. doi:10.1093/jxb/erv166.

[14] Chen JW, Liu W-J, Hu D-X, Wang X, Balamurugan S, Alimujiang A, Yang W-D, Liu J-S, Li H-Y. Identification of a malonyl CoA-acyl carrier protein transacylase and its regulatory role in fatty acid biosynthesis in oleaginous microalga *Nannochloropsis oceanica. Biotech Appl Biochem*. 2016. doi: 10.1002/bab.1531.

[15] Lei A, Chen H, Shen G, Hu Z, Chen L, Wang J. Expression of fatty acid synthesis genes and fatty acid accumulation in *Haematococcus pluvialis* under different stressors. *Biotechnol Biofuels*. 2012; 5: 18. doi: 10.1186/1754-6834-5-18.

[16] Morales-Sánchez D, Kyndtb J, Ogdenc K, Martinez A. Toward an understanding of lipid and starch accumulation in microalgae: a proteomic study of *Neochloris oleoabundans* cultivated under N-limited heterotrophic conditions. *Algal Res*. 2016; 20: 22-34. doi: 10.1016/j.algal.2016.09.006.

[17] Gu H, Jinkerson RE, Davies FK, Sisson LA, Schneider PE, Posewitz MC. Modulation of medium-chain fatty acid synthesis in *Synechococcus sp*. PCC 7002 by replacing FabH with a *Chaetoceros* Ketoacyl-ACP synthase. *Front Plant Sci*. 2016; 7: 690. doi: 10.3389/fpls.2016.00690.

[18] Peng KT, Zheng CN, Xue J, Chen XY, Yang WD, Liu JS, Bai W, Li HY. Delta 5 fatty acid desaturase upregulates the synthesis of polyunsaturated fatty acids in the marine diatom *Phaeodactylum tricornutum*. *J Agric Food Chem*. 2014; 62: 8773-8776. doi: 10.1021/jf5031086.

[19] Petrie JR, Liu Q, Mackenzie AM, Shrestha P, Mansour MP, Robert SS, Frampton DF, Blackburn SI, Nichols PD, Singh SP. Isolation and characterisation of a high-efficiency desaturase and elongases from microalgae for transgenic LC-PUFA production. *Mar Biotechnol*. 2010; 12: 430-438. doi: 10.1007/s10126-009-9230-1.

[20] Pereira SL, Leonard AE, Huang YS, Chuang LT, Mukerji P. Identification of two novel microalgal enzymes involved in the conversion of the omega3-fatty acid, eicosapentaenoic acid, into docosahexaenoic acid. *Biochem J*. 2004; 384: 357-366.

[21] Cook O, Hildebrand M. Enhancing LC-PUFA production in *Thalassiosira pseudonana* by overexpressing the endogenous fatty acid elongase genes. *J Appl Phycol*. 2016; 28: 897-905.

[22] Hu Q, Sommerfeld M, Jarvis E, Ghirardi M, Posewitz M, Seibert M, Darzins A. Microalgal triacylglycerols as feedstocks for biofuel production: perspectives and advances. *Plant J*. 2008; 54: 621-639. doi: 10.1111/j.1365-313X.2008.03492.x.

[23] Liu B, Benning C. Lipid metabolism in microalgae distinguishes itself. *Curr Opin Biotechnol*. 2013; 24: 300-309. doi: 10.1016/j.copbio.2012.08.008.

[24] Damiani MC, Popovich CA, Constenla D, Leonardi PI. Lipid analysis in *Haematococcus pluvialis* to assess its potential use as a biodiesel feedstock. *Bioresour Technol*. 2010; 101: 3801-3807. doi: 10.1016/j.biortech.2009.12.136.

[25] Popovich CA, Damiani C, Constenla D, Leonardi PI. Lipid quality of *Skeletonema costatum* and *Navicula gregaria* from South Atlantic Coast (Argentina): evaluation of its suitability as biodiesel feedstock. *J Appl Phycol*. 2012; 24: 1-10. doi:10.1007/s10811-010-9639-y.

[26] Popovich CA, Damiani MC, Constenla D, Martínez AM, Freije H, Giovanardi M, Pancaldi S, Leonardi PI. *Neochloris oleoabundans* grown in enriched natural seawater for biodiesel feedstock: evaluation of its growth and biochemical composition. *Bioresour Technol*. 2012; 114: 287-293. doi: 10.1016/j.biortech.2012.02.121.

[27] Bongiovani N, Popovich C, Martínez A, Freije H, Constenla D, Leonardi PI. *In vivo* measurements to estimate culture status and neutral lipid accumulation in *Nannochloropsis oculata* CCALA 978: implications for biodiesel oil studies. *Algological Studies*. 2013; 142: 3-16 doi: 10.1127/1864-1318/2013/0104.

[28] Martín L, Popovich C, Martinez A, Damiani C, Leonardi PI. Oil assessment of *Halamphora coffeaeformis* diatom growing in a hybrid two-stage system for biodiesel production. *Renewable Energy*. 2016; 92: 127-135. doi: 10.1016/j.renene.2016.01.078

[29] Rismani-Yazdi H, Haznedaroglu BZ, Hsin C, Peccia J. Transcriptomic analysis of the oleaginous microalga *Neochloris oleoabundans* reveals metabolic insights into triacylglyceride accumulation. *Biotechnol Biofuels*. 2012; 5: 74. doi: 10.1186/1754-6834-5-74.

[30] Yu ET, Zendejas FJ, Lane PD, Gaucher S, Simmons BA, Lane TW. Triacylglycerol accumulation and profiling in the model diatoms *Thalassiosira pseudonana* and *Phaeodactylum tricornutum* (Bacillariophyceae) during starvation. *J Appl Phycol*. 2009; 21: 669-681. doi: 10.1007/s10811-008-9400-y.

[31] Guarnieri MT, Nag A, Smolinski SL, Darzins A, Seibert M, et al. Examination of triacylglycerol biosynthetic pathways via de novo transcriptomic and proteomic analyses in an unsequenced microalga. PLoS One. 2011; 6: e25851. doi: 10.1371/journal.pone.0025851.

[32] Pal D, Khozin-Goldberg I, Cohen Z, Boussiba S. The effect of light, salinity, and nitrogen availability on lipid production by *Nannochloropsis* sp. *Appl Microbiol Biotechnol*. 2011; 90: 1429-1441. doi:10.1007/s00253-011-3170-1.

[33] Tang H, Chen M, Garcia ME, Abunasser N, Ng KY, Salley SO. Culture of microalgae *Chlorella minutissima* for biodiesel feedstock production. *Biotechnol Bioeng*. 2011; 108: 2280-2287. doi: 10.1002/bit.23160.

[34] Breuer G, Lamers PP, Martens DE, Draaisma RB, Wijffels RH. Effect of light intensity, pH, and temperature on triacylglycerol (TAG) accumulation induced by nitrogen starvation in *Scenedesmus obliquus*. *Bioresour Technol*. 2013; 143: 1-9. doi: 10.1016/j.biortech.2013.05.105.

[35] Gong Y, Guo X, Wan X, Liang Z, Jiang M. Triacylglycerol accumulation and change in fatty acid content of four marine oleaginous microalgae under nutrient limitation and at different culture ages. *J Basic Microbiol*. 2013; 53: 29-36. doi: 10.1002/jobm.201100487.

[36] Yu SY, Li H, Tong M, Ouyang LL, Zhou ZG. Identification of a Δ6 fatty acid elongase gene for arachidonic acid biosynthesis localized to the endoplasmic reticulum in the green microalga *Myrmecia incisa* Reisigl. *Gene*. 2012; 493: 219-227. doi: 10.1016/j.gene.2011.11.053.

[37] Aguirre AM, Bassi A. Investigation of biomass concentration, lipid production, and cellulose content in Chlorella vulgaris cultures using response surface methodology. *Biotechnol Bioeng*. 2013; 110: 2114-2122. doi: 10.1002/bit.24871.

[38] Simionato D, Block MA, La Rocca N, Jouhet J, Maréchal E, Finazzi G, Morosinotto T. The response of *Nannochloropsis gaditana* to nitrogen starvation includes de novo biosynthesis of triacylglycerols, a decrease of chloroplast galactolipids, and reorganization of the photosynthetic apparatus. *Eukaryot Cell*. 2013; 12: 665-676. doi: 10.1128/EC.00363-12.

[39] La Russa M, Bogen C, Uhmeyer A, Doebbe A, Filippone E, Kruse O, Mussgnug JH. Functional analysis of three type-2 DGAT homologue genes for triacylglycerol production in the green microalga *Chlamydomonas reinhardtii*. *J Biotechnol*. 2012; 162: 13-20. doi: 10.1016/j.jbiotec.2012.04.006.

[40] Roleda MY, Slocombe SP, Leakey RJ, Day JG, Bell EM, Stanley MS. Effects of temperature and nutrient regimes on biomass and lipid production by six oleaginous microalgae in batch culture employing a two-phase cultivation strategy. *Bioresour Technol*. 2013; 129: 439-449. doi: 10.1016/j.biortech.2012.11.043.

[41] Adarme-Vega TC, Lim DKY, Timmins M, Vernen F, Li Y, Schenk PM. Microalgal biofactories: a promising approach towards sustainable omega-3 fatty acid production. *Microb Cell Fact*. 2012; 11: 96. doi: 10.1186/1475-2859-11-96.

[42] Coleman RA, Lee DP. Enzymes of triacylglycerol synthesis and their regulation. *Prog Lipid Res*. 2004; 43: 134-176. doi: 10.1016/S0163-7827(03)00051-1.

[43] Courchesne NMD, Parisien A, Wang B, Lan CQ. Enhancement of lipid production using biochemical, genetic and transcription factor engineering approaches. *J Biotechnol*. 2009; 141: 31-41. doi: 10.1016/j.jbiotec.2009.02.018.

[44] Xu J, Zheng Z, Zou J. A membrane-bound glycerol-3-phosphate acyltransferase from *Thalassiosira pseudonana* regulates acyl composition of glycerolipids. *Botany*. 2009; 87: 544-551. doi: 10.1139/B08-145.

[45] Misra N, Panda PK, Parida BK. Agrigenomics for microalgal biofuel production: an overview of various bioinformatics resources and recent studies to link OMICS to bioenergy and bioeconomy. *OMICS*. 2013; 17: 537-549. doi: 10.1089/omi.2013.0025.

[46] Chapman KD, Ohlrogge JB. Compartmentation of triacylglycerol accumulation in plants. *J Biol Chem*. 2012; 287: 2288-2294. doi: 10.1074/jbc.R111.290072.

[47] Gwak Y, Hwang YS, Wang B, Kim M, Jeong J, Lee CG, Hu Q, Han D, Jin E. Comparative analyses of lipidomes and transcriptomes reveal a concerted action of multiple defensive systems against photooxidative stress in *Haematococcus pluvialis*. *J Exp Bot*. 2014; 65: 4317-4334. doi: 10.1093/jxb/eru206.

[48] Yamaoka Y, Achard D, Jang S, Legéret B, Kamisuki S, Ko D, Schulz-Raffelt M, Kim Y, Song WY, Nishida I, Li-Beisson Y, Lee Y. Identification of a *Chlamydomonas* plastidial 2-lysophosphatidic acid acyltransferase and its use to engineer microalgae with increased oil content. Plant Biotechnol J. 2016; 14: 2158-2167. doi: 10.1111/pbi.12572.

[49] Deng XD, Cai JJ, Fei XW. Involvement of phosphatidate phosphatase in the biosynthesis of triacylglycerols in *Chlamydomonas reinhardtii*. *J Zhejiang Univ Sci B*. 2013; 14: 1121-1131. doi: 10.1631/jzus.B1300180.

[50] Csaki LS, Dwyer JR, Fong LG, Tontonoz P, Young SG, Reue K. Lipins, lipinopathies, and the modulation of cellular lipid storage and signaling. *Progr Lipid Res*. 2013; 52: 305-316. doi: 10.1016/j.plipres.2013.04.001.

[51] Radakovits R, Jinkerson RE, Darzins A, Posewitz MC. Genetic engineering of algae for enhanced biofuel production. *Eukaryot Cell*. 2010; 9: 486-501. doi: 10.1128/EC.00364-09.

[52] Taylor DC, Katavic V, Zou J, MacKenzie SL, Keller WA, An J, Friesen W, Barton DL, Pedersen KK, Giblin EM, et al. Field testing of transgenic rapeseed cv. Hero transformed

with a yeast sn-2 acyltransferase results in increased oil content, erucic acid content and seed yield. *Mol Breed*. 2002; 8: 317-322. DOI: 10.1023/A:1015234401080.

[53] Lardizabal KD, Thompson GA, Hawkins D. Diacylglycerol acyl transferase proteins. In Official Gazette of the United States Patent and Trademark Office Patents Edited by: Office TUSPaT. USA, 2006.

[54] Lung SC, Weselake RJ. Diacylglycerol acyltransferase: a key mediator of plant triacylglycerol synthesis. *Lipids*. 2006; 41: 1073-1088.

[55] Chen JE, Smith AG. A look at diacylglycerol acyltransferases (DGATs) in algae. *J Biotechnol*. 2012; 162: 28-39. doi: 10.1016/j.jbiotec.2012.05.009.

[56] Wagner M, Hoppe K, Czabany T, Heilmann M, Daumb G, Feussner I, Fulda M. Identification and characterization of an acyl-CoA:diacylglycerol acyltransferase 2 (DGAT2) gene from the microalga *O. tauri*. *Plant Physiol Biochem*. 2010; 48: 407-416. doi: 10.1016/j.plaphy.2010.03.008.

[57] Xu J, Kazachkov M, Jia Y, Zheng Z, Zou J. Expression of a type 2 diacylglycerol acyltransferase from *Thalassiosira pseudonana* in yeast leads to incorporation of docosahexaenoic acid β-oxidation intermediates into triacylglycerol. *FEBS J*. 2013; 280: 6162-6172. doi: 10.1111/febs.12537.

[58] Fan J, Cui Y, Wan M, Wang W, Li Y. Lipid accumulation and biosynthesis genes response of the oleaginous *Chlorella pyrenoidosa* under three nutrition stressors. *Biotechnol Biofuels*. 2014; 7: 17. doi: 10.1186/1754-6834-7-17.

[59] Xiao Y, Zhang J, Cui J, Feng Y, Cui Q. Metabolic profiles of *Nannochloropsis oceanica* IMET1 under nitrogen-deficiency stress. *Bioresour Technol*. 2013; 130: 731-738. doi: 10.1016/j.biortech.2012.11.116.

[60] Miller R, Wu G, Deshpande RR, Vieler A, Gärtner K, Li X, Moellering ER, Zäuner S, Cornish AJ, Liu B, Bullard B, Sears BB, Kuo M-H, Hegg EL, Shachar-Hill Y, Shiu S-H, Benning C. Changes in transcript abundance in *Chlamydomonas reinhardtii* following nitrogen deprivation predict diversion of metabolism. *Plant Physiol*. 2010; 154: 1737-1752. doi: 10.1104/pp.110.165159.

[61] Msanne J, Xu D, Konda AR, Casas-Mollano JA, Awada T, Cahoon EB, Cerutti H. Metabolic and gene expression changes triggered by nitrogen deprivation in the photoautotrophically grown microalgae *Chlamydomonas reinhardtii* and *Coccomyxa* sp. C-169. *Phytochemistry*. 2012; 75: 50-59. doi: 10.1016/j.phytochem.2011.12.007.

[62] Boyle NR, Page MD, Liu B, Blaby IK, Casero D, Kropat J, Cokus SJ, Hong-Hermesdorf A, Shaw J, Karpowicz SJ, Gallaher SD, Johnson S, Benning C, Pellegrini M, Grossman A, Merchant SS. Three acyltransferases and nitrogen-responsive regulator are implicated in nitrogen starvation-induced triacylglycerol accumulation in *Chlamydomonas*. *J Biol Chem*. 2012; 287: 15811-1525. doi: 10.1074/jbc.M111.334052.

[63] Guihéneuf F, Leu S, Zarka A, Khozin-Goldberg I, Khalilov I, Boussiba S. Cloning and molecular characterization of a novel acyl-CoA:diacylglycerol acyltransferase 1-like

gene (PtDGAT1) from the diatom *Phaeodactylum tricornutum. FEBS J.* 2011: 278: 3651-3666. doi: 10.1111/j.1742-4658.2011.08284.x.

[64] Niu YF, Zhang MH, Li DW, Yang WD, Liu JS, Bai WB, Li HY. Improvement of neutral lipid and polyunsaturated fatty acid biosynthesis by overexpressing a type 2 diacylglycerol acyltransferase in marine diatom *Phaeodactylum tricornutum. Mar Drugs.* 2013; 11: 4558-4569. doi: 10.3390/md11114558.

[65] Dahlqvist A, Stahl U, Lenman M, Banas A, Lee M, Sandager L, Ronne H, Stymne S. Phospholipid:diacylglycerol acyltransferase: an enzyme that catalyzes the acyl-CoA-independent formation of triacylglycerol in yeast and plants. *Proc Natl Acad Sci USA.* 2000; 97: 6487-6492. doi: 10.1073/pnas.120067297.

[66] Yoon K, Han D, Li Y, Sommerfeld M, Hu Q. Phospholipid:diacylglycerol acyltransferase is a multifunctional enzyme involved in membrane lipid turnover and degradation while synthesizing triacylglycerol in the unicellular green microalga *Chlamydomonas reinhardtii. Plant Cell.* 2012; 24: 3708-3724. doi: 10.1105/tpc.112.100701.

[67] La Russa M, Bogen C, Uhmeyer A, Doebbe A, Filippone E, Kruse O, Mussgnug JH. Functional analysis of three type-2 DGAT homologue genes for triacylglycerol production in the green microalga *Chlamydomonas reinhardtii. J Biotechnol.* 2012; 162: 13-20. doi: 10.1016/j.jbiotec.2012.04.006.

[68] Moellering ER, Benning C. RNA interference silencing of a major lipid droplet protein affects lipid droplet size in *Chlamydomonas reinhardtii. Eukaryot Cell.* 2010; 9: 97-106. doi: 10.1128/EC.00203-09.

[69] Nguyen HM, Baudet M, Cuiné S, Adriano JM, Barthe D, Billon E, Bruley C, Beisson F, Peltier G, Ferro M, Li-Beisson Y. Proteomic profiling of oil bodies isolated from the unicellular green microalga *Chlamydomonas reinhardtii*: with focus on proteins involved in lipid metabolism. *Proteomics.* 2011; 11: 4266-4273. doi: 10.1002/pmic.201100114.

[70] Kato N, Dong T, Bailey M, Lum T, Ingram D. Triacylglycerol mobilization is suppressed by brefeldin A in *Chlamydomonas reinhardtii. Plant Cell Physiol.* 2013; 54: 1585-1599. doi:10.1093/pcp/pct103.

[71] Merchant SS, Kropat J, Liu B, Shaw J, Warakanont J. TAG, you're it! *Chlamydomonas* as a reference organism for understanding algal triacylglycerol accumulation. *Curr Opin Biotechnol.* 2012; 23: 352-363. doi: 10.1016/j.copbio.2011.12.001.

[72] Li X, Benning C, Kuoc M-H. Rapid Triacylglycerol Turnover in *Chlamydomonas reinhardtii* Requires a Lipase with Broad Substrate. *Eukaryot Cell.* 2012; 11: 1451-1462. doi: 10.1128/EC.00268-12.

[73] Li X, Moellering ER, Liu B, Johnny C, Fedewa M, Sears BB, Kuo MH, Benning C. A galactoglycerolipid lipase is required for triacylglycerol accumulation and survival following nitrogen deprivation in *Chlamydomonas reinhardtii. Plant Cell.* 2012; 24: 4670-4686. doi: 10.1105/tpc.112.105106.

[74] Guarnieri MT, Pienkos PT. Algal omics: unlocking bioproduct diversity in algae cell factories. *Photosynth Res.* 2014; 123: 255-263. doi: 10.1007/s11120-014-9989-4.

[75] Jamers A, Blust R, De Coen W. Omics in algae: paving the way for a systems biological understanding of algal stress phenomena? *Aquat Toxicol*. 2009; 92: 114-121. doi: 10.1016/j.aquatox.2009.02.012.

[76] Rismani-Yazdi H, Haznedaroglu BZ, Bibby K, Peccia J. Transcriptome sequencing and annotation of the microalgae *Dunaliella tertiolecta*: pathway description and gene discovery for production of next-generation biofuels. *BMC Genomics*. 2011; 12: 148. doi: 10.1186/1471-2164-12-148.

[77] Guarnieri MT, Nag A, Yang S, Pienkos PT. Proteomic analysis of *Chlorella vulgaris*: potential targets for enhanced lipid accumulation. *J Proteomics*. 2013; 93: 245-253. doi: 10.1016/j.jprot.2013.05.025.

[78] Lv H, Qu G, Qi X, Lu L, Tian C, Ma Y. Transcriptome analysis of *Chlamydomonas reinhardtii* during the process of lipid accumulation. *Genomics*. 2013; 101: 229-237. doi: 10.1016/j.ygeno.2013.01.004.

[79] Wase N, Black PN, Stanley BA, DiRusso CC. Integrated quantitative analysis of nitrogen stress response in *Chlamydomonas reinhardtii* using metabolite and protein profiling. *J Proteome Res*. 2014; 13: 1373-1396. doi: 10.1021/pr400952z.

[80] Valledor L, Furuhashi T, Recuenco-Muñoz L, Wienkoop S, Weckwerth W. System-level network analysis of nitrogen starvation and recovery in *Chlamydomonas reinhardtii* reveals potential new targets for increased lipid accumulation. *Biotechnol Biofuels*. 2014; 7: 171. doi: 10.1186/s13068-014-0171-1.

[81] Yang ZK, Ma YH, Zheng JW, Yang WD, Liu JS, Li HY. Proteomics to reveal metabolic network shifts towards lipid accumulation following nitrogen deprivation in the diatom *Phaeodactylum tricornutum*. *J Appl Phycol*. 2014; 26: 73-82. doi: 10.1007/s10811-013-0050-3.

[82] Li Y, Han F, Xu H, Mu J, Chen D, Feng B, Zeng H. Potential lipid accumulation and growth characteristic of the green alga *Chlorella* with combination cultivation mode of nitrogen (N) and phosphorus (P). *Bioresour Technol*. 2014; 174: 24-32. doi: 10.1016/j.biortech.2014.09.142.

[83] Garnier M, Carrier G, Rogniaux H, Nicolau E, Bougaran G, Saint-Jean B, Cadoret JP. Comparative proteomics reveals proteins impacted by nitrogen deprivation in wild-type and high lipid-accumulating mutant strains of *Tisochrysis lutea*. *J Proteomics*. 2014; 105: 107-120. doi: 10.1016/j.jprot.2014.02.022.

[84] Guihéneuf F, Khan A, Tran L-S. Genetic Engineering: a promising tool to engender physiological, biochemical, and molecular stress resilience in green microalgae. *Front Plant Sci*. 2016; 7: 400 doi: 10.3389/fpls.2016.00400.

[85] Roesler K, Shintani D, Savage L, Boddupalli S, Ohlrogge J. Targeting of the Arabidopsis homomeric acetyl-coenzyme A carboxylase to plastids of rapeseeds. Plant Physiol. 1997; 113:75-81.

[86] Dehesh K, Tai H, Edwards P, Byrne J, Jaworski JG. Overexpression of 3-ketoacyl-acyl-carrier protein synthase IIIs in plants reduces the rate of lipid synthesis. Plant Physiol. 2001; 125:1103-14.

[87] Xue J, Wang L, Zhang L, Balamurugan S, Li D-W, Zeng H, Yang W-D, Liu J-S, Li H-Y. The pivotal role of malic enzyme in enhancing oil accumulation in green microalga *Chlorella pyrenoidosa. Microb Cell Fact*. 2016; 15: 120. doi: 10.1186/s12934-016-0519-2.

[88] Iskandarov U, Sitnik S, Shtaida N, Cohen SD,Stefan, Khozin-Goldberg L, Cohen Z, Boussiba S. Cloning and characterization of a GPAT-like gene from the microalga *Lobosphaera incisa* (Trebouxiophyceae): overexpression in *Chlamydomonas reinhardtii* enhances TAG production. *J Appl Phycol*. 2016; 28: 907-919. doi:10.1007/s10811-015-0634-1.

[89] Niu Y-F, Wang X, Hu D-X, Balamurugan S, Li D-W, Yang W-D, Liu J-S, Li H-Y. Molecular characterization of a glycerol-3-phosphate acyltransferase reveals key features essential for triacylglycerol production in *Phaeodactylum tricornutum. Biotechnol Biofuels*. 2016; 9: 60. doi: 10.1186/s13068-016-0478-1.

[90] Hsieh H-J, Su C-H, Chien L-J. Accumulation of Lipid Production in *Chlorella minutissima* by triacylglycerol biosynthesis-related genes cloned from *Saccharomyces cerevisiae* and *Yarrowia lipolytica. J Microbiol*. 2012; 50: 526-534. doi: 10.1007/s12275-012-2041-5.

[91] Yamaoka Y, Achard D, Jang S, Legéret B, Kamisuki S, Ko D, Schulz-Raffelt M, Kim Y, Song W-Y, Nishida I, Li-Beisson Y, Lee Y. Identification of a *Chlamydomonas* plastidial 2-lysophosphatidic acid acyltransferase and its use to engineer microalgae with increased oil content. *Plant Biotechnol J*. 2016; 14: 2158-2167. doi: 10.1111/pbi.12572

[92] Li DW, Cen SY, Liu YH, Balamurugan S, Zheng XY, Alimujiang A, Yang WD, Liu JS, Li HY. A type 2 diacylglycerol acyltransferase accelerates the triacylglycerol biosynthesis in heterokont oleaginous microalga *Nannochloropsis oceanica. J Biotechnol*. 2016; 229: 65-71. doi: 10.1016/j.jbiotec.2016.05.005.

[93] Chen C-X, Sun Z, Cao H-S, Fang F-L, Ouyang L-L, Zhou Z-G. Identification and characterization of three genes encoding acyl-CoA: diacylglycerol acyltransferase (DGAT) from the microalga *Myrmecia incisa. Algal Res*. 2015; 12: 280-288. doi: 10.1016/j.algal.2015.09.007.

[94] Chungjatupornchai W, Watcharawipas A. Diacylglycerol acyltransferase type 2 cDNA from the oleaginous microalga *Neochloris oleoabundans*: cloning and functional characterization. *J Appl Phycol*. 2015; 27: 1499. doi:10.1007/s10811-014-0448-6

[95] Deng X-D, Gu B, Li Y-J, Hu X-W, Guo J-C, Fei X-W. The roles of acyl-CoA:diacylglycerol acyltransferase 2 genes in the biosynthesis of triacylglycerols by the green algae *Chlamydomonas reinhardtii. Mol Plant*. 2012; 5: 945-947. doi: 10.1631/jzus.B1300180.

[96] Murphy D. The biogenesis and functions of lipid bodies in animals, plants and microorganisms. *Prog Lipid Res*. 2001; 40, 325-438.

[97] Farese RV Jr, Walther TC. Lipid droplets finally get a little R-E-S-P-E-C-T. *Cell*. 2009; 139: 855-860. doi: 10.1016/j.cell.2009.11.005.

[98] Fujimoto T, Parton RG. Not just fat: the structure and function of the lipid droplet. *Cold Spring Harb Perspect Biol*. 2011; 3: a004838. doi: 10.1101/cshperspect.a004838.

[99] Vieler A, Brubaker SB, Vick B, Benning C. A lipid droplet protein of *Nannochloropsis* with functions partially analogous to plant oleosins. *Plant Physiol.* 2012; 158: 1562-1569. doi: 10.1104/pp.111.193029

[100] Kimmel AR, Brasaemle DL, McAndrews-Hil M, Sztalryd C, Londos C. Adoption of PERILIPIN as a unifying nomenclature for the mammalian PAT-family of intracellular lipid storage droplet proteins. *J Lipid Res.* 2010; 51: 468-471. doi: 10.1194/jlr.R000034.

[101] Subramanian V, Rothenberg A, Gomez C, Cohen AW, Garcia A, Bhattacharyya S, Shapiro L, Dolios G, Wang R, Lisanti MP, et al. Perilipin A mediates the reversible binding of CGI-58 to lipid droplets in 3T3-L1 adipocytes. *J Biol Chem.* 2004; 279: 42062-42071. doi: 10.1074/jbc.M407462200.

[102] Jolivet P, Roux E, D'Andrea S, Davanture M, Negroni L, Zivy M, Chardot T. Protein composition of oil bodies in *Arabidopsis thaliana* ecotype WS. *Plant Physiol Biochem.* 2004; 42: 501-509. doi: 10.1016/j.plaphy.2004.04.006.

[103] Peled E, Leu S, Zarka A, Weiss M, Pick U, Khozin-Goldberg I, Boussiba S. Isolation of a novel oil globule protein from the green alga *Haematococcus pluvialis* (Chlorophyceae). *Lipids.* 2011; 46: 851-861. doi: 10.1007/s11745-011-3579-4.

[104] Ouyang LL, Chen SH, Li Y, Zhou ZG. Transcriptome analysis reveals unique C4-like photosynthesis and oil body formation in an arachidonic acid-rich microalga *Myrmecia incisa* Reisigl H4301. *BMC Genomics.* 12013; 4: 396. DOI: 10.1186/1471-2164-14-396.

[105] Nojima D, Yoshino T, Maeda Y, Tanaka M, Nemoto M, Tanaka T. Proteomics analysis of oil body-associated proteins in the oleaginous diatom. *J Proteome Res.* 2013; 12: 5293-5301.

[106] Maeda Y, Sunaga Y, Yoshino T, Tanaka T. Oleosome-associated protein of the oleaginous diatom *Fistulifera solaris* contains an endoplasmic reticulum-targeting signal sequence. *Mar Drugs.* 2014; 12: 3892-3903. doi: 10.3390/md12073892.

[107] Kim S, Kim H, Ko D, Yamaoka Y, Otsuru M,Kawai-Yamada M, Toshiki Ishikawa T, Oh H-M, Nishida I, Li-Beisson Y, Lee Y. Rapid Induction of Lipid Droplets in *Chlamydomonas reinhardtii* and *Chlorella vulgaris* by Brefeldin A. *PLoS One* 2013; 8: e81978. doi: 10.1371/journal.pone.0081978.

[108] Goodson C, Roth R, Wang ZT, Goodenough U. Structural correlates of cytoplasmic and chloroplast lipid body synthesis in *Chlamydomonas reinhardtii* and stimulation of lipid body production with acetate boost. *Eukariot Cell.* 2011; 10: 1592-1606. doi: 10.1128/EC.05242-11.

[109] Solovchenko AE. Physiological role of neutral lipid accumulation in eukaryotic microalgae under stresses. *Russian J Plant Physiol.* 2012; 59: 167-176. doi:10.1134/S1021443712020161.

[110] Trentacoste EM, Shrestha RP, Smith SR, Glé C, Hartmann AC, Hildebrand M, Gerwick WH. Metabolic engineering of lipid catabolism increases microalgal lipid accumulation

without compromising growth. *Proc Natl Acad Sci USA*. 2013; 110: 19748-19753. doi: 10.1073/pnas.1309299110.

[111] Barnech Bielsa G, Popovich CA, Rodríguez MC, Martín LA, Matulewicz MC, Leonardi PI. Simultaneous production assessment of triacylglycerols for biodiesel and exopolysaccharides as valuable co-products in *Navicula cincta. Algal Res*. 2016; 15: 120-128. doi: 10.1016/j.algal.2016.01.013.

[112] Ho S-H, Huang S-W, Chen C-Y, Hasunuma T, Kondo A, Chang J-S. Bioethanol production using carbohydrate-rich microalgae biomass as feedstock. *Bioresour Technol*. 2013; 135: 191-198. Doi: 10.1016/j.biortech.2012.10.015.

[113] Kumar VB, Pulidindi IN, Kinel-Tahan Y, Yehoshua Y, Gedanken A. Evaluation of the potential of chlorella vulgaris for bioethanol production. *Energy Fuels*. 2016; 30: 3161-3166. doi: 10.1021/acs.energyfuels.6b00253.

[114] Trabelsi L, Chaieb O, Mnari A, Abid-Essafi S, Aleya L. Partial characterization and antioxidant and antiproliferative activities of the aqueous extracellular polysaccharides from the thermophilic microalgae *Graesiella sp. BMC Complement Altern Med*. 2016; 16: 210. doi: 10.1186/s12906-016-1198-6.

[115] Scodelaro Bilbao PG, Damiani C, Salvador GA, Leonardi PI. *Haematococcus pluvialis* as a source of fatty acids and phytosterols: potential nutritional and biological implications. *J Appl Phycol*. 2016; 28: 3283. doi:10.1007/s10811-016-0899-z.

[116] Francavilla M, Trotta P, Luque R. Phytosterols from *Dunaliella tertiolecta* and *Dunaliella salina*: a potentially novel industrial application. *Bioresour Technol*. 2010; 101: 4144-4150. doi: 10.1016/j.biortech.2009.12.139.

[117] Ryckebosch E, Bruneel C, Termote-Verhalle R, Goiris K, Muylaert K, Foubert I. Nutritional evaluation of microalgae oils rich in omega-3 long chain polyunsaturated fatty acids as an alternative for fish oil. *Food Chem*. 2014; 160: 393-400. doi: 10.1016/j.foodchem.2014.03.087.

[118] Waghmare AG, Manoj K. Salve MK, Jean Guy LeBlanc JG, Shalini S. Arya SS Concentration and characterization of microalgae proteins from *Chlorella pyrenoidosa. Biores Bioproc*. 2016; 3:16. doi: 10.1186/s40643-016-0094-8.

[119] Tibbetts SM, Whitney CG, MacPherson MJ, Bhatti S, Banskota AH, Stefanova R, McGinn PJ. Biochemical characterization of microalgal biomass from freshwater species isolated in Alberta, Canada for animal feed applications. *Algal Res*. 2015; 11: 435-447. doi: 10.1016/j.algal.2014.11.011.

[120] Bong SC, Loh SP. A study of fatty acid composition and tocopherol content of lipid extracted from marine microalgae, *Nannochloropsis oculata* and *Tetraselmis suecica*, using solvent extraction and supercritical fluid extraction. *Int Food Res J*. 2013; 20: 721-729.

[121] Safafar H, van Wagenen J, Møller P, Jacobsen C. Carotenoids, phenolic compounds and tocopherols contribute to the antioxidative properties of some microalgae species grown on industrial wastewater. *Mar Drugs*. 2015; 13: 7339-7356. doi: 10.3390/md13127069.

[122] Shah MR, Liang Y, Cheng JJ, Daroch M. Astaxanthin-producing green microalga *Haematococcus pluvialis*: from single cell to high value commercial products. *Front Plant Sci*. 2016; 7: 531. doi: 10.3389/fpls.2016.00531.

[123] Fu W, Paglia G, Magnúsdóttir M, Steinarsdóttir EA, Gudmundsson S, Palsson BO, Andrésson OS, Brynjólfsson S. Effects of abiotic stressors on lutein production in the green microalga *Dunaliella salina*. *Microb Cell Fact*. 2014; 13: 3. doi: 10.1186/1475-2859-13-3.

[124] Borodina A, Ladygina LV. The Effect of cultivation conditions on accumulation of carotenoids in *Phaeodactylum tricornutum* bohl. (Bacillariophyta). *Int J Algae*. 2013; 15: 274-284. doi: 10.1615/InterJAlgae.v15.i3.70

[125] Borowitzka MA. Carotenoid production using microorganisms. In Cohen Z, Ratledge C (eds). Single Cells Oils: Microbial and Algal Oils. AOCS Press, Urbana. 2010; 225-240.

Permissions

All chapters in this book were first published in FA, by InTech Open; hereby published with permission under the Creative Commons Attribution License or equivalent. Every chapter published in this book has been scrutinized by our experts. Their significance has been extensively debated. The topics covered herein carry significant findings which will fuel the growth of the discipline. They may even be implemented as practical applications or may be referred to as a beginning point for another development.

The contributors of this book come from diverse backgrounds, making this book a truly international effort. This book will bring forth new frontiers with its revolutionizing research information and detailed analysis of the nascent developments around the world.

We would like to thank all the contributing authors for lending their expertise to make the book truly unique. They have played a crucial role in the development of this book. Without their invaluable contributions this book wouldn't have been possible. They have made vital efforts to compile up to date information on the varied aspects of this subject to make this book a valuable addition to the collection of many professionals and students.

This book was conceptualized with the vision of imparting up-to-date information and advanced data in this field. To ensure the same, a matchless editorial board was set up. Every individual on the board went through rigorous rounds of assessment to prove their worth. After which they invested a large part of their time researching and compiling the most relevant data for our readers.

The editorial board has been involved in producing this book since its inception. They have spent rigorous hours researching and exploring the diverse topics which have resulted in the successful publishing of this book. They have passed on their knowledge of decades through this book. To expedite this challenging task, the publisher supported the team at every step. A small team of assistant editors was also appointed to further simplify the editing procedure and attain best results for the readers.

Apart from the editorial board, the designing team has also invested a significant amount of their time in understanding the subject and creating the most relevant covers. They scrutinized every image to scout for the most suitable representation of the subject and create an appropriate cover for the book.

The publishing team has been an ardent support to the editorial, designing and production team. Their endless efforts to recruit the best for this project, has resulted in the accomplishment of this book. They are a veteran in the field of academics and their pool of knowledge is as vast as their experience in printing. Their expertise and guidance has proved useful at every step. Their uncompromising quality standards have made this book an exceptional effort. Their encouragement from time to time has been an inspiration for everyone.

The publisher and the editorial board hope that this book will prove to be a valuable piece of knowledge for researchers, students, practitioners and scholars across the globe.

List of Contributors

Kieran N. Kilcawley and David T. Mannion
Teagasc Food Research Centre, Co. Cork, Ireland

Jiangjiang Zhu
Key Laboratory of Sichuan Province for Qinghai-Tibetan Plateau, Animal Genetic Resources Reservation and Exploitation, Chengdu, P.R. China
Key Laboratory of State Ethnic Affairs Commission and Ministry of Education for Animal Genetics & Breeding, Chengdu, Sichuan, P.R. China
College of Animal Science and Technology, Northwest A&F University, Yangling, Shaanxi, P.R. China

Jun Luo
College of Animal Science and Technology, Northwest A&F University, Yangling, Shaanxi, P.R. China

Livia de Oliveira Ruiz Moreti, Letícia Nishi and Rosângela Bergamasco
State University of Maringa, Maringá-PR, Brasil

Rosa Maria Ribeiro
Cesumar Institute of Science, Technology and Innovation (ICETI), Maringá-PR, Brazil

Marcos Vinicius Antunes de Lemos, Inaê Cristina Regatieri, Fabieli Louise Braga Feitosa and Fernando Baldi
São Paulo State University, Jaboticabal, Brazil

Angelica S.C. Pereira
University of São Paulo, Pirassununga, Brazil

Oğuz Taşbozan and Mahmut Ali Gökçe
Cukurova University, Fisheries Faculty, Department of Aquaculture, Adana, Turkey

Sergio Montserrat-de la Paz, Rocio Abia, Sergio Lopez and Francisco JG Muriana
Laboratory of Cellular and Molecular Nutrition, Spanish National Research Council (CSIC), Seville, Spain

Beatriz Bermudez
Department of Cell Biology, University of Seville, Seville, Spain

Jubie Selvaraj
Department of Pharmaceutical Chemistry, JSS College of Pharmacy (A Constituent Institution of JSS University-Mysuru), Rock Lands, Udhagamandalam, Tamil Nadu, India

Siobhan Simpson and Catrin S. Rutland
Faculty of Medicine, School of Veterinary Medicine and Science, University of Nottingham, UK

Alison Mostyn
Faculty of Medicine, School of Veterinary Medicine and Science, University of Nottingham, UK
Faculty of Medicine, School of Health Sciences, University of Nottingham, UK

Mohammad Salah Abaza and Mohammad Afzal
Department of Biological Sciences, Faculty of Science, Kuwait University, Kuwait

Aneela Afzal
Advanced Imaging Center, Oregon Health Sciences University, Portland, USA

Afaf El-Ansary
Central Laboratory, King Saud University, Riyadh, Saudi Arabia
Autism Research and Treatment Center, King Saud University, Riyadh, Saudi Arabia

Hanan Qasem
Autism Research and Treatment Center, King Saud University, Riyadh, Saudi Arabia

Katalin Nagy and Ioana-Daria Tiuca
University of Medicine and Pharmacy "I. Hațieganu", Cluj-Napoca, Romania

Ana M.M. Gonçalves
IMAR (Marine and Environmental Research Centre) & MARE (Marine and Environmental Sciences Centre), Faculty of Sciences and Technology, University of Coimbra, Coimbra, Portugal
Department of Biology and CESAM, University of Aveiro, Aveiro, Portugal

João C. Marques
IMAR (Marine and Environmental Research Centre) & MARE (Marine and Environmental Sciences Centre), Faculty of Sciences and Technology, University of Coimbra, Coimbra, Portugal

Fernando Gonçalves
Department of Biology and CESAM, University of Aveiro, Aveiro, Portugal

Paola Scodelaro Bilbao
Laboratorio de Estudios Básicos y Biotecnológicos en Algas (LEBBA), Centro de Recursos Naturales Renovables de la Zona Semiárida (CERZOS), Consejo Nacional de Investigaciones Científicas y Técnicas (CONICET), Camino de La Carrindanga, Bahía Blanca, Argentina
Universidad Nacional del Sur (UNS), Departamento de Biología, Bioquímica y Farmacia, San Juan, Bahía Blanca, Argentina
Instituto de Investigaciones Bioquímicas de Bahía Blanca (INIBIBB), Consejo Nacional de Investigaciones Científicas y Técnicas (CONICET), Camino de La Carrindanga, Bahía Blanca, Argentina

Gabriela A. Salvador
Universidad Nacional del Sur (UNS), Departamento de Biología, Bioquímica y Farmacia, San Juan, Bahía Blanca, Argentina
Instituto de Investigaciones Bioquímicas de Bahía Blanca (INIBIBB), Consejo Nacional de Investigaciones Científicas y Técnicas (CONICET), Camino de La Carrindanga, Bahía Blanca, Argentina

Patricia I. Leonardi
Laboratorio de Estudios Básicos y Biotecnológicos en Algas (LEBBA), Centro de Recursos Naturales Renovables de la Zona Semiárida (CERZOS), Consejo Nacional de Investigaciones Científicas y Técnicas (CONICET), Camino de La Carrindanga, Bahía Blanca, Argentina
Universidad Nacional del Sur (UNS), Departamento de Biología, Bioquímica y Farmacia, San Juan, Bahía Blanca, Argentina

Index

www.ingramcontent.com/pod-product-compliance
Lightning Source LLC
LaVergne TN
LVHW060556210926
848485LV00005B/1324